PYRETHRUM

The Natural Insecticide

Contributors

NORMAN B. AKESSON
WILLIAM F. BARTHEL
HOWARD BOND
G. S. BURDEN
GEORGE CAMOUGIS
JOHN E. CASIDA
JOHN J. DeFEO
MICHAEL ELLIOTT
H. B. GILLENWATER
STAFFORD W. HEAD
NORMAN F. JANES
G. D. GLYNNE JONES
W. HARRY LANGE
ROBERT L. LYON
KAREN MAUGER
GEORGE A. McLAUGHLIN
RAYMOND P. MISKUS
JOSEPH B. MOORE
EMIL M. MRAK
RICHARD E. PILLMORE
CARROLL N. SMITH
CLARA H. WILLIAMS
IZURU YAMAMOTO

PYRETHRUM

The Natural Insecticide

Edited by

JOHN E. CASIDA

Division of Entomology and Parasitology

College of Agricultural Sciences

University of California

Berkeley, California

ACADEMIC PRESS New York and London 1973

A Subsidiary of Harcourt Brace Jovanovich, Publishers

ACADEMIC PRESS, INC.
111 Fifth Avenue, New York, New York 10003

United Kingdom Edition published by
ACADEMIC PRESS, INC. (LONDON) LTD.
24/28 Oval Road, London NW1

LIBRARY OF CONGRESS CATALOG CARD NUMBER: 73–9419

PRINTED IN THE UNITED STATES OF AMERICA

Contents

Chapter 2 **Pyrethrum Production**

G. D. GLYNNE JONES

Part II **CHEMISTRY**

Chapter 3 **Composition of Pyrethrum Extract and Analysis of Pyrethrins**

STAFFORD W. HEAD

Chapter 4 **Chemistry of the Natural Pyrethrins**

MICHAEL ELLIOTT AND NORMAN F. JANES

Part IV USE AREAS

List of Contributors

Numbers in parentheses indicate the pages on which the authors' contributions begin.

Norman B. Akesson (261), Department of Agricultural Engineering, College of Agricultural and Environmental Sciences, University of California, Davis, California

William F. Barthel (123), Center for Disease Control, Health Services and Mental Health Administration, Public Health Service, United States Department of Health, Education, and Welfare, Atlanta, Georgia

Howard Bond (177), Department of Medicinal Chemistry, College of Pharmacy, University of Rhode Island, Kingston, Rhode Island

G. S. Burden (243), Insects Affecting Man and Animals Research Laboratory, United States Department of Agriculture, Gainesville, Florida

George Camougis (211), New England Research, Inc., Worcester, Massachusetts

John E. Casida (101), Division of Entomology and Parasitology, College of Agricultural Sciences, University of California, Berkeley, California

John J. DeFeo (177), Department of Pharmacology and Toxicology, College of Pharmacy, University of Rhode Island, Kingston, Rhode Island

Michael Elliott (55), Department of Insecticides and Fungicides, Rothamsted Experimental Station, Harpenden, Herts, England

H. B. Gillenwater (243), Stored-Product Insect Research and Development Laboratory, United States Department of Agriculture, Savannah, Georgia

Stafford W. Head (25), The Pyrethrum Marketing Board, Nakuru, Kenya, East Africa

Norman F. Janes (55), Department of Insecticides and Fungicides, Rothamsted Experimental Station, Harpenden, Herts, England

G. D. Glynne Jones (17), The Pyrethrum Marketing Board, Nakuru, Kenya, East Africa

W. Harry Lange (261), Department of Entomology, College of Agricultural and Environmental Sciences, University of California, Davis, California

Robert L. Lyon (281), Pacific Southwest Forest and Range Experiment Station, Forest Service, United States Department of Agriculture, Berkeley, California

Karen Mauger (177), Department of Medicinal Chemistry, College of Pharmacy, University of Rhode Island, Kingston, Rhode Island

George A. McLaughlin (3), McLaughlin Gormley King Company, Minneapolis, Minnesota

Raymond P. Miskus (281), Pacific Southwest Forest and Range Experiment Station, Forest Service, United States Department of Agriculture, Berkeley, California

Joseph B. Moore (293), McLaughlin Gormley King Company, Minneapolis, Minnesota

Emil M. Mrak (307), University of California, Davis, California

Richard E. Pillmore (143), Bureau of Sport Fisheries and Wildlife, Fish and Wildlife Service, United States Department of the Interior, Denver Federal Center, Denver, Colorado

Carroll N. Smith (225), 317 N.W. 32nd St., Gainesville, Florida

Clara H. Williams (167), Toxicology Branch, Pesticides Tolerances Division, Office of Pesticides Programs, Environmental Protection Agency, Washington, D.C.

Izuru Yamamoto (195), Department of Agricultural Chemistry, Laboratory of Pesticide and Bio-Organic Chemistry, Tokyo University of Agriculture, Tokyo, Japan

Participants

The symposium participants are shown in a photograph taken in August, 1972, when the "Symposium on Pyrethrum the Natural Insecticide" was held. Although Dr. Carroll Smith was at the symposium, he was absent at the time the photograph was taken; he is shown in an insert photograph. Dr. G. F. MacLeod of the Division of Agricultural Sciences, Office of the President, University of California, Berkeley, California, is also included on the photograph. Dr. MacLeod was the guest speaker at a banquet for the symposium participants.

The contributors are as follows (left to right):

FIRST ROW: Emil M. Mrak, Clara H. Williams, John E. Casida, Richard E. Pillmore, and John J. DeFeo

SECOND ROW: George McLaughlin, Hagen B. Gillenwater, Robert L. Lyon, Joseph B. Moore, William F. Barthel, and Michael Elliott

THIRD ROW: George Camougis, Stafford W. Head, Izuru Yamamoto, G. F. MacLeod, G. D. Glynne Jones, and Norman B. Akesson

Preface

The use of chemical toxicants or insecticides to control pest insects has proved to be one of the most controversial and emotional subjects in recent years. It is becoming increasingly evident that in addition to effectiveness in insect control an ideal or even an acceptable insecticide chemical must have an unusual set of biological and chemical properties and economic features. Of greatest importance, it must not be toxic on a short- or long-term basis to humans exposed to the compound as it is applied or to residues resulting from its use. The compound must have a physiological specificity or use pattern that spares other beneficial organisms including not only useful insects (predators, parasites, pollinators) but all exposed biota, particularly those directly contacted by the insecticide in the form and manner applied. The residues of the compound must dissipate, by physical or chemical means, at a rate so that hazardous levels do not accumulate in an individual organism or on passing through a food chain. The insecticide must also be susceptible to good formulation, compatible with other agents with which it is normally combined, available in adequate supply, and must be favorably priced relative to its effectiveness. This book attempts to define the extent to which one of the insecticides in common use fits in with our current concepts of an ideal insect control agent. This insecticide is pyrethrum.

Pyrethrum is a product of nature and not of a chemical factory. This means that its availability is determined by the crop acreage, the yields, and the pyrethrins content. Pyrethrum production is an important farming practice in some parts of the world, especially Kenya and Tanzania, helping to support their economies. This insecticide, as a natural material, fits in with the concepts of organic farming. This marvelous gift of nature has helped man control pestiferous insects for more than a century and possibly for many centuries. Thus, it has one of the longest histories of safety in use. However, the fact that pyrethrum is a natural material with a long history of use does not exempt it from reevaluation on the same standards applied to newer synthetic organic insecticides. Also, it does not exempt it from the challenge of man's imagination and ingenuity in attempting to improve upon it by altering its structure through preparation of analogs. Pyrethrum provides an excellent example of a natural combination of materials that is very difficult to reproduce, either chemically or in its useful biological properties, in the organic chemistry laboratory. In many desirable features, pyrethrum remains unsurpassed by any type of synthetic organic insecticide and even by the best synthetic analogs or pyrethroids.

The chapters comprising this work deal with all facets of pyrethrum, ranging from the culture of pyrethrum plants to the use of pyrethrum extract to control crop pests, from the chemical and biological stability of the pyrethrins to their persistence as residues, and from the action of pyrethrum on the insect nervous system to the toxicity of pyrethrum constituents to mammals. All contributors are specialists in their areas.

This book is based on papers presented at the "International Symposium on Recent Advances with Pyrethrum the Natural Insecticide" held in conjunction with the American Institute of Biological Sciences Silver Anniversary Meeting at the University of Minnesota, Minneapolis, August, 1972. The symposium was coorganized by the editor and Professor Emil Mrak, Chancellor Emeritus of the University of California, Davis. It was sponsored by McLaughlin Gormley King Co. and the Pyrethrum Board of Kenya.

I gratefully acknowledge the help of Jane Chamberlain and Gladys Moffett of the Division of Entomology and Parasitology, University of California, Berkeley, for assistance in manuscript reviews and revisions and Dr. Joseph B. Moore, Senior Staff Vice President of McLaughlin Gormley King Co., for frequent encouragement and advice on organizing the symposium and preparing this book. Special thanks are due the Governing Board of McLaughlin Gormley King Co. for committing their Company to the importance of assembling all published and unpublished information on pyrethrum in a form available to all interested parties.

John E. Casida

Note on Nomenclature

Pyrethrum flowers yield "pyrethrum extract" of which the insecticidal constituents are collectively "the pyrethrins" or "the natural pyrethrins." The six insecticidal esters, or "the pyrethrins," are shown in Fig. 1, page 26. The three chrysanthemates shown there are referred to as the "pyrethrins I" and the three pyrethrates as the "pyrethrins II." Pyrethrin I refers specifically to the chrysanthemate whereas "pyrethrins I and II" is a mixture of the chrysanthemate and pyrethrate. The same relationships are involved with cinerin I, "cinerins I and II," jasmolin I and "jasmolins I and II." These terms occur so frequently throughout this book that the quotation marks or italic designations are usually deleted for convenience.

Part I

INTRODUCTION

Chapter 1

History of Pyrethrum

GEORGE A. McLAUGHLIN

I. Introduction

Early discovery of the utility of "insect flowers" (pyrethrum) may well have been accidental. Pyrethrum flowers are in the genus *Chrysanthemum*, the commercially important species being *cinerariaefolium* (Treviranus) Boccone. The casual observer may think it an ordinary field daisy, which the flower head resembles. However, the plants are readily distinguishable. Commercial use of pyrethrum probably originated in Persia. Secrecy surrounding its production makes it difficult to pinpoint the actual date of recognition of its insecticidal activity. We do know that the species *C. cinerariaefolium* was produced in Dalmatia about 1840, replacing a Persian species. A story of interest has it that a German woman of Dubrovnik, Dalmatia, who picked the flowers for beauty, threw them into a corner after they withered. Several weeks later, they were found surrounded by dead insects. Apparently she associated death of the insects with the insecticidal property of the plants.

Although accounts of the discovery of the toxicity to insects vary, it is fairly certain that pyrethrum was in use in Europe as an insecticide more than a century ago and in Persia considerably earlier.

Gnadinger (1936) reports early importation, about 1860, into the United States of "insect powder," later as flowers called "insect flowers," to be powdered in this country. Consumption increased from about 600,000 lb in 1885 to 3,000,000 lb in 1919.

II. Early History of MGK and Insect Flowers

It is interesting to note that MGK, incorporated in 1908 as McLaughlin Gormley King Company, an importer and miller of botanical materials, was dealing with insect flowers in 1915. When, in 1919, importers began to extract the insect flowers with kerosene, the resulting liquid sprays proved to be more effective and useful than the powders.

Also in 1919, the founder of MGK Company, Mr. Alexander McLaughlin, and his associates decided that they should conduct investigations into specialization relating to some of the botanicals they were handling, which included pyrethrum. Their first step was to employ Mr. Charles B. Gnadinger, a chemist, enticing him away from the local office of the Food and Drug Administration.

No study of the history of pyrethrum can be complete without recognition of the brilliant work and persistence that Gnadinger showed after he read the early experiments of Staudinger and Ruzicka (1924). Their important scientific studies of pyrethrum were made in the period 1910 to 1916 but no publication of the work was made until 1924 (Staudinger and Ruzicka, 1924; Staudinger *et al.*, 1924). Staudinger and Ruzicka separated and partially identified the two primary active principles of pyrethrum, pyrethrin I and pyrethrin II, and in the long series of papers published in 1924, they created a truly scientific foundation for the present knowledge about pyrethrum.

Gnadinger studied Staudinger and Ruzicka's conclusions and assay method, as did Tattersfield and others, in the period 1925 to 1935. Gnadinger and Corl (1929) published an assay method that proved to be more practicable and accurate than any formerly available. This method was immediately applied to studies of the content of pyrethrins of several commercial varieties, made possible the manufacture of standardized and concentrated pyrethrum extract for the preparation of insecticides of uniform effectiveness, and opened many other avenues of investigation that a good practical and accurate first method of assay provides.

Gnadinger was a tireless and brilliant worker on pyrethrum. On his own time, only modestly paid for by the company, he wrote, edited, proofread, and prepared three texts, still available today in many reference libraries but now out of print. These were *Pyrethrum Flowers*, the original edition, September 1933; the second edition, 1936, which combined the work of the two editions; and finally the Supplement to the second edition, in 1945, in total almost 1000 pages of text and references. Gnadinger retired in 1950 having been briefly president of MGK, and his personally selected younger associates, Mr. Carroll A. Clark, president of MGK from 1958 to 1969, followed by his retirement at the end of 1971, and Dr. Joseph B. Moore, vice president of MGK, who is undoubtedly the most widely known commercial research worker in pyrethrum in the United States today, are both intensely interested, as I am, in bringing together the research findings in recent years on pyrethrum into a reference book which will be generally available. This goal is achieved, in part, by publication of this volume, based on a Symposium on *Pyrethrum: The Natural Insecticide*. It is clear from the following chapters in this volume that pyrethrum, with such a long history of effective use, continues to be a suitable modern insecticide even with the changing views on the appropriateness of various chemicals for insect control.

III. Origin and Development of Pyrethrum Production in Kenya

In the history of pyrethrum, world wars seem to have affected very substantially the location of sources of pyrethrum. Obviously after 1914 it became impossible to import Dalmatian pyrethrum flowers, and growers in Japan took over the market in the United States. Production of pyrethrum in Japan soared to 71,800 acres yielding 28,000,000 lb by 1935. Following Gnadinger's pyrethrum publications, growers in Kenya, interested since 1928 in cultivating the crop, made contact with MGK and we were pleased to observe that the content of pyrethrins of the African flowers was far superior to that of either Japanese or Dalmatian flowers. In addition, yields from African flowers were much higher per acre than from any other part of the world. MGK purchased the first commercially produced Kenya pyrethrum (4 tons) from the farm of D. H. Pell-Smith in December 1933. After this early contact, MGK aggressively encouraged Kenya pyrethrum cultivation and export. In 1936 (Gnadinger, 1936–1945a), the company offered for the first time anywhere a higher price based on active principle content determined by Gnadinger's chemical assay.

Pioneers of pyrethrum growing in Kenya enthusiastically organized growing of the crop, forming the Kenya Pyrethrum Growers Association early in 1933. The Kenya Farmers Association, a large marketing cooperative, then graded and baled the flowers, also tending to export and routing. Excellent and dependable handling permitted rapid growth of the project. At the request of the growers, the *Sale of Pyrethrum Ordinance* was created in 1935, and a Pyrethrum Board appointed (Gnadinger, 1936–1945b). The ordinance required that all pyrethrum to be grown must be licensed and that all produced must be sold to a single central agency, insuring careful inspection and grading, and quality maintenance. Licenses were issued to planters and the entire origination and marketing was kept under tight control. Many changes in rules and practices took place in the almost 40 intervening years, and entirely new ordinances were created in 1938, in 1956, and again in 1964. Today, after independence, the Kenya pyrethrum industry is the most completely organized grower system, following much of the English law and business philosophy rather than the American style.

IV. East African Sources and World War II

Consequently, in the years between 1935 and 1941, Kenya pyrethrum replaced the Japanese product in the market because of its superiority. Of course, World War II completely shut off supply from Japan and greatly encouraged the production of pyrethrum in Africa, especially Kenya, Tanganyika, and the eastern portion of the Belgian Congo. After World War II, Japan never regained any large proportion of pyrethrum export capability.

At the same time that Gnadinger's assay method was beginning to be used internationally, before World War II, Tattersfield, Hobson, and other workers at Rothamsted Experimental Station, Harpenden, England had been following entirely different procedures to develop accurate, practical assay methods. Many other researchers became involved, including H. A. Seil (1934), using an acid method. Pyrethrum was declared a strategic war material for malaria control on June 13, 1942 (Knowlson, 1942). Despite the fact that the Seil method gave somewhat higher results than other available acid methods, it was selected as the official method when the War Production Board took over importation and control of pyrethrum. All available pyrethrum then (December 1942) was allocated only for the production of aerosol bombs for use by the armed forces, with one exception among domestic uses: the production on a limited scale of a patented, stabilized pyrethrum dust concentrate, used widely in victory gardens. Biological methods of assay had been introduced several years earlier but the chemical assay methods proved to be more accurate.

Although experimental pyrethrum growing projects were established in many regions of the United States and in many countries elsewhere, the most ambitious project apparently was that initiated at the Colorado Experiment Station, Fort Collins, in 1933 (Gnadinger *et al.,* 1933). Climatic conditions there seemed to favor growth of healthy plants and led to a joint project of MGK and Colorado that continued until 1936. Many interesting new facts about pyrethrum plants were discovered and revealed in the literature during this time. Results were sufficiently encouraging to permit the purchase by MGK of 600 acres of suitable land near Pagosa Springs, Colorado in 1938. The company supported the commercial growing experiment thereafter except for some minor interest taken by the United States Navy, which hoped to eliminate the need for ships bottoms required during the war for the importation of pyrethrum flowers from Kenya. In 1939, Gnadinger, in charge of the research from the start, employed Dr. J. B. Moore, first as assistant on a project at Wenatchee, Washington, and then, in 1942, as manager of the Colorado project. The final consensus in the company in 1947 was that we had not produced the flowers on a commercially successful basis so the project was closed out. Dr. Moore has retained his lively interest in all aspects of pyrethrum ever since, much to the benefit of MGK.

Certainly, one of the most important moments in the history of pyrethrum was July 29, 1941, when L. D. Goodhue and W. N. Sullivan applied for patent on a mixture of a liquified gas and a parasiticide, an aerosol. A patent was granted (Goodhue and Sullivan, 1943). These men, with J. H. Fales, had been experimenting with smokes as insecticide aerosols for several years.

As the active principles of pyrethrum were then known as among the most effective insecticides, weight for weight, I believe the workers theorized that creating very small, light particles of concentrated solution floating in air for a long time might improve results and reduce the cost of pyrethrins used per unit of space. The first aerosol to accomplish this on a large scale, as mentioned before, was the wartime bomb, a high-pressure device which gave particles probably of 10 to 30 μm in size. These proved to be more effective and economical than oil sprays, which delivered particles 100 μm in diameter. In 1945, DDT was officially added and pyrethrins reduced, conserving scarce supplies of pyrethrum and reducing cost.

V. Postwar Scarcity of Supplies

So many of these bombs ordered by the military became surplus after the war and such a great commercial flow of DDT surged into the insecticide market that pyrethrum was dealt a fierce double blow, almost a fatal one. World consumption of pyrethrum declined, growers turned to other crops, and the situation for those dependent on pyrethrum looked bleak, to say the least. Kenya pyrethrum production declined from 7400 tons in 1945 to 1560 tons in 1948.

However, work went on to make the bomb commercially more acceptable for household use. Reduced gas pressures without loss of small particle size, development of new and better valves and liners for tin can containers, automatic filling machines, mixtures of propellants, all surfaced in 1948 and 1949. Acceptance by the public was instant except for those few products which gave slow effect and produced too little knockdown. Pyrethrum stepped back into the picture to supply the missing element.

Concurrently, a patent was granted to Dr. Herman Wachs (1949) of U.S. Industrial Chemicals, Inc., for the first truly effective, commercially viable synergist, piperonyl butoxide. Whereas there had been many synergists offered and used before, and it was not an entirely new thought for pyrethrum users, piperonyl butoxide became by all odds the most widely used pyrethrum cost-reducing substance within a few years. It is today, long after the United States patent has expired, still a requirement in most pyrethrum insecticides.

At about the same time, another cost-reducing synergist, *N*-octyl bicycloheptene dicarboximide, an entirely different type, became commercially available. Combinations of the two synergists with pyrethrum helped to reduce the cost and improve the quality of pyrethrum aerosols and some sprays. Without these and other similar steps, it is doubtful that pyrethrum used alone could have faced the competitive markets in the 1950's.

It became clear that greater production was needed. The rapid United States development of pyrethrum-containing "beer-can" aerosols in 1949–1950, and the cost-reducing economics of synergists failed at first to come through loud and clear to the growers. From the low point of production in 1948–1949, 1950 production only went up to 2175 tons. Great fears are expressed in the records lest the takeover by synthetic chemicals reduce the market for pyrethrum. Rapid origination of scores of synthetic compounds following DDT caused tremors throughout the pyrethrum growing world. As late as 1956, production in Kenya had not yet reached 3500 tons. The demand continued to grow so that an intense shortage of supplies was felt, especially in the United States.

Strangely enough, it became apparent that aerosol sales did not substitute for oil-spray sales among household and industrial users. Shortage of pyrethrum led to increased use of synthetics, such as isobutyl undecylenamide, isobornyl thiocyanoacetate and β-butoxy thiocyanodiethyl ether, instead of pyrethrum in sprays, but they were not widely used in aerosols. As is often typical in such a period of shortage, one or two large buyers of pyrethrum virtually cornered the market in the United States, which did not appeal to the Kenya growers' organization. In the years prior to independence of Kenya, not more than a few hundred growers, owning large, diversified farms, supplied 80% of the pyrethrum crop.

In 1951, Kenya pyrethrum officials expressed interest (private exchanges) in local domestic extraction of the flowers to produce a concentrated product, superior for storage and export. Baled whole pyrethrum flowers, the only kind available theretofore, were expensive to ship and pyrethrins lost during storage or in transit were expensive. With the help of refiners and machinery manufacturers, the first 3000-ton crude extract plant was completed in 1956. Thereafter, a continuous program of process development and improvement made the importation of whole baled flowers, and the United States plants to extract them, obsolete.

The history of insect resistance to chemical insecticides is old, going back to 1916 with resistance of citrus red scale to cyanide fumigation (Porter, 1952). When DDT became ubiquitous, insect resistance soon became apparent. The most spectacular example is said to be resistance of the housefly to DDT, an area of use in which pyrethrum figures prominently. Although DDT and other chlorinated and organophosphorus products had replaced pyrethrum in agriculture as well as some other uses, because of great cost advantages—the bloom eventually came off the rose. Pyrethrum became even more in demand for quick killing effect and the inability of insects to develop significant resistance under normal use conditions.

Production of pyrethrum in countries other than Kenya, affected by the same postwar factors, slumped after World War II but began to build up again in the 1950's. Tanganyika resumed production as did eastern Congo for a time, and production continued to increase after Tanganyikan independence, as did that of Kenya, but the Congo's production virtually disappeared. Growing of pyrethrum in Ecuador was initiated in 1960 but it was clear as early as 1964 that cost of production there could not improve on East Africa's. Many other sources were tried, New Guinea, Peru, Brazil, Japan, Rhodesia, Rwanda, each contributing a few score tons to meet world requirements, as detailed by Glynne Jones (Chapter 2 in this volume), but none were commercially successful. None was as well organized for production as Kenya, either. From 1959 to 1960 Kenya's production rose from about 4100 tons to 6500 tons and up to 9100 tons in 1961.

After independence in Kenya and Tanganyika, both sources became stabilized and most reliable shippers. As detailed in Chapter 2, this volume, 85,000 farmers now grow pyrethrum in small plots as a cash crop in Kenya alone.

Furthermore, advanced analytical competence in Kenya, using sophisticated new gas–liquid chromatographic methods, encouraged sampling of every field in Kenya and the selection of higher-content individual plants for splits. Farmers and their cooperative's representatives could be informed quickly of the real monetary value of their production.

In the late 1950's and early 1960's, more completely refined pyrethrum extract produced in the United States reduced staining complaints arising from the use of aerosols in households and eliminated the allergens. Industry also devised the first successful water-based aerosol, powered with liquefied petroleum gas, a lower-cost propellant. This last was a spectacular and unexpected bonus in the production of aerosols.

Development of intermittent aerosol application began also in the 1960's. Metered valves electrically actuated to release bursts of about 100 mg every 15 min proved very popular in industrial and food processing plants. Pyrethrum, piperonyl butoxide, and *N*-octyl bicycloheptene dicarboximide are all cleared for use in food processing plants as food additives, since May 19, 1967, and are therefore the active materials of choice for use in intermittent sprayers.

All these improvements in the techniques, in Kenya and in the United States, and the development of an international aerosol industry raised the demand for pyrethrum. However, production outstripped demand and Kenya seriously overproduced in 1967–1968 with 11,000 tons of flowers.

VI. Pyrethrum Recognized as a Modern Insecticide

Toxicologists had been aware for several years of traces of DDT and other chlorinated pesticides in soil, water, birds, and some farm animals. Tolerances for residues on food had been initiated in 1955. In the early 1960's, more sophisticated methods of analysis became available, allowing detection of tenths of a part per billion instead of parts per million of insecticide. It is well said that the modern history of pyrethrum started with the publication of Rachel Carson's *Silent Spring* in 1962. Her mention of pyrethrins refers to them as preferable, a "less toxic chemical" than the synthetic ones, such as DDT, lindane, phosphates, and the like. However, many people expecting the use of pyrethrum to surge upward following the emotional acceptance by the public and politicians of her point of view, were to be disappointed. The most obvious immediate effect was the general reduction in the use of all pesticides. Although the importation of pyrethrum into the United States did not rise greatly, except for the years 1966 and 1967 when there was substantial overproduction in East Africa, there was a gradual rise in the consumption of pyrethrum.

State and federal administrators enacted a host of stringent new regulations about pesticides. Naturally, pyrethrum was one of these, as were synergists. Although pyrethrum had been used for more than 100 years without evidence of dangerous effects to man or animals, such a history, although unique, was not sufficient to obtain labeling which made its safety obvious to the buyer, usually an uninformed woman bereft of advisors and help in a modern supermarket. The industry in the United States, poorly organized to present the case for "nontoxic" pyrethrum, struggled hard and eventually successfully, to gain Interpretation 24 (Agricultural Research Service, 1965), permitting suitably formulated mixtures of pyrethrum and synergists to be registered for sale bearing the affirmative inscription "Nontoxic to humans and pets." Such aerosols and sprays showed marked sales improvements at once, probably having some additive effect on pyrethrum consumption. Industry mailed 20,800 questionnaires to people with children and received 4300 replies, more than 20% return, the overwhelming proportion heavily favoring the need for distinctive "nontoxic" labeling.

In August 1972, the Environmental Protection Agency published a proposed new labeling requirement withdrawing Interpretation No. 24, a great disappointment. The industry, and especially the consumer, will be affected unfavorably should this requirement go into effect. The consumer needs a clear, affirmative statement on the label enabling her to differentiate among the household insecticide products on the shelf those which she can use, even misuse, without danger to herself, her family, or the environment. This right of the consumer should and must be given the dominant position in any proposed changes in regulations.

The observations by pyrethrum growers in east Africa of unsold portions of overproduced flowers coincided with announcements of some new synthetic pyrethrin-like compounds called *pyrethroids*. It has been estimated that some 200 synthetic organic insecticides had made their appearance, and disappearance, over the years prior to 1969. These were supposed to replace pyrethrum but had never quite succeeded. However, the official weight of the United States Department of Agriculture, Entomology Research Division, was brought to bear, revealing to the Kenya pyrethrum growers confidential, then unpublished, test results on new synthetic chemicals compared to pyrethrins, which shook the confidence of growers. In May 1968, the Minister of Agriculture of Kenya, The

Honorable Bruce McKenzie, visited pyrethrum buyers in the United States, asking for opinions from pyrethrum buyers as to a suitable course of action to support and extend the pyrethrum market. Reaction was, I believe, universally consistent—reduce prices. This was done to the extent of about 15% in October 1968. The outlook to pyrethrum growers must have looked dark at that point. Some errors were apparently made in communications with grower's cooperatives. The effect of these events was to discourage many growers, who replaced pyrethrum plots with potatoes or some other crop. Drought and other unfavorable conditions exaggerated the reduction of the crop and, once again, in 1970 the industry learned that all of the oversupply had been used up. There was as desperate a shortage again in 1970–1971 as there had been in the early 1950's. During this period of scarcity, no large pyrethrum buyer was permitted to corner the market.

An aggressive and obviously competent farmer, Mr. A. F. Brown, related in 1965 some results of plant selection experiments he had conducted over several years on his large farm on the slopes of Mount Kilimanjaro in Tanzania (Brown, 1965). At about the same time, industry in the United States perceived an opportunity to suggest combining plant selection with the more extensive project of plant breeding to develop controlled seed, offering to make available the consulting services of a renowned plant geneticist, Dr. H. E. Brewbaker, who visited Kenya in 1965 to discuss such a project. Not all the steps he recommended were taken but plant selection was entered into in a big way, as described later in this volume. The results have been very good, perhaps have improved quality, and at the least have preserved it. Shortage of supply is presently being relieved, with Kenya producing 15,000 tons in the year ending September 30, 1972 and possibly raising its sights to 16,000–18,000 tons next year.

Funds have been provided by industry in the United States and the Kenya Pyrethrum Board since 1965 for the purpose of exploring unknown aspects of the fate of pyrethrum and synergists in the environment and mammals. Interest has been aroused in state and federal agencies, especially experiment stations, to promote and explore the subject scientifically, in some cases with matching funds. It is estimated that these sources have provided a total of about one-half million dollars for such purposes since 1965.

Two separate geographical areas have undertaken commercial production of pyrethrum in recent years. The first to do so is Rwanda, a nation of East Africa just east of Lake Kivu and west of Lake Victoria, neighboring on Burundi and Uganda. An American, Mrs. Rosamond Carr, a pyrethrum grower resident in Rwanda for years, was effective with others in interesting the United Nations Industrial Development Organization to finance and construct a substantial pyrethrum extraction plant (Anonymous, 1970) in Ruhengeri, which operated initially in 1971. The other new source, in an initial stage but apparently privately financed, is in Rhodesia.

Extensive commercial development of DDT and other synthetic insecticides following World War II killed the modest market previously established for agricultural uses of pyrethrum. The newer compounds provided insect control of the first order at such low cost per acre per application as to stop pyrethrum in its tracks. Even use for truck-crop insect control with stabilized dusts, and for home garden products in which nontoxicity, widely advertised for years, was important disappeared because of relative cost. After all, what could compete with insect controls costing as little as 35¢ to 50¢ per acre per application! Certainly not pyrethrum. The only remaining broad markets for which no substitutes were at hand were household aerosols and sprays and some industrial application.

Twenty years later, after insect resistance, setting of tolerances, stringent "bans" by environmental enthusiasts, and general "putting down" of the use of such chemical pesticides had decimated the general pesticide industry, a new and efficient method of applying pyrethrum formulations for agricultural and forest insect control has appeared. It has been referred to generally as ULV—ultralow volume. To some extent the machinery required does the same thing as the aerosol does in the household products—without heat, it mechanically reduces particle size to as little as 5 μm in diameter and it is flexible. Costs of area application with pyrethrum are beginning to compete with other permitted products, especially for control of mosquitoes, tabanids, exposed forms of Lepidoptera, roaches, and flies. This new technique is opening up a whole modern use for pyrethrum, denied it heretofore because of its cost. Additional information on this technique can be found in other chapters in this volume. Several other publications are presently in print (Preiss, 1973).

The list of pyrethrum and synergist exemptions from tolerances and accepted tolerances for them, under the Federal Food, Drug, and Cosmetic Act of 1954, is given by Moore (Chapter 16 in this volume). Under the Food Additives Amendment of 1958 to the 1954 Act, tolerances were established for almost every suitable use. Protocols were obtained for all this work many years ago, and the costs were paid by industry. These are the broadest uses permitted for any active principle or formulations and represent many thousands of man-hours of careful scientific testing at a cost to industry of many hundreds of thousands of dollars. None of these tests were considered necessary until 1955.

VII. Pyrethrum in the Future

Some work remains to be done—carcinogenic studies and inhalation data on prolonged use of intermittent sprayers. Although such work has been started, it is too early to reach conclusions but the outlook is for favorable findings on these substances as in the massive amount of earlier work that has been done. All the findings will be made available to the proper United States authorities, and those in the World Health Organization where required, for inclusion in the Codex Alimentarius.

Those of us who have been in this kind of work for 40 years or more, as laymen working with technologists, many of whom are truly scientists, have seen the pesticide industry accused of preying on the environment and consumers without sufficient public attention being drawn to its accomplishments. In contrast, I offer you the pyrethrum industry, a small one, but long-lived and proud of its accomplishments. It has no monopoly; it is characterized by rugged competition, not only among its members but with uncontrollable natural elements. The numbers of farmers, most of them now black, probably exceeds 85,000 in Kenya alone plus those in Tanzania, Rwanda, Congo, New Guinea, Ecuador, Brazil, Japan, and elsewhere. It is often their only cash crop and keeps whole families busy, and earning, with handpicking for as many as 10 months of the year. The commercial manufacturing, marketing, packaging, and research must occupy several more thousands of people. It is all these people who have kept the faith for pyrethrum who have provided funds for proving that this commodity is as safe to use as it was thought to be in 1930.

The pyrethrum industry has borne its own losses and used its relatively small gains to support continuing research designed to further evaluate the usefulness and safety of its products. We have recognized, and responded to, our responsibilities for social progress and environmental protection. The efficacy and safety of pyrethrum in protecting man from insect annoyance and insect-transmitted diseases, and in protecting his stored food from insect attack, have been established over many generations. I am thankful that I have had the opportunity to spend my business lifetime in constructive efforts to make this useful insecticide more generally available in forms of ever-increasing effectiveness and lower cost.

References

Agricultural Research Service. (1965). Part 362. Regulations for the Enforcement of the Federal Insecticide, Fungicide, and Rodenticide Act, Interpretation 24. *Fed. Regist.* **30,** 1728.

Anonymous. (1970). Contract award to build pilot pyrethrum plant. U.N. Development Program, pp. 1–3, March.

Brown, A. F. (1965). A pyrethrum improvement programme. *Pyrethrum Post* **8**(1), 8–10.

Gnadinger, C. B. (1936–1945a). "Pyrethrum Flowers," Supplement to 2nd ed., pp. 400–404. McLaughlin Gormley King, Minneapolis, Minnesota.

Gnadinger, C. B. (1936–1945b). "Pyrethrum Flowers," Supplement to 2nd ed., p. 397. McLaughlin Gormley King, Minneapolis, Minnesota.

Gnadinger, C. B., and Corl, C. S. (1929). Pyrethrum flowers. I. The quantitative determination of the active principles. *J. Amer. Chem. Soc.* **51,** 3054–64.

Gnadinger, C. B., Evans, L. E., and Corl, C. S. (1933). Pyrethrum investigation in Colorado. I. *Colo. Agr. Exp. Sta. Bull.* **401,** 1–19.

Goodhue, L. D., and Sullivan, W. N. (1943). Parasiticidal aerosols suitable for combating flies, roaches, mosquitoes, etc. U.S. Patent 2,321,023.

Knowlson, J. S. (1942). Pyrethrum. General preference order M-179. *Fed. Regist.* **7,** 4479.

Porter, B. A. (1952). Insects are harder to kill. *Yearb. Agr. (U.S. Dep. Agr.)*, p. 318.

Preiss, F. J. (1973). Mosquito adulticiding with synergized pyrethrins. *Soap Chem. Spec.* **49**(5), 72–74, 106.

Seil, H. A. (1934). Estimation of pyrethrins. *Soap* **10**(5), 89, 91, and 111.

Staudinger, H., and Ruzicka, L. (1924). Insektentötende stoffe. I–VI and VIII–X. *Helv. Chim. Acta* **7,** 177–201, 201–211, 212–235, 236–244, 245–259, 377–390, 406–441, 442–448, and 448–458.

Staudinger, H., Muntwyler, O., Ruzicka, L., and Seibt, S. (1924). Insektentötende stoffe. VII. *Helv. Chim. Acta* **7,** 390–406.

Wachs, H. (1949). Methylenedioxyphenol derivatives. U.S. Patent 2,485,681.

Chapter 2

Pyrethrum Production

G. D. GLYNNE JONES

I. Introduction

Gnadinger's account (1936) of the early development of the pyrethrum industry has not been improved upon. He established that the present commercially cultivated crop originated from temperate climatic areas and that the species *Chrysanthemum cinerariaefolium* came from Dalmatia (now Yugoslavia). It is probable that the early Chinese records of pyrethrum refer to other species.

II. World Production of Pyrethrum and Prices for Pyrethrum Products

World production of pyrethrum is normally measured in metric tons of dry flowers and the pyrethrins content varies from 0.8% (w/w) to over 2%. Table I summarizes the amounts produced in leading growing countries. At present, the world's leading producers are Kenya, Tanzania, Ecuador, Rwanda, and Japan, with smaller amounts being grown in New Guinea, Brazil, Zaire, Indonesia, India, the U.S.S.R., Taiwan, Rhodesia, Yugoslavia, and South Africa. The first four of these countries export over 95% of their production, largely in the form of extract, but

TABLE I

Comparative Pyrethrum Production of Different Countries During 1967–1972

Country	Metric tons on a yearly basis (some overlap of dates up to 6 months) 1967–1968	1968–1969	1969–1970	1970–1971	1971–1972
Kenya	11,059	7300	5909	9747	14,400[a]
Tanzania	5102	4757	2416	2665	4300
Ecuador	1609	1744	1457	1241	1100[a]
Japan	950	838	700	600	380[a]
Rwanda	120[a]	200	640	800	1000[a]
Others	700[a]	620[a]	600[a]	580[a]	600[a]
Total	19,540	15,459	11,722	15,633	21,780

[a] Estimated.

Kenya also exports over 3000 tons of ground flowers for the world's mosquito coil market. Kenya is also a major producer of fully refined extract suitable for direct inclusion in aerosols.

World production in 1967–1968 was approximately 2500 tons in excess of demand and the existence of unsold stocks coupled with considerable publicity on the development of new synthetic pyrethroids led to a general decline in world production during the next 2 years. By late 1969, there was an acute world shortage and this coincided with much stronger governmental action in the United States, Canada, and Europe against persistent insecticides that were alleged to pollute the environment. These changes have resulted in an increase in demand for safe, nonpersistent insecticides. In the latter half of 1971 and the first part of 1972, world production slightly in excess of 20,000 tons has proved insufficient to meet demand and a tentative figure of 23,000 tons is given as the present demand with a projected progressive annual increase in the region of 10% per annum.

Over the past 2 years interest has renewed in the use of pyrethrum in agriculture, horticulture, and forestry and the existence of reliable ultralow volume application equipment has made it possible to apply very small dosages of pyrethrins per acre to achieve a high kill of a range of insect pests. If, as is likely, this usage continues to expand, then this forecast of demand may prove to be a serious underestimate. In Kenya there are no problems in effecting a further expansion of the crop and probably this statement also applies to Tanzania and Rwanda.

World prices for pyrethrum products have remained almost static for over 10 years but progressive monetary inflation has also affected all the pyrethrum producing countries. Considerable experimental data has been accumulated to show that increases in production of pyrethrins in the field can be obtained and these higher yields can be sustained without any significant additional costs in cultivation, picking, etc.

Processing factories in Kenya are owned by the Pyrethrum Board, which is organized on a cooperative basis with all profits being returned to the producer. The Board also recognizes that, in order to maintain confidence in the future of supplies, large stockpiles of extract must exist so as to insure the continued orderly marketing of this valuable insecticide.

III. Growing of Pyrethrum

During the past 10 years, the growing of pyrethrum in Kenya has undergone a major change. Until 1962, the majority of pyrethrum was produced on farms with land in excess of 50 acres and individual fields in excess of 30 acres were not uncommon. Following Kenya's becoming independent and the gradual change in land ownership, over 90% of pyrethrum now comes from small plot holders with less than 10 acres of land, with pyrethrum plots averaging in size from $\frac{1}{4}$ to 1 acre. The individual plot holders are organized into cooperative societies and there are now over 151 such societies, with memberships ranging from 300 to 1500. A few large-scale farms still exist and generally their level of productivity is very much higher than before 1963.

There are now over 85,000 families engaged in growing pyrethrum and in many cases a cash income obtained from the dry flowers on a monthly basis is a vital part of their economy.

IV. Plant Improvement Programs

Pyrethrum can be produced from clones or from seed. A pyrethrum clone is a group of plants obtained by the continuous vegetative propagation of a single plant; seed is produced by the hybridization of two or more clones and is less uniform but it has been shown that, if the right selection of parents is made, seed can be planted to give both a high content of 1.7% or over and a high yield of flowers. However, subsequent vegetative propagation from seedlings must be restricted because of the observed inverse correlation between content of pyrethrins and vigor, with the tendency of the more vigorous plants with lower content of pyrethrins to produce more splits and, therefore, dominate the subsequent plant population. The Pyrethrum Board, therefore, regards clone production in nurseries and seed production to be complimentary to its general development program.

In 1963 the Pyrethrum Board of Kenya and the Ministry of Agriculture recognized that it was insufficient to organize agronomic research on pyrethrum without insuring that the newly developed clones and seed were made available to the increasing numbers of small farmers, many of whom had the minimum of formal education. This has posed a very difficult problem in agricultural extension as the new circumstances are quite different from when there were only a few hundred large-scale farmers. It was also realized that the future of the industry was dependent on increasing productivity both in terms of yields per acre and the content of pyrethrins in the individual flower.

The first step was to develop methods of vegetative propagation. Fortunately, at this time a farmer in Tanzania, Mr. A. F. Brown (1965), pioneered methods of large-scale vegetative propagation and his techniques were adapted for use in Kenya. It was also accepted that local technical resources were insufficient to institute such a large crop-improvement policy and the Netherlands Government, having appreciated the importance of this problem, gave generous aid in the form of plant breeders, agriculturists, and young volunteers with technical training. The joint collaboration of the Pyrethrum Board and the Ministry of Agriculture and the technical assistance from Holland have resulted in the establishment of five large-scale irrigated nurseries for vegetative propagation covering 400 acres, in addition to many smaller nurseries and fields producing high-quality seed. All clones being propagated have a pyrethrins content in excess of 1.6% (present crop average 1.3%) and a yield potential at least twice the present average for the country. Many commercial fields are now producing flowers with over 2% pyrethrins, with yields well in excess of 800 lb per acre. This improvement program has taken 2 years longer to produce direct results in the field than was originally anticipated. The Pyrethrum Board has increased its own field extension staff and during 1972 over six million plants, each with a minimum splitting potential of times five, have been distributed to growers and 3200 lb of high-content seed have also been grown and harvested.

Other countries, such as Tanzania and Rwanda, have adopted parallel policies of plant improvement and recently we have heard of attempts in Rhodesia to grow pyrethrum under irrigation as an intensive cash crop. (In Kenya, irrigation is restricted to the nurseries for effecting rapid propagation of high content clones.)

Many new techniques of plant breeding have been applied to pyrethrum improvement, e.g., the inducement of polyploidy and the crossing of diploids with tetraploids to produce triploids, but we still rely on the older and more laborious methods of mass selection and controlled hybridization of proven clones. We do, however, attempt to update these older techniques and, for example, the yield of viable seed following hybridization has more than doubled in the past 4 years.

We have every reason to anticipate that our plant breeders will continue to produce improved hybrids and select better clones. We are constantly improving the methods of distribution of these materials to the grower, and provided these positive actions continue we believe that the future of the industry will be assured.

Further details and a number of relevant papers on the methods of pyrethrum plant improvement have been published in the journal *Pyrethrum Post.*

V. Factors Affecting Pyrethrum Production

The bulk of the world's production is now grown on land near the equator from 6000 to 8000 ft with a rainfall of 35 to 60 in, evenly spread throughout at least 7 months of the year. Under these conditions flowering continues from 7 to 11 months each year. Prolonged frost conditions and continued waterlogging can be very detrimental.

We are particularly conscious of the effects of extremes of climate on production. In Kenya, every 10 to 11 years there is a serious drought and a less serious one every 3 to 5 years. In the Highlands in Kenya there are several different climatic areas and every attempt is being made to spread the growing of the crop among these different areas so as to minimize the effects of climate on production.

The use of phosphatic fertilizers has been shown to increase the yields of dried flowers on a variety of soils (Kroll, 1962) but experiments with other macro- and micronutrients have always produced inconclusive results. Recently it has been shown that some of the new clones do appear to respond to nitrogen and this interesting development is being closely investigated.

The main insect pests that attack pyrethrum are thrips but outbreaks are only occasional and are relatively easily controlled (Bullock, 1961). Root knot eelworms are commonly found in most fields but it has been difficult to prove that their existence significantly affects yields of dried flowers. Every attempt is made to free nurseries from eelworm infestations but chemical control on a field scale is not practicable.

VI. Summary

Kenya produces more than half of the world's pyrethrum, a little over 20,000 metric tons of dried flowers per year. Tanzania, Ecuador, and Rwanda also contribute important amounts. Pyrethrum production in Kenya comes mostly from small plot holders and more than 85,000 families are engaged in growing pyrethrum. A plant improvement program involving improved hybrids and better clones has resulted in many commercial fields with over 2% pyrethrins content and yields well in excess of 800 lb per acre. Production is sometimes adversely affected by drought, prolonged frost conditions, and waterlogging. Certain types of fertilizers are useful in improving yields. Insect and eelworm pests are not serious problems in pyrethrum production. Production of pyrethrum flowers can be expanded to meet the present increasing world requirements without causing any drastic price increases in the end products.

References

Brown, A. F. (1965). A pyrethrum improvement programme. *Pyrethrum Post* 8(1), 8–10.

Bullock, J. A. (1961). The pests of pyrethrum in Kenya. *Pyrethrum Post* 6(2), 22–24.

Gnadinger, C. B. (1936). "Pyrethrum Flowers," 2nd ed. McLaughlin Gormley King, Minneapolis, Minnesota.

Kroll, U. (1962). The improvement of pyrethrum yields through the application of fertilizers. *Pyrethrum Post* 6(3), 32–33.

Part II

CHEMISTRY

Chapter 3

Composition of Pyrethrum Extract and Analysis of Pyrethrins

STAFFORD W. HEAD

I. Introduction

The insecticidally active constituents of pyrethrum extract, the pyrethrins, are esters, formed by a combination of two acids, chrysanthemic acid and pyrethric acid, and three alcohols, pyrethrolone, cinerolone, and jasmololone. The esters of chrysanthemic acid are called pyrethrin I, cinerin I, and jasmolin I, respectively, and are together known as the pyrethrins I fraction, whereas the esters of pyrethric acid

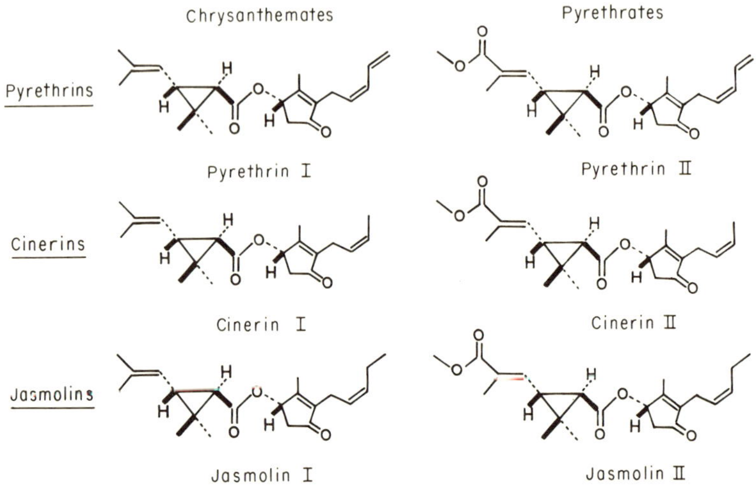

Fig. 1. Structures of the six insecticidal constituents of pyrethrum extract.

are called pyrethrin II, cinerin II, and jasmolin II, these representing the *pyrethrins II* fraction. Their structures are shown in Fig. 1. These six components together account for the kill and knockdown properties of pyrethrum extract and it is unlikely that any additional components will be found that add significantly in these areas of biological activity. It is not clear, however, whether the known pyrethrins account for the repellent properties of the extract. The stereochemical configuration of these compounds, which is so important to insecticidal activity, is discussed by Elliott and Janes (this volume, Chapter 4).

The pyrethrum plant of commerce is *Chrysanthemum cinerariaefolium* (Vis) and although many other members of the family Compositae have been screened for activity, no other significant source has been established (Gnadinger, 1936, 1945; Contant, 1963). The botanical structure of pyrethrum has been described by Chandler (1951, 1954). The flower head is typical of the Compositae and is a collection of small flowers (florets) set on a slightly convex receptacle, the under surface of which is covered by the scales of the involucre. The disc florets, which are sited over the center of the receptacle, are surrounded by an outer ring of ray florets. Both disc and ray florets have ribbed achenes, which are seated upon the receptacle. A cross-section of mature flower head is shown in Fig. 2a. The structures of the disc florets are bisexual whereas the ray florets are unisexual (female), having no stamens. Each disc floret (Fig. 2c) has a yellow tubular corolla with a small calyx at its base. Five stamens, rising from the base of the inner surface of

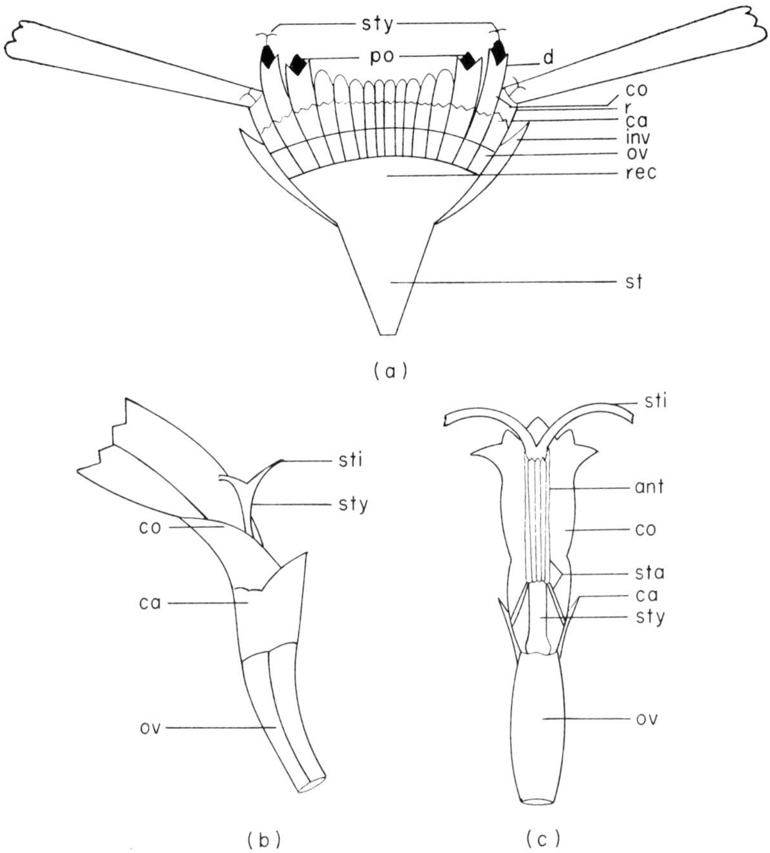

Fig. 2. The structure of the pyrethrum flower head. (*a*), inflorescence; (*b*), ray floret; (c), disc floret; sty, style; po, pollen; d, disc floret; co, corolla; r, ray floret; ca, calyx; inv, involucre; ov, ovary; rec, receptacle; st, stalk; ant, anthers; sti, stigma; sta, stamens. (Head, 1969b).

the corolla, each terminate in an elongated anther; the anthers join together at their outer edges to form a cylinder. At the center of the floret stands the style, which rises from the ovary. The style terminates in a bilobed stigma and is located within the cylinder formed by the anthers. In the unopened floret, the two lobes of the stigma are vertical and the inner receptive surfaces are oppressed. As the corolla opens, the style elongates and when the stigma is exposed the two lobes open outward, ready for pollination. The outer disc florets open first and the development proceeds progressively across the disc toward the center. The ray florets (Fig. 2b) differ in basic structure from the disc florets in that the corolla is enlarged, forming the petals, and the stamens are absent. More than 90% of the pyrethrins in the flower head are found within the achenes. Chandler (1951, 1954) suggests that these are located in the oil glands on the outer wall of the achene and also in the inner secretory ducts. However, recent work (Brewer, 1972) indicates that the major concentration of pyrethrins is within the secretory ducts (Fig. 3a and b).

Early studies on the distribution of the pyrethrins in the pyrethrum plant and their rate of development in the flower head have been reviewed by Gnadinger (1936). A more recent investigation (Head, 1966b) has confirmed the presence of the pyrethrins throughout the

TABLE I

Weights of Flower Heads Before and After Drying at Different Stages of Development[a]

Stage of flower development						
Stage number	Description	Approximate time for development (days)	Average fresh weight (mg)	Average dry weight (mg)	Total Pyrethrins (mg)	Pyrethrins dry wt. (%)
1	Well-developed closed buds	0	178	52	0.40	0.76
2	Ray florets vertical	12	418	119	1.78	1.49
3	Ray florets horizontal; first row of disc florets open	16	480	126	2.48	1.97
4	Approximately three rows of disc florets open	19	615	164	3.45	2.10
5	All disc florets open; fully mature	21	689	195	3.89	2.00
6	Early overblown condition; color of disc florets diminishing but ray florets still intact	31	716	253	3.92	1.55
7	Late overblown condition; little color remaining in disc, florets dried out	43	666	347	4.02	1.16
8	Disc florets fallen, stems dry $\frac{1}{2}$ in below head—suitable for collection for seed	60	321	280	[b]	[b]

[a] Clone 1708 (Head, 1966b).
[b] Not analyzed.

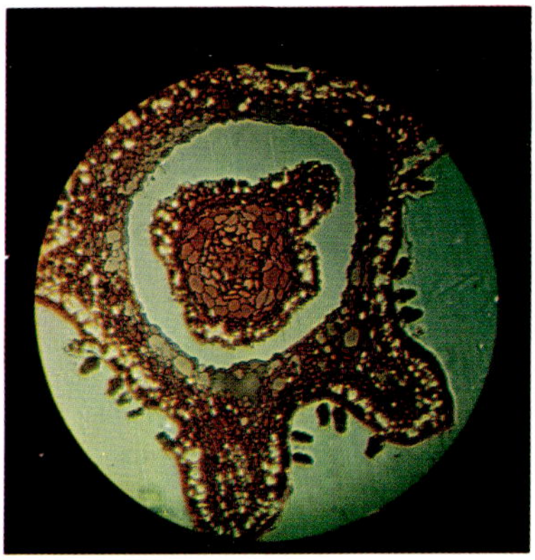

(a)

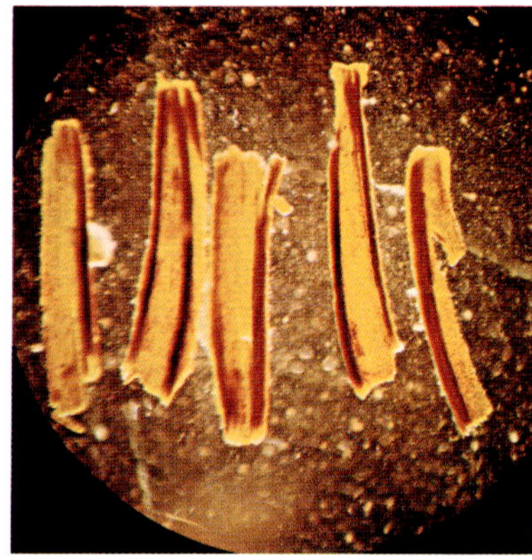

(b)

(c)

Fig. 3. Color photographs of different stages or portions of the flower head. (*a*) Transverse section of an achene. The large gray cells toward the inner wall of the achene contain pyrethrins (Courtesy of Jan Brewer). (*b*) Strips of the achene wall stained with 2,4-dinitrophenylhydrazine to show the location of pyrethrins in the secretory ducts (Courtesy of Jan Brewer). (*c*) Stages of development of the flower head numbered to correspond with Table I.

whole of the plant, the greatest concentration being located in the flower head. The development of a typical flower head is shown in Fig. 3c and the pyrethrins concentration and average flower head weight at each stage of development are recorded in Table I. Although the weight of pyrethrins in a single flower head increases with maturity, the pyrethrins content (percent pyrethrins w/w in the dry flowers) is at an optimum at stage 4. In practice, only flower heads with horizontal petals (stages 3 to 5, inclusive) are harvested (Head, 1963a). The interval between picking is regulated so as to keep to a minimum the number of overblown flowers (stages 6 to 7) which tend to reduce the plants' flower-bearing capabilities and hence the yield of pyrethrins.

Flowers at harvest contain approximately 80% water and are dried to contain around 10% moisture, either in the sun or by mechanical driers. Liquid extracts of the ground flowers are usually prepared using countercurrent extraction procedures with a light petroleum solvent. The primary extract yields, after concentration, a dark viscous oleoresin containing approximately 30% pyrethrins (oleoresin concentrate). This is sometimes diluted with mineral oil and standardized to contain 25% pyrethrins (oleoresin extract). The demands of the aerosol industry have resulted in the development of various refining processes that convert the dark viscous oleoresin concentrate into a decolorized and dewaxed refined extract. The initial step in the refining process is usually the extraction of oleoresin concentrate with methanol. The methanol extract is further purified and decolorized with charcoal to give the final product. In addition to their use in extract production, pyrethrum flowers are finely ground for use in the manufacture of mosquito coils. The chemistry, analysis, and mode of action of the pyrethrins have been the subject of a recent review (Matsui and Yamamoto, 1971).

II. Composition of Pyrethrum Extract

The relative proportion of the pyrethrins components in extracts from different plants can vary (Head, 1967a) but commercial pyrethrum extract, which is derived from many different plants, is relatively constant in composition. The mean ratio of pyrethrins I to pyrethrins II for Kenyan refined extract is a little over unity, although individual production runs show considerable variation (see Table II). Analysis of pyrethrins by the Association of Official Agricultural Chemists (AOAC) method gives a measure of pyrethrins I and pyrethrins II, whereas gas–liquid chromatography (GLC) gives the cinerin, jasmolin, and pyrethrin composition of each fraction. Analyses of a typical oleoresin concentrate are recorded in Table III (Head, 1969a).

TABLE II

Variation in Pyrethrins I to Pyrethrins II Ratio in Refined Pyrethrum Extract of Kenya Origin

Range py I/py II	Number
0.90–0.95	4
0.95–1.05	46
1.05–1.15	34
1.15–1.25	14
1.25–1.35	5
Mean py I/py II 1.08	103

TABLE III

Composition of the Pyrethrins in a Typical Pyrethrum Oleoresin[a]

Component	%	Group	Total
Cinerin I	2.2%	pyrethrins I 14.8%	total pyrethrins 30%
Jasmolin I	1.2%		
Pyrethrin I	11.4%		
Cinerin II	3.5%	pyrethrins II 15.2%	
Jasmolin II	1.2%		
Pyrethrin II	10.5%		

[a] Head (1969a).

A. Chemical Components

1. *Sesquiterpenes*

Several sesquiterpenes have been identified in pyrethrum flowers. Pyrethrosin (chrysanthin) was first isolated in 1937 by Rose and Haller and has been the subject of a detailed investigation by Barton and his co-workers (Barton and de Mayo, 1957; Barton *et al.,* 1960). The isolation of three new sesquiterpenoid lactones, chrysanin, dihydro-β-cyclopyrethrosin, and chrysanolide, has been described recently (Doskotch *et al.,* 1971). Another investigation reports the presence of β-cyclopyrethrosin (Doskotch and El-Faraly, 1969) and mentions the presence of sesamin (a synergist for pyrethrins) in alcohol extracts from pyrethrum flowers. The relative proportion of these materials in pyrethrum flowers is unknown but pyrethrosin is likely to predominate. Pyrethrosin has low solubility in light petroleum and it is probable that only low levels occur in pyrethrum extract.

2. *Flavonoids*

A recent investigation by Glennie and Harborne (1972) has shown the presence of the 7-glucosides and 7-glucuronides of apigenin, luteolin, and quercetin and also the 3,6-dimethyl ether and 3,6,4-trimethyl ether of quercetagetin. Of these compounds, only the two methyl ethers were detected in low concentration in pyrethrum extract.

3. *Triterpenols and Sterols*

Pyrethrol was first isolated from pyrethrum flowers by Fujitani in 1909. Its presence in pyrethrum extract was noted by Staudinger and Ruzicka (1924) but it was not until 1966 that Herz and Mirrington established its identity as taraxasterol, since confirmed in two further publications (Elliott and Stephenson, 1966; Castille, 1966).

Fukushi in 1952 investigated the composition of the unsaponifiable matter present in a waxlike residue from commercial extract production. He isolated a substance that he thought to be ceryl alcohol and a sterol that was identified as β-amyrin. Recent investigations (Stevenson, 1972) have shown, in addition to β-amyrin and taraxasterol, the presence of two additional triterpenols and four sterols. The mixed triterpenol and sterol fraction represents some 5% of the oleoresin concentrate and approximately 3% of the refined concentrate.

4. *Alkanes*

The alkane fraction from pyrethrum extract was examined in 1955 by Wanless, King, and Ritter. The mixed alkanes, which were isolated from a factory residue, were separated by column chromatography, and the fractions were examined by x-ray diffraction and mass spectrometry. All members of the series from n-C_{24} to n-C_{36} were reported and low-weight recoveries indicated the possible presence of high members of the series. The major components found were n-heptacosane, n-nonacosane, and n-hentriacontane. The presence of n-nonacosane and n-hentriacontane in pyrethrum waxes was also reported by Fukushi (1952). Gas–liquid chromatographic analysis of the alkane fraction (Head, 1969a,b) shows two series of components, one containing all the members of the series from n-C_{23} to n-C_{31} and the other containing a series of unknown components with high molecular weights. The total alkane content of pyrethrum oleoresin concentrate is about 4% by weight and this is reduced to 0.2% in refined extract. The composition of the normal alkane series, which together represents the major part of the fraction, is shown in Table IV.

5. *Fatty Acids*

Ripert, in 1934, claimed the presence of a number of free acids in an ether extract of pyrethrum flowers, including chrysanthemic, chrysan-

TABLE IV

Distribution of the *n*-Alkanes in the Alkane Fraction from Pyrethrum Oleoresin Concentrate[a]

Component	Percent of total
n-$C_{23}H_{48}$	3.0
n-$C_{24}H_{50}$	1.6
n-$C_{25}H_{52}$	25.3
n-$C_{26}H_{54}$	1.2
n-$C_{27}H_{56}$	15.5
n-$C_{28}H_{58}$	3.4
n-$C_{29}H_{60}$	40.6
n-$C_{30}H_{62}$	2.3
n-$C_{31}H_{64}$	7.0

[a] Head (1969a).

themum dicarboxylic, isovaleric, caproic, lauric, palmitic, oleic, and linoleic acids. He stated that all these acids were also present in the combined state. In 1937, Acree and LaForge reported the presence of palmitic and linoleic acids combined with pyrethrolone, and the presence of tiglic acid has also been suggested (Katsuda and Tikamoto, 1958).

A recent investigation (Head, 1968a) has shown that combined fatty acids represent up to 40% by weight of the pyrethrum oleoresin concentrate, approximately one half of this being non-hydroxy fatty acids. The composition of the latter as given by GLC analysis is recorded in Table V, where it can be seen that palmitic, stearic, oleic, linoleic, and linolenic acids together constitute up to 90% of this acid fraction. The composition of the remaining acids, which appear to be hydroxy fatty acids, has not as yet been investigated.

In refined pyrethrum extract the combined fatty acids are reduced to around 20% of the extract composition, of which some 12% are non-hydroxy fatty acids. The proportions of the major non-hydroxy fatty acids present in oleoresin and refined extract are shown in Table VI.

6. *Carotenoids*

The presence of β-carotene in pyrethrum extract was reported by Subbaratnam and Parsameswaran Pillay (1947), and the use of residues from the manufacture of refined pyrethrum extract as a source of xanthophyll for addition to chicken feed has also been described (Head, 1966c; Plebluda *et al.*, 1967). Recent work (Head, 1969a,b) has shown that up to 90% of the carotenoid fraction is related to cis and trans isomers of the xanthophyll lutein. These occur mainly as diesters but are also

TABLE V

Fatty Acid Composition of a Commercial Pyrethrum Extract Containing 30% Pyrethrins[a]

Common name	Code[b]	Percent acid (w/w)	Percent distribution
Myristic	14:0	0.25	1.2
Pentadecanoic	15 (unsat)	0.02	0.1
	15 (br)	<0.01	—
	15:0	0.05	0.25
Palmitic	16 (unsat)	0.04	0.2
	16 (br)	0.01	0.05
	16:0	4.64	22.8
Heptadecanoic	17 (unsat)	0.12	0.6
	17 (br)	0.02	0.1
	17:0	0.13	0.65
Stearic	18 (br)	0.04	0.2
	18:0	1.71	8.4
Oleic	18:1	2.09	10.3
Linoleic	18:2	6.62	32.5
Linolenic	18:3	2.94	14.5
Nonadecanoic	19 (unsat)	0.05	0.25
	19 (br)	0.01	0.05
	19:0	0.03	0.15
Arachidic	20 (unsat)	0.07	0.35
	20:0	0.34	1.7
Heneicosanoic	21:0	0.04	0.2
Behenic	22:0	0.37	1.8
Tricosanoic	23:0	0.07	0.35
Lignoceric	24:0	0.36	1.8
Pentacosanoic	25:0	0.05	0.25
Hexacosanoic	26:0	0.15	0.7
Heptacosanoic	27:0	0.02	0.1
Octacosanoic	28:0	0.07	0.35
Nonacosanoic	29:0	<0.01	—
Triacontanoic	30:0	0.02	0.1
Total		20.34	

[a] Head (1968b).
[b] Br = branched; unsat = unsaturated.

present as monoesters and the free diols. The carotenoid hydrocarbons α- and β-carotene represent only 1% of the carotenoid fraction. The analysis of an oleoresin concentrate is recorded in Table VII. The carotenoid content of the concentrate is between 0.8 and 0.9%, expressed in terms of free xanthophyll, or between 1.3 and 1.5%, expressed as the naturally occurring esters. In refined extract the carotenoid level is reduced to

TABLE VI

Comparison of the Major Non-Hydroxy Fatty Acids Present in Oleoresin and Refined Extract

Sample	Approximate total fatty acids (%)	Distribution (%)				
		Palmitic	Stearic	Oleic	Linoleic	Linolenic
Oleoresin concentrate 30% py	20	25.8	9.5	11.6	36.7	16.4
Refined concentrate 55% py	12	30.0	6.7	15.0	36.7	11.6

TABLE VII

Carotenoid Analysis of Pyrethrum Oleoresin Concentrate[a]

Carotenoid	Wt. %	Percent of total
α- and β-Carotene	0.01	1.2
"Neurosporene-like"	0.02	2.4
Lutein diesters[b]	0.57	69.5
"Flavoxanthin-like"	0.05	6.1
Lutein monoesters[b]	0.14	17.1
Lutein	0.03	3.7
Total carotenoids	0.82	

[a] Head (1969a).
[b] Expressed as the free xanthophyll.

approximately 0.04% w/w. Chlorophyll or its breakdown products are also present in oleoresin extract (Head, 1969a,b) but absent in refined material.

The known components of pyrethrum oleoresin and refined extract are summarized in Table VIII.

B. Isolation of Pyrethrins

Many methods for the separation of pyrethrins by column chromatography have been described, but most of this work was carried out before the advent of GLC with the result that the jasmolins were not detected. Separation by absorption chromatography either on alumina (Lord *et al.*, 1952) or on a silica gel–plaster of Paris mixture (Spickett, 1957) is of value in obtaining concentrates of pyrethrins I and pyrethrins II.

TABLE VIII

The Known Components of Pyrethrum Oleoresin and Refined Extract

Component	Approximate percent composition, w/w	
	Oleoresin	Refined
Total pyrethrins	30–35	50–55
Non-hydroxy fatty acids	20	12
Hydroxy fatty acids	20	8
Alkanes	4	0.4
Carotenoids (as free xanthophylls)	0.8	0.04
Sterols and triterpenols	5	3
Sesquiterpenes	trace	unknown
Flavonoids	trace	unknown
Chlorophylls	0.1	nil
Ceryl alcohol	unknown	unknown

Purified extracts should be employed; commercial refined concentrates containing approximately 50% pyrethrins or concentrates obtained by the nitromethane method (Barthel *et al.*, 1944) are both suitable. Pure pyrethrin I can be isolated from the pyrethrins I concentrate using a displacement chromatographic method (Godin *et al.*, 1966). This method, however, fails to give either pure cinerin I or jasmolin I. Jasmolin II, cinerin II, and pyrethrin II can be isolated from the pyrethrins II fraction by repeated chromatography on a silica gel–plaster of Paris column (Godin *et al.*, 1965). Mixtures of jasmolin I and pyrethrin I or cinerin I and jasmolin I can be separated with good yield by preparative GLC (Beevor *et al.*, 1965; Godin *et al.*, 1966).

A simpler method that allows the separation of all six constituents has been described by Rickett (1972) in which liquid–liquid partition chromatography is employed. Nitromethane supported in Celite is used as the stationary phase with hexane and carbon tetrachloride–hexane (1:3) being used to elute the pyrethrins I and pyrethrins II components, respectively. A typical separation for a column loaded with more than 270 mg total pyrethrins is shown in Fig. 4. The separation of small amounts of the individual esters can be achieved by thin-layer chromatography (TLC), using a continuous development technique (Stahl and Pfeifle, 1966) or multiple development (Abe and Fujita, 1971).

When large quantities of pyrethrin I and pyrethrin II are required, reconstitution from chrysanthemic and pyrethric acid chlorides and pyrethrolone should be considered (Elliott and Janes, 1969). These com-

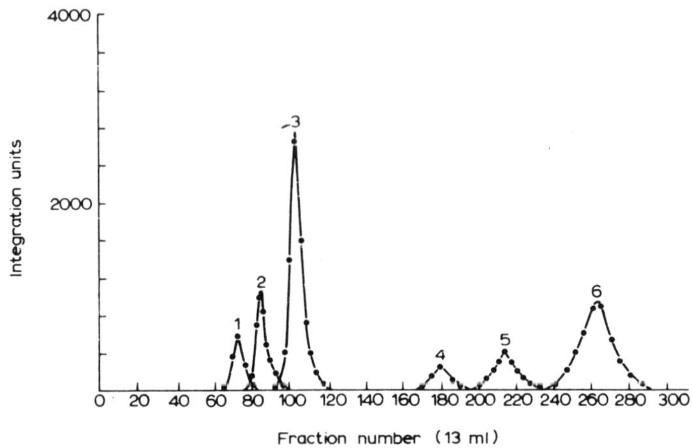

Fig. 4. Liquid–liquid partition chromatographic separation of 270 mg total pyrethrins on a 1000 × 38 mm column using GLC for analyzing the content of eluate fractions. 1, Jasmolin I; 2, cinerin I; 3, pyrethrin I; 4, jasmolin II; 5, cinerin II; 6, pyrethrin II. [Reproduced from F. E. Rickett, Preparative scale separation of pyrethrins by liquid–liquid partition chromatography, *J. Chromatogr.* **66** (1972) 356–360, Fig. 3.]

ponents are best obtained by isolation from pyrethrum extract and methods avoiding structural alteration are well established (Elliott, 1964).

C. Physical Data and Storage of the Pyrethrins

The pyrethrins are all clear, viscous, high-boiling gums. Their boiling points, refractive indices, and ultraviolet (UV) absorption data are recorded in Table IX. Optical rotation data for the pyrethrins and cinerins are available (Elliott, 1960) but vary considerably, depending on the method of isolation. Elliott (1961) has detailed the differences between the infrared (IR) spectra of the cinerins and pyrethrins, the major points being that cinerin I and pyrethrin I show a single absorption peak at 8.7 μm, whereas cinerin II and pyrethrin II show two peaks of the same intensity at 8.5 and 8.7 μm (differences in the acid moiety). Also pyrethrin I and pyrethrin II differ from cinerin I and cinerin II by possessing an additional sharp peak at 11.0 μm [—CH_2 out-of-plane deformation in the conjugated vinyl (—CH=CH_2) group]. The IR spectra of the jasmolins are quoted as being very similar to those of the cinerins (Godin *et al.*, 1966). Nuclear magnetic resonance spectra (Bramwell *et al.*, 1969) are discussed by Elliott and Janes in this volume (Chapter 4).

TABLE IX

Physical Constants of the Pyrethrins

Name	Formula	Molecular weight	Boiling point (°C)[a]	n_D^{20a}	λ_{max} (nm)	ϵ
Pyrethrin I	$C_{21}H_{28}O_3$	328.4	146–148 at 2×10^{-3} mm Hg	1.5237	222.5–223[b]	38,650
Pyrethrin II	$C_{22}H_{28}O_5$	372.4	196–198 at 7×10^{-3} mm Hg	1.5370	227–228[b]	47,350
Cinerin I	$C_{20}H_{28}O_3$	316.4	136–138 at 8×10^{-3} mm Hg	1.5064	220[b]	21,000
Cinerin II	$C_{21}H_{28}O_5$	360.4	182–184 at 1×10^{-3} mm Hg	1.5183	229[b]	28,300
Jasmolin I	$C_{21}H_{30}O_3$	330.4	—	—	219[c]	21,500
Jasmolin II	$C_{22}H_{30}O_5$	374.4	—	—	229[c]	22,900

[a] Data on reconstituted components (Elliott, 1961).

[b] λ_{max} and ϵ in *n*-hexane, summarizing data of Ward (1953) and Sawicki and Thain (1961) carried out on material isolated by column chromatography.

[c] λ_{max} and ϵ in ethanol (Godin *et al.*, 1966).

The stability of pyrethroid concentrates and of the pure esters has been reviewed by Godin (1968). As the pyrethrins approach 100% purity their stability appears to decrease, pyrethrins I and II being more unstable than the cinerins and jasmolins. Godin recommends that individual pyrethrins be stored at about −25°C in tightly stoppered glass containers that are wrapped in tinfoil and sealed in polythene. Chlorinated hydrocarbons, alcohols, and ethers are not suitable solvents and the compounds are best stored in acetone at a concentration of 10 mg/ml. The stored esters can be checked for polymer formation by TLC (Stahl, 1960) and if necessary should be purified by column chromatography on alumina using benzene as the solvent (Elliott and Janes, 1969).

III. Analysis of Pyrethrins

There can be no doubt that the successful exploitation of pyrethrum in the 1930's was closely related to the development of analytical methods for the active constituents that gave reasonable concordance with biological activity. Early methods have been reviewed by Gnadinger (1936) and were of three types: (*a*) the so-called acid methods, in which the pyrethrins are saponified followed by isolation of the chrysanthemum acids, which are estimated either together or separately; (*b*) isolation of the pyrethrins semicarbazones followed by estimation by nitrogen determination; and (*c*) reduction methods, utilizing alkaline copper solutions or potassium ferricyanide. Methods (*b*) and (*c*) give a measure of total pyrethrins, whereas the acid methods can be adapted to give estimates of what were then considered to be pyrethrin I and pyrethrin II. Only the acid methods (in refined forms) survive and these are applied in commerce to the analysis of both extract and flowers.

A. Commercial Methods of Analysis

All the acid methods rely on a separation of the chrysanthemum acids from fatty acids through the solubility of their barium salts. In the Seil method (1934) chrysanthemic acid is separated by steam distillation and chrysanthemum dicarboxylic acid is extracted with ether, both acids being determined by alkalimetry. Loss of chrysanthemic acid occurs during distillation because of its partial conversion to a nonvolatile saturated hydroxy acid, and although a modification to reduce this loss has been described (Mitchell *et al.*, 1948) the method is now rarely employed. The present Japanese official method (1940, 1942) is a modified form of Seil assay.

Wilcoxon (1936) investigated the reaction of Denigès reagent (acidic mercuric sulfate) with chrysanthemic acid, a reaction that had previously been reported by Seil (1934) and also Audiffren (1934). The reduction to mercurous sulfate, which is accompanied by a series of characteristic color changes, was estimated by iodometry. The reaction, further modified by Holaday (1938), is the basis of the pyrethrins I determination in todays' so-called mercury reduction method. The first official method of the AOAC based on the mercury reduction procedure was adopted in 1940 and, with a revision of the pyrethrin I factor in 1945, remained substantially unchanged until the appearance of the tenth edition in 1965. Methods developed and adopted in Europe [Pyrethrum Board of Kenya (PBK) method (1954); Tanganyika Extract Co. Ltd. (TECO) method (1954); British Pharmaceutical Codex method (1954); British Veterinary Codex method (1953)] using the same principle gave results for pyrethrins I some 20% higher than those recorded by the AOAC method. The major difference between the methods is at the acidification stage after the precipitation of insoluble barium salts. The AOAC method employs sulfuric acid with the subsequent separation of barium sulfate, which is filtered, whereas the PBK method utilizes hydrochloric acid, which gives a substantially clear solution. Examination of the AOAC procedure (Mitchell, 1953; Beckley and Hopkins, 1954) shows that some loss of chrysanthemic acid occurs by occlusion on a precipitate of insoluble acids removed during filtration of the barium sulfate. There was considerable

TABLE X

Comparative Pyrethrins I Analysis[a]

Determination: Isolation method:	GLC AOAC	GLC PBK	Denigès AOAC	Denigès PBK
	10.3	11.1	11.6	14.2
	10.6	11.0		

[a] Head (1968a).

discussion on the validity of the respective methods and for a brief period the AOAC seventh edition was modified to allow the use of hydrochloric acid (Mitchell, 1953). It is now well established that the PBK procedure definitely gives high results. Brierley and Brown (1962) examined the acidic material determined in the PBK procedure as pyrethrins I and estimated that the true figure could not exceed 75% of that normally recorded by the method. Although there is loss of chrysanthemic acid in the AOAC procedure, this is more than balanced by inclusion of extraneous material reacting with Denigès reagent. A recent comparison (Head, 1968a) of chrysanthemic acid recovered by the AOAC and PBK isolation procedures using GLC for estimation is recorded in Table X.

In the AOAC tenth edition the procedure for pyrethrins II is modified so that only water-soluble acidic material is estimated. This additional step is very effective in the analysis of crude extracts in which results can be up to 4% lower (McClellan, 1963). However, for refined extract, in which the difference is reduced, the inclusion of this modification can give slightly higher results, presumably because of absorption of carbon dioxide during the filtration step. The true figure for pyrethrins II is likely to be at least 10% lower than that recorded by the mercury reduction methods (Byrne *et al.*, 1965; Brierley and Brown, 1964; Head, 1963b; Abe and Fujita, 1971). A detailed comparison of the mercury reduction methods has been carried out by Shah (1970).

B. False Pyrethrins

Material registered as pyrethrins by the mercury reduction methods that are biologically inactive are known as *false pyrethrins*. The mercury reduction methods include a dewaxing step for those extracts not giving a clear solution in light petroleum but no account is taken of the free acid content which certainly contributes to errors in the analysis. *False pyrethrins* are best removed by chromatographic purification on alumina

(Brown *et al.*, 1954) but for satisfactory results particular care should be taken in grading the alumina, and it is wise to check for complete elution of pyrethrins II by TLC. The level of *false pyrethrins* in oleoresin concentrate registered by this method can vary considerably depending on the nature of the flowers and the conditions of extraction but is usually around 10%, whereas refined extracts contain between 3 and 5%.

It should be noted that the mercury reduction procedure has an inherent error in the figure recorded for pyrethrins II. A highly purified sample of allethrin, a synthetic homolog of cinerin I, analyzed by the AOAC, seventh edition method, recorded a figure of 8% in the pyrethrins II determination (Moore, 1954); similar figures have been observed for pure pyrethrin I (S. W. Head, unpublished work). It has not been established whether the modifications to the pyrethrins II procedure set out in the AOAC tenth edition help to eliminate this deficiency.

C. 2,4-Dinitrophenylhydrazine (DNP) Methods

A method based on the determination of the ketonic group in the alcohol moiety through 2,4-dinitrophenylhydrazone formation was first suggested by Moore (1956), improved by Smith (1959), and subsequently modified to give a technique suitable for routine operation (Head, 1963b). The 2,4-dinitrophenylhydrazones of pyrethrins I and pyrethrins II were prepared under standardized conditions and chromatographed on alumina to give two easily discernible bands that were collected separately and determined from their spectral absorption at 377 nm. The method gave lower results than the mercury reduction method, and also a different ratio of pyrethrins I to pyrethrins II, but was considered to give a better measure of the active constituents. The possibility of alcoholysis of the pyrethrins II dinitrophenylhydrazones to give equivalent quantities of the methyl ethers, which chromatograph with the pyrethrins I derivatives, cast some doubt on the pyrethrins I–pyrethrins II ratios given by the method. When a longer column was used (Smith, 1960) it was possible to apply the method to the determination of pyrethrin I, cinerin I, pyrethrin II, and cinerin II. This work was carried out before the discovery of the jasmolins and their behavior in this four-component separation is unknown.

D. The Ethylenediamine (EDA) Method

This method was first developed for the analysis of allethrin (Hogsett *et al.*, 1953) and has been applied for many years to the analysis of pyrethrins, especially in process control (Wachs and Hanley, 1967). In the method, pyrethrum extract was refluxed with ethylenediamine with the liberation of pyrethric and chrysanthemic acids which were determined by nonaqueous titration with sodium methylate. Free acids in the pyrethrum extract were removed by chromatography on alumina prior to the analysis and the esters of fatty acids did not react. The method was quick and reproducible but in its simple form gave only a figure for total pyrethrins. Wachs and Hanley (1969) overcame this difficulty with an alkoxyl method for determining the pyrethrins II components. When used in conjunction with the EDA method both pyrethrins I and pyrethrins II could be determined. Unfortunately, no comparative studies between this method and the mercury reduction procedures have been given but it is known that for refined extract the EDA method gives a result some 10% lower than the AOAC method.

E. Comparison of Results Given by Different Analytical Methods

Results given by the mercury reduction assays, the EDA method, and the DNP method are compared in Table XI. These are compiled from data on the 1970 world standard pyrethrum extract (Head, 1970), which was a blend of refined extract from five sources standardized in several different laboratories. One can judge the difficulties experienced with the mercury reduction methods from the range of results recorded. The ratio between the AOAC and PBK methods can vary with both the source and quality of the extracts. The AOAC result for refined extract is between 92 and 93% of the PBK figure and for oleoresin this level is reduced to around 89–90%. The lower DNP and EDA figures are likely to reflect the true level of pyrethrins in the extract. This has been confirmed by applying more specific measuring methods to the acids isolated by the techniques employed in the mercury reduction methods (Byrne *et al.*, 1965; Mitchell *et al.*, 1963; Head, 1968a; Abe and Fujita, 1971).

F. Spectrophotometric Methods

The UV spectrophotometric method of Beckley (1950) and the modified procedure of Shukis *et al.* (1951) have been widely employed for the rapid assay of pyrethrins. The method is applied to advantage to the analysis of pyrethrum flowers both for payment to growers and also for individual plant selection in crop improvement programs. It is now unusual to extract the pyrethrins from the ground flowers in

TABLE XI

Comparison of the Results Given by Different Analytical Methods, Compiled from Analytical Data on the 1970 World Standard Pyrethrum Extract[a]

Method	Number of laboratories	Pyrethrins I		Pyrethrins II		Total pyrethrins		Total pyrethrins relative to PBK
		Range (%)	Mean (%)	Range (%)	Mean (%)	Range (%)	Mean (%)	
PBK	5	12.8–13.5	13.3	9.3–10.4	9.7	22.4–23.6	23.0	100
AOAC 8th ed.	5	11.3–12.1	11.5	9.8–10.6	10.0	21.1–22.2	21.5	93
AOAC 9th ed.	5	11.0–11.8	11.4	9.3–10.4	10.1	20.5–22.0	21.5	93
AOAC 10th ed.	7	10.9–11.7	11.3	9.0–10.6	9.8	20.1–21.9	21.1	92
DNP	1	—	10.5	—	8.0	—	18.5	80
EDA	2	—	—	—	—	19.2–19.5	19.4	84

[a] Head (1970).

a Soxhlet apparatus and extraction at room temperature by agitation with light petroleum is preferred. The latter method removes approximately 95% of the total pyrethrins present in the sample and the reproducibility, although not as good as the Soxhlet technique, is adequate. After suitable dilution the optical density of the extract is determined at 225 nm. The method is normally calibrated against the AOAC mercury reduction assay and gives good linear regression lines. However, discrepancies can occur in samples that contain unusual amounts of extraneous materials which absorb at 225 nm or have an unusual ratio of cinerins to pyrethrins (Ward and Newham, 1962). Similarly, variations of up to 10% are shown in the apparent specific absorption of pyrethrins present in different grades of extracts. This variation is greatly reduced if the sample is initially chromatographed on alumina (Ward and Newham, 1962).

Infrared spectroscopy has been employed for the analysis of extracts using the carbonyl band at 5.80 μm for quantitation (McTaggart *et al.*, 1958).

Several colorimetric methods have been described and each has proved valuable in specific applications. The sulfur color (Cueto and Dale, 1953; Levy and Estrada, 1954; Brown and Wood, 1962) and DNP method (Head, 1964) in its rapid form determine total pyrethrins and are unaffected by synergists. The phosphoric acid or red color method (Williams *et al.*, 1956) determines only the pyrethrins and not the cinerins or jasmolins. The use of Denigès reagent (Schreiber and McClellan, 1954), which determines pyrethrins I, has also been described. All the colorimetric methods suffer the same deficiency as the UV spectrophotometric method in that the apparent response to pyrethrins in different grades of extract varies because of interference by extraneous materials.

G. Thin Layer Chromatography

Thin-layer chromatography is a valuable tool for the detection of pyrethrins. The method of Stahl (1960), in which pyrethrins I and pyrethrins II are separated on silica gel using ethyl acetate in light petroleum (1:4) for development, is generally employed. A typical separation achieved by this method is shown in Fig. 5. Stahl utilized the technique to demonstrate the rapid photolysis of pyrethrins and detected oxidation products. Thin-layer chromatography methods using continuous development (Stahl and Pfeifle, 1966) and multiple development (Abe and Fujita, 1971) have been applied to the difficult separation of the cinerin, jasmolin, and pyrethrin components. In theory, pyrethrins

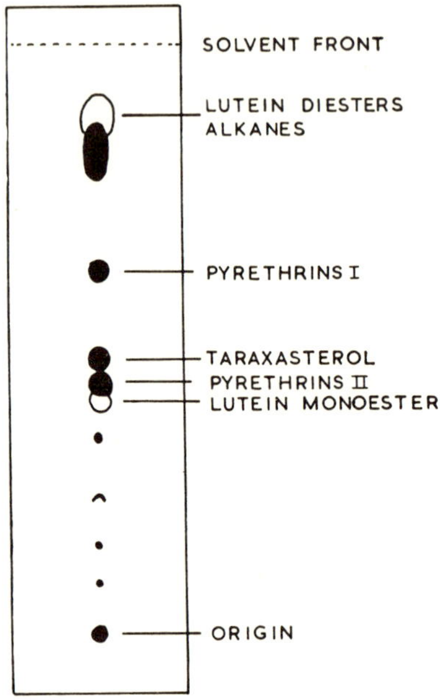

Fig. 5. Thin-layer chromatogram of pyrethrum extract on silica gel G. Solvent system, ethyl acetate in light petroleum (1:4). [Reproduced from S. W. Head, The composition of pyrethrum extract, *Pyrethrum Post* **10**(2) (1969) 17–21, Fig. 1.]

I and pyrethrins II could be isolated by TLC and determined by any of the colorimetric methods mentioned above. The use of the phosphoric acid method has been described in this application (Furmanec *et al.*, 1969) and also the more general vanallin–sulfuric acid color reaction (Tyihak, 1966).

H. Gas–Liquid Chromatography

The separation of pyrethrins by GLC was first achieved by Donegan *et al.* (1962) and has proved to be the most valuable advance in pyrethrins analysis in recent years. The pyrethrins respond to the electron-capture detector and the sensitivity is such as to provide a method suitable for residue determination. The relative response of the six components to the electron-capture detector appears to be proportional to that given by the flame-ionization detector (Head, 1966a, 1967b). Degradation of the pyrethrin II component is best avoided by employing all-glass systems, short columns, and column packing with low levels of stationary phase. Many stationary phases have been employed but neopentylglycol succinate has proved to be of particular value. A recent publication (Rickett, 1972) indicates that modified silicone stationary phase OV225 may give superior results. When using flame ionization for detection, dibutyl phthalate and dicyclohexyl phthalate are suitable as internal standards. It is usual to employ a refined pyrethrum extract of known AOAC analysis as a standard using the total pyrethrins I and pyrethrins II response for the calculation. The cinerin–jasmolin–pyrethrin composition can then be estimated by simple proportion.

Abe and Fujita (1971) have attempted to analyze pyrethrum extract using the separated pyrethrins as standards and have observed differences in response to the flame-ionization detector between the pure components. Their results were considerably lower than the AOAC method, the main difference being in the figure for pyrethrins II.

I. High-Speed Liquid Chromatography

It is almost certain that recent advances in liquid chromatography will prove to be valuable in pyrethrins analysis. The technique offers hopes of high resolution without the possibility of pyrethrins degradation so often experienced with GLC techniques. A separation on alumina (Fig. 6) detecting the eluent at 225 nm in a 1-mm flow-through cell gives an indication of the potential of the method. In this system, quantitation can be achieved either by area comparison with a standard or by UV measurements on the collected fractions (Cornelius, 1954).

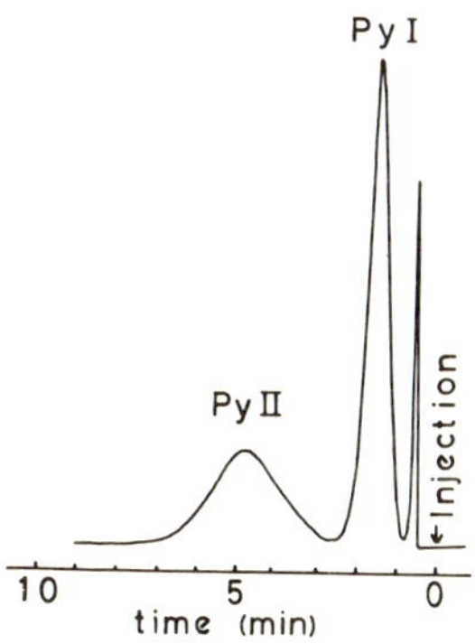

Fig. 6. Separation of pyrethrins on alumina. Column, 300 mm by 2.8 mm i.d. packed with alumina (neutral, grade 3, 170–200 mesh). Solvent, acetone in *n*-hexane (2.5:97.5 by volume). Flow rate, 3.5 ml/min. Sample, 5 μl of *n*-hexane solution containing 20 mg pyrethrins per milliliter. Detection by UV absorption at 225 nm using a 1-mm flow-through cell.

J. Analysis of Pyrethrins in Formulations

Pyrethrins are commonly formulated together with a synergist as dusts, oil- or water-based sprays, or mosquito coils in which finely ground pyrethrum flowers and an organic filler are bound with a water-soluble swelling gum.

The only method of assay generally applicable to such formulations is GLC, using electron capture for detection. The method requires a solution of pyrethrins in either hexane or acetone containing approximately 2 ppm pyrethrins for injection into the gas chromatograph. Oil-based formulations usually only require dilution. When analyzing water-based formulations, dilutions should be carried out in acetone and water removed at the final dilution step by vacuum treatment at 40°C. The pyrethrins can usually be extracted from dusts in a Soxhlet apparatus with acetone or *n*-butanol. The latter must be employed for the analysis of mosquito coils. The GLC method can also detect the presence of certain chlorinated hydrocarbons and retention values relative to pyrethrin I are given in Table XII. Under the conditions described piperonyl butoxide and other common synergists are not detected. Gas–liquid chromatography with flame ionization is also of value but is somewhat limited

TABLE XII

Retention Characteristics of the Pyrethrins and Some Other Insecticides (Pyrethrin I = 100)[a]

Insecticide	Relative retention
Lindane	16
Allethrin	35
Dieldrin	44
Cinerin I	53
Jasmolin I	69
Pyrethrin I	100
p,p'-DDT	123
Tetramethrin	299
Methoxychlor	334
Cinerin II	425
Jasmolin II	525
Pyrethrin II	765

[a] Conditions of chromatography: Instrument, Aerograph 204 with electron-capture (EC) detector; glass column 75 cm by 2 mm i.d. packed with 2% neopentylglycol succinate on 80–100 mesh Chromosorb WHMDS. Detector temp., 185°C; injection temperature, 180°C. Nitrogen flow rate, 60 ml/min. Range E.C. 10, attenuation ×2. Column temperature, 150°C; data for cinerin II, jasmolin II, and pyrethrin II obtained at 180°C.

in that (*a*) the method requires solutions for injection at fairly high pyrethrins concentration and (*b*) other components can interfere with the pyrethrins response. However, the method can detect synergists and is therefore of particular value in checking concentrates (Bevenue *et al.*, 1970). Methods estimating total pyrethrins from the pyrethrin I response are open to doubt because of possible variation in the pyrethrins I to pyrethrins II ratio. The DNP method in both its full and shortened form can be of value in formulation analysis but the mercury reduction methods are rarely satisfactory, the presence of emulsifiers causing particular difficulties.

K. Analysis of Pyrethrins in the Presence of Synthetic Pyrethroids

There are three major synthetic pyrethroids presently available commerically that could be found blended with natural pyrethrins, namely, allethrin, resmethrin, and tetramethrin (Fig. 7). These are all esters of chrysanthemic acid that may be in the (±)-cis,trans form or for improved biological activity have the (+)-trans structure (the isomer shown in Fig. 7). Thin-layer chromatography separation of these pyrethroids on silica gel using ethyl acetate–light petroleum (1:9) for development gives a clear separation of resmethrin (R_f 0.69) from pyrethrins I (R_f 0.42) and pyrethrins II (R_f 0.19). Allethrin is coincident with pyrethrins I but tetramethrin (R_f 0.26), although clearly separated from the pyrethrins, merges with the taraxasterol (R_f 0.27) present in the natural product.

The pyrethrins, allethrin, and tetramethrin all react to the EDA method, the latter giving recoveries in excess of 100%. Technical resmethrin gives a low response to the EDA method but the response

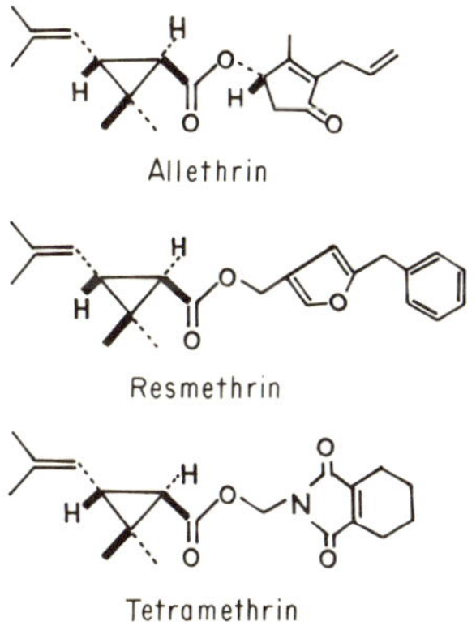

Fig. 7. Structures of three synthetic pyrethroid insecticide chemicals.

is sufficient to render the method unsuitable for the direct assay of pyrethrins–resmethrin mixtures. Only allethrin and pyrethrins react in the DNP method and this can therefore be of value for pyrethrins assay in samples admixed with resmethrin or tetramethrin. All three synthetic materials react in the mercury reduction assay to give an apparent pyrethrins I content. Resmethrin can be separated from the pyrethrins by careful chromatography on alumina, the separation being monitored by TLC. After the resmethrin has been separated the pyrethrins may be eluted in the normal way and determined by mercury reduction assay.

The pyrethrins and allethrin and tetramethrin respond to the electron-capture detector and can be easily separated on the neopentylglycol succinate column (Table XII). Resmethrin is also separated, being eluted after pyrethrin I, but is not detected, at least at low concentration, by the electron-capture detector.

IV. Summary

The pyrethrum plant of commerce, *Chrysanthemum cinerariaefolium*, contains six closely related insecticidally active esters collectively termed pyrethrins which are mainly located in the flower head. The nature and development of the flower head are discussed and related to practical recommendations for harvesting.

Components identified either in the flower head or commercial extracts include sesquiterpenes, flavonoids, triterpenols and sterols, alkanes, fatty acids, and carotenoids. Physical data for the individual pyrethrins are tabulated and methods for their isolation reviewed. The results given by present commercial mercury reduction methods of analysis are compared with those obtained by the 2,4-dinitrophenylhydrazone and ethylenediamine methods.

Methods of analysis requiring standardized extract for calibration, i.e., colorimetric, UV, TLC, and GLC methods, are reviewed. The suitability of these methods for the analysis of pyrethrins in formulations and in the presence of synthetic pyrethroids is discussed.

References

Abe, Y., and Fujita, Y. (1971). Studies on pyrethroidal compounds. Part 1. The quantitative determination of natural pyrethrins by gas-liquid chromatography. *J. Agr. Chem. Soc. Jap.* **45**, 22–28.

Acree, F., Jr., and LaForge, F. B. (1937). Constituents of pyrethrum flowers. X. Identification of the fatty acids combined with pyrethrolone. *J. Org. Chem.* **2**, 308–313.

Association of Official Agricultural Chemists. (1940). "Official and Tentative Methods of Analysis," 5th ed., pp. 66–67. Ass. Offic. Agr. Chem., Washington, D.C.

Association of Official Agricultural Chemists. (1945). Revision of method for pyrethrin I. *J. Ass. Offic. Agr. Chem.* **28**, 72.

Association of Official Agricultural Chemists. (1960). "Official Methods of Analysis," 9th ed. pp. 41–43. Ass. Offic. Agr. Chem., Washington, D.C.

Association of Official Agricultural Chemists. (1965). "Official Methods of Analysis," 10th ed., pp. 50–52. Ass. Offic. Agr. Chem., Washington, D.C.

Audiffren, M. (1934). Colour reaction of chrysanthemum monocarboxylic acid. *J. Pharm. Chim.* [8] **19**, 535–536.

Barthel, W. F., Haller, H. L., and LaForge, F. B. (1944). Pyrethrins for aerosols. *Soap* **20**, 121.

Barton, D. H. R., and de Mayo, P. (1957). Sesquiterpenoids. Part VIII. The constitution of pyrethrosin. *J. Chem. Soc., London* pp. 150–158.

Barton, D. H. R., Bockman, O. C., and de Mayo, P. (1960). Sesquiterpenoids. Part XII. Further investigations on the chemistry of pyrethrosin. *J. Chem. Soc., London* pp. 2263–2271.

Beckley, V. A. (1950). The spectrophotometric estimation of the pyrethrins. *Pyrethrum Post* **2**(1), 23–24.

Beckley, V. A., and Hopkins, J. (1954). Determination of pyrethrin I. *Soap Sanit. Chem.* **30**, 141.

Beevor, P. S., Godin, P. J., and Snarey, M. (1965). Jasmolin I, cinerin I, and a new method for isolating research quantities of pyrethroids. *Chem. Ind. (London)*, pp. 1342–1343.

Bevenue, A., Kawano, Y., and DeLano, F. (1970). Analytical studies of pyrethrin formulations by gas chromatography. *J. Chromatogr.* **50**, 49–58.

Bramwell, A. F., Crombie, L., Hemesley, P., Pattenden, G., Elliott, M., and Janes, N. F. (1969). Nuclear magnetic resonance spectra of the natural pyrethrins and related compounds. *Tetrahedron Lett.* **25**, 1727–1741.

Brewer, J. (1972). Personal communication. Government Research Station, Molo, Kenya.

Brierley, A., and Brown, N. C. (1962). Estimation of pyrethrins. *Soap Chem. Spec.* **38**, 105.

Brierley, A., and Brown, N. C. (1964). Estimation of pyrethrin II. *Soap Chem. Spec.* **40**, 149.

British Pharmaceutical Codex. (1954). The Pharmaceutical Press, London.

British Veterinary Codex. (1953). Supplement 1959, p. 53. The Pharmaceutical Press, London.

Brown, N. C., and Wood, M. C. (1962). Applications of the "sulphur colour" test in the estimation of pyrethrins. *Pyrethrum Post* **6**(4), 11–15.

Brown, N. C., Phipers, R. F., and Singleton, K. G. (1954). The estimation of "false" pyrethrins in pyrethrum extracts. *Pyrethrum Post* **3**(3), 3–5.

Byrne, J. H. N., Mitchell, W., and Tresadern, F. H. (1965). An infrared spectrophotometric method for the determination of "Pyrethrin II." *Analyst* **90**, 362–365.

Castille, A. (1966). Contribution à l'etude du pyrethrol. *An. Real. Acad. Farm.* **32**, 121–130.

Chandler, S. E. (1951). Botanical aspects of pyrethrum. General considerations; the seat of the active principles. *Pyrethrum* Post **2**(3), 1.

Chandler, S. E. (1954). Botanical aspects of pyrethrum II. Further observations. *Pyrethrum Post* 3(3), 6.

Contant, R. B. (1963). Possible use of chrysanthemum species in the genetic improvement of pyrethrum (*C. cinerariaefolium* Vis.). *Proc. East Afr. Acad.* **1**, 85–92.

Cornelius, J. A. (1954). A chromatographic procedure for the determination of "Pyrethrins" in pyrethrum extracts. *Analyst* **79**, No. 940, 458–460.

Cueto, C., and Dale, W. E. (1953). Colorimetric determination of pyrethrins, allethrin and furethrin. *Anal. Chem.* **25**, 1367–1369.

Donegan, L., Godin, P. J., and Thain, E. M. (1962). The separation and estimation of the insecticidal constituents of pyrethrum extract by gas-liquid chromatographic analysis. *Chem. Ind.* (*London*) **31**, 1420.

Doskotch, R. W., and El-Faraly, F. S. (1969). Isolation and characterization of (+)-sesamin and β-cyclopyrethrosin from pyrethrum flowers. *Can. J. Chem.* **47**, 1139–1142.

Doskotch, R. W., El-Faraly, F. S., and Hafford, C. D. (1971). Sesquiterpene lactones from pyrethrum flowers. *Can. J. Chem.* **49**, 2103.

Elliott, M. (1960). Pyrethrolone and related compounds. *Chem. Ind.* (*London*) pp. 1142–1146.

Elliott, M. (1961). The pyrethrins and related compounds. Part II. Infra-red spectra of the pyrethrins and of other constituents of pyrethrum extract. *J. Appl. Chem.* **11**, 19–23.

Elliott, M. (1964). The pyrethrins and related compounds. Part V. Purification of (+)-pyrethrolone as the monohydrate, and the nature of "Pyrethrolone-C." *J. Chem. Soc., London,* pp. 5225–5228.

Elliott, M., and Janes, N. F. (1969). Pyrethrin II and related esters obtained by reconstitution. *Chem. Ind.* (*London*) pp. 270–271.

Elliott, M., and Stephenson, H. (1966). "Pyrethrol"-taraxasterol—a note. *Pyrethrum Post* **8**(3), 27–28.

Fujitani, J. (1909). Chemistry and pharmacology of insect powder. *Arch. Exp. Pathol. Pharmakol.* **61**, 47–75.

Fukushi, S. (1952). Studies on the components of the unsaponifiable matter of the wax of *Chrysanthemum cinerariaefolium. J. Agr. Chem. Soc. Jap.* **26**, 1–2.

Furmanec, D., Schilling, F. A. E., and Brown, B. B. (1969). A rapid colorimetric assay for pyrethrins. *Pyrethrum Post* **10**, 21–23.

Glennie, C. W., and Harborne, J. B. (1972). Flavonoids of pyrethrum. *Pyrethrum Post* **11**(3), 82–84.

Gnadinger, C. B. (1936). "Pyrethrum Flowers," 2nd ed. McLaughlin Gormley King, Minneapolis, Minnesota.

Gnadinger, C. B. (1945). "Pyrethrum Flowers, supplement to 2nd ed. (1936–1945). McLaughlin Gormley King, Minneapolis, Minnesota.

Godin, P. J. (1968). The stability of pyrethroid concentrates and pure esters. *Pyrethrum Post* 9(4), 17.

Godin, P. J., Stevenson, J. H., and Sawicki, R. M. (1965). The insecticidal activity of jasmolin II and its isolation from pyrethrum (*Chrysanthemum cinerariaefolium* Vis). *J. Econ. Entomol.* **58**, 548–551.

Godin, P. J., Sleeman, R. J., Snarey, M., and Thain, E. M. (1966). The jasmolins, new insecticidally active constituents of *Chrysanthemum cinerariaefolium* Vis. *J. Chem. Soc., C* pp. 332–334.

Head, S. W. (1963a). An examination of the effect of picking methods on the pyrethrins content of dry pyrethrum flowers. *Pyrethrum Post* **7**(2), 3–9.

Head, S. W. (1963b). Determination of pyrethrins. *Soap Chem. Spec.* **39**, 97.

Head, S. W. (1964). Rapid method for the determination of "total pyrethrins" by spectroscopy of their 2,4-dinitrophenylhydrazones. *J. Sci. Food Agr.* **15**, 390–395.

Head, S. W. (1966a). The quantitative determination of pyrethrins by gas-liquid chromatography. Part I. Detection by electron capture. *Pyrethrum Post* **8**(4), 3–7.

Head, S. W. (1966b). A study of the insecticidal constituents in *Chrysanthemum cinerariaefolium*. (1) Their development in the flower head. (2) Their distribution in the plant. *Pyrethrum Post* **8**(4), 32–37.

Head, S. W. (1966c). The effect of a residue from the decolourisation of pyrethrum extract on yolk colour. *Poultry Sci.* **45**, 1269–1275.

Head, S. W. (1967a). A study of the insecticidal constituents of *Chrysanthemum cinerariaefolium* (3). Their composition in different pyrethrum clones. *Pyrethrum Post* **9**(2), 3–7.

Head, S. W. (1967b). The quantitative determination of "pyrethrins" by gas-liquid chromatography. Part II. Detection by hydrogen flame ionisation. *Pyrethrum Post* **9**(1), 12–17.

Head, S. W. (1968a). The quantitative determination of "pyrethrins" by gas-liquid chromatography. Part III. Through chromatography of the methyl esters of chrysanthemic acid and chrysanthemum dicarboxylic acid. *Pyrethrum Post* **9**(4), 31–36.

Head, S. W. (1968b). Fatty acid composition of extract from pyrethrum flowers (*Chrysanthemum cinerariaefolium*). *J. Agr. Food Chem.* **16**, 762–765.

Head, S. W. (1969a). The composition of pyrethrum extract. *Pyrethrum Post* **10**(2), 17–21.

Head, S. W. (1969b). Constituents of extracts from pyrethrum flowers; paraffins, fatty acids and carotenoids. Ph.D. Thesis, University of East Africa.

Head, S. W. (1970). World standard pyrethrum extract. *Pyrethrum Post* **10**(3), 27–31.

Herz, W., and Mirrington, R. N. (1966). Identification of pyrethrol with taraxasterol. *J. Pharm. Sci.* **55**, 104.

Hogsett, J. N., Kacy, H. W., and Johnson, J. B. (1953). Analysis of commercial allethrin. Determination of allethrin content by reaction with ethylenediamine. *Anal. Chem.* **25**, 1207–1211.

Holaday, D. A. (1938). Determination of pyrethrin I. *Ind. Eng. Chem., Anal. Ed.* **10**, 5–6.

Japanese Official Method for the Assay of Pyrethrins. (1940). *J. Agr. Chem. Soc. Jap.* **16**, 389.

Japanese Official Method for the Assay of Pyrethrins. (1942). *J. Agr. Chem. Soc. Jap.* **18**, 766.

Katsuda, Y., and Tikamoto, T. (1958). Studies on the degradation of pyrethrins. *Botyu-Kagaku* **23**, 60–63.

Levy, L. W., and Estrada, R. J. (1954). A colorimetric method for the determination of pyrethrins. *J. Agr. Food Chem.* **2**, 629–632.

Lord, K. A., Ward, J., Cornelius, J. A., and Jarvis, M. W. (1952). Chromatographic separation of the pyrethrins. *J. Sci. Food Agr.* **9**, 419–426.

McClellan, D. B. (1963). Analytical methods for determination of pyrethrins. *J. Ass. Offic. Agr. Chem.* **46**, 664–667.

McTaggart, N. G., Thornton, E., and Harford, A. D. (1958). The determination of pyrethrins and other insecticidal compounds by infra-red spectrometry. *Pyrethrum Post* **4**(4), 12–15.

Matsui, M., and Yamamoto, I. (1971). Pyrethroids. *In* "Naturally Occurring Insecticides" (M. Jacobson and D. G. Crosby, eds.), pp. 3–70. Dekker, New York.

Mitchell, W. (1953). Note on recent changes in the A.O.A.C. method for the assay of pyrethrins. *J. Sci. Food Agr.* **5**, 246–248.

Mitchell, W., Tresadern, F. H., and Wood, S. A. (1948). Analysis of pyrethrum flowers by the Seil and Wilcoxon-Holaday methods. *Analyst* **73**, 484–494.

Mitchell, W., Byrne, J. H. N., and Tresadern, F. H. (1963). Pyrethrum analysis: An infra-red method. *Analyst* **88**, 538–543.

Moore, B. P. (1954). The assay of "pyrethrin" and allethrin concentrates with 2:4-dinitrophenylhydrazine. *J. Sci. Food Agr.* **5**, 500–504.

Moore, B. P. (1956). Notes on the 2:4-dinitrophenylhydrazine method for pyrethrum assay. *J. Sci. Food Agr.* **7**, 740–744.

Plebluda, H. J., Trenton, N. J., Wertz, P. F., and Hoffman, W. H. (1967). Edible xanthophyll pigment; a by-product of pyrethrin manufacture. U.S. patent 3,333,962; *Chem.* Abstr. **65**, 19233c (1966).

Pyrethrum Board of Kenya. (P.B.K.). (1954). "Determination of Pyrethrins." Nakuru, Kenya.

Rickett, F. E. (1972). Preparative scale separation of pyrethrins by liquid-liquid partition chromatography. *J. Chromatogr.* **66**, 356–360.

Ripert, J. (1934). New method of analysis for products containing extracts of pyrethrum. *Ann. Fals. Fraudes* **27**, 580–595.

Rose, W. G., and Haller, H. L. (1937). Constituents of pyrethrum flowers. XI. Chrysanthin. *J. Org. Chem.* **2**, 484–488.

Sawicki, R. M., and Thain, E. M. (1961). Chemical and biological examination of commercial pyrethrum extracts for insecticidal constituents. *J. Sci. Food Agr.* **12**, 137–145.

Schreiber, A. A., and McClellan, D. B. (1954). Estimation of micro quantities of pyrethroids. *Anal. Chem.* **26**, 604–607.

Seil, H. A. (1934). Estimation of pyrethrins. *Soap* **10**, 89.

Shah, V. M. (1970). A comparison of the mercury reduction methods of pyrethrum analysis. *Pyrethrum Post* **10**(4), 27–35.

Shukis, A. J., Cristi, D., and Wachs, H. (1951). Spectrophotometric evaluation of pyrethrum. *Soap Sanit. Chem.* **27**, 124.

Smith, H. J. (1959). A modified 2,4-dinitrophenylhydrazone method for the determination of "Pyrethrin I" and "Pyrethrin II." *J. Sci. Food Agr.* **10**, 260–267.

Smith, H. J. (1960). Detection and estimation of the biologically active constituents of pyrethrum. *J. Sci. Food Agr.* **11**, 172–176.

Spickett, R. G. W. (1957). A method for the separation of the constituents of pyrethrum extract. *Chem. Ind.* (*London*) p. 561.

Stahl, E. (1960). Inactivation of pyrethrins at the site of action. *Arch. Pharm.* (*Weinheim*) **293**, 531–537.

Stahl, E., and Pfeifle, J. (1966). Thin layer and gas chromatography of pyrethrum-based insecticides. *Pyrethrum Post* **8**(4), 8–9.

Staudinger, H., and Ruzicka, L. (1924). Synthesis of the pyrethrins. *Helv. Chim. Acta* **7**, 448–458.

Stevenson, D. S. (1972). Application of liquid-gel chromatography to the analytical characterisation of pyrethrum extract. *Pyrethrum Post* **11**(3), 90–93.

Subbaratnam, A. V., and Parsameswaran Pillay, P. (1947). Carotenoid pigments of pyrethrum flowers. *J. Sci. Ind. Res., Sect. B* **6**, 100–102.

Tanganyika Extract Co. Ltd. (T.E.C.O.). (1954). "Method for the determination of 'Pyrethrins' (extract)."

Tyihak, E. (1966). Thin layer chromatographic investigation of pyrethrins. *Herba Hung.* **5**, 306–313.

Wachs, H., and Hanley, A. (1967). Pyrethrum analysis by E.D.A. method. *Soap Chem. Spec.* **43**, 78.

Wachs, H., and Hanley, A. (1969). Determining pyrethrins I:II ratios. *Soap Chem. Spec.* **45**, 107.

Wanless, G. G., King, W. H., and Ritter, J. J. (1955). Hydrocarbons in pyrethrum cuticle wax. *Biochem. J.* **95**, 684–690.

Ward, J. (1953). Separation of the "pyrethrins" by displacement chromatography. *Chem. Ind.* (*London*) pp. 586–587.

Ward, J., and Newham, G. (1962). A study of the spectrophotometric determination of the pyrethrins. *Pyrethrum Post* **6**(3), 34–38.

Wilcoxon, F. (1936). The determination of pyrethrin I. *Contrib. Boyce Thompson Inst.* **8**, 175–181.

Williams, H. L., Dale, W. E., and Sweeney, J. P. (1956). A new colorimetric method for pyrethrins. *J. Ass. Offic. Agr. Chem.* **39**, 872.

Chapter 4

Chemistry of the Natural Pyrethrins

MICHAEL ELLIOTT and NORMAN F. JANES

I. Introduction

This review covers developments in the chemistry of the natural pyrethrins during the 10 years since a previous survey (Crombie and Elliott, 1961). Results from physical methods of structure determination and analysis are emphasized. Other reviews (Elliott, 1969; Matsui and Yamamoto, 1971) consider particular aspects of the field.

The optical and geometrical isomerism shown by the natural esters and related synthetic compounds and a system of nomenclature have been fully described (Crombie and Elliott, 1961).

II. Structure of the Natural Esters

The insecticidal activity in fresh pyrethrum flowers is not diminished by drying at temperatures up to 80°C (Godin *et al.*, 1965). Extraction of such dried flowers by the usual process (Head, in Chapter 3 of this volume) and subsequent partition into nitromethane (Barthel *et al.*, 1944) separates all the active from the inactive constituents; the fraction insoluble in nitromethane neither increases nor diminishes the potency of the active insecticidal constituents (Sawicki and Thain, 1961).

Examination of the insecticidal fraction by column chromatography (Ward, 1953; Stephenson, 1960) and more recently by thin-layer chromatography (TLC) (Stahl and Pfeifle, 1966; Copius-Peereboom, 1969), analytical gas–liquid chromatography (GLC) (Donegan *et al.*, 1962; Head, 1964, 1966, 1967), preparative gas–liquid chromatography (Beevor *et al.*, 1965; King and Paisley, 1969), and liquid–liquid chromatography (Rickett, 1972) reveals six related compounds. These recent techniques have replaced earlier methods for examining or characterizing the constituents, such as those using the 2,4-dinitrophenylhydrazones (Smith, 1960), which are difficult to separate and crystallize; however, the semicarbazones are still important in isolating the natural rethrolones (Section III,B).

The insecticidal constituents were recognized as esters early in the history of the chemical investigations (Fujitani, 1909; Yamamoto, 1919, 1923; Staudinger and Ruzicka, 1924). Some of the physical properties of these compounds are discussed by Head (Chapter 3 in this volume) and they are further fully characterized by ultraviolet (UV) (Crombie and Elliott, 1961), infrared (IR) (Elliott, 1961), nuclear magnetic resonance (NMR) (Bramwell *et al.*, 1969) and mass (King and Paisley, 1969) spectral data. The strongest and most direct evidence for the

TABLE I

Chemical Shifts and Coupling Constants in NMR Spectra of Natural Pyrethrins

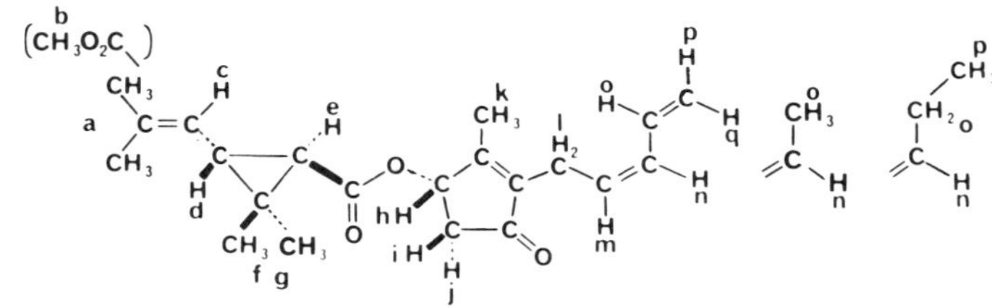

Ester	a	b	c	d	e	f	g	h	i	j	k	l	m	n	o	p	q
	Allocation																
Pyrethrin I	8.29	—	5.07	7.9[a]	8.62	8.88	8.76	4.31	7.12	7.81	7.97	6.88	4.7[a]	3.9	3.2	4.8[a]	4.8[a]
Cinerin I	8.27	—	5.07	7.9[a]	8.59	8.86	8.73	4.32	7.11	7.81	7.95	7.02	~4.5 (m)		8.28		
Jasmolin I	8.28	—	5.06	7.9[a]	8.59	8.85	8.73	4.33	7.13	7.8[a]	7.95	7.0[a]	~4.5 (m)		7.85	9.04	
Pyrethrin II	8.07	6.25	3.52	7.79	8.27	8.77	8.70	4.31	7.10	7.81	7.97	6.88	4.7[a]	3.9	3.28	4.8[a]	4.8[a]
Cinerin II	8.07	6.26	3.47	7.8[a]	8.26	8.77	8.70	4.29	7.10	7.80	7.97	7.02	~4.5 (m)		8.28		
Jasmolin II	8.05	6.25	3.50	7.8[a]	8.25	8.74	8.65	4.31	7.10	7.8[a]	7.95	7.01	~4.5 (m)		7.85	9.04	

Ester	J_{ac}	J_{cd}	J_{de}	J_{hi}	J_{hj}	J_{ij}	J_{lm}	J_{mn}	J_{no}	J_{op}	J_{oq}
	Coupling constant										
Pyrethrins I and II		8 ("I"s)[c]						10	10	10	16
Cinerins I and II	1.5	9.5 ("II"s)[d]	5.5	6	2	18	7	[b]	6		
Jasmolins I and II								[b]	[b]	7	

[a] Peak partially obscured. τ value quoted is approximate.
[b] Appropriate areas complex and not analyzed.
[c] For example, chrysanthemates.
[d] For example, pyrethrates.

structure of the six compounds now rests on their mass spectral molecular weights combined with detailed analysis of their NMR spectra.

The NMR spectrum of each of the six esters is a combination of the spectra of the acid and alcohol from which it is constituted, with changes only in the chemical shifts of the protons near the site of esterification (Bramwell *et al.*, 1969). Table I shows the chemical shift values of the various protons and most of the coupling constants thus established which confirm the structures accepted for the six esters. Chemical degradation had left the nature of the side chains of the alcohols in dispute. However, each proton in the pentadienyl side chains of pyrethrin I and pyrethrin II gives a separate signal with the chemical shift and multiplicity predicted, as do the terminal methyl and ethyl groups in the cinerins and jasmolins, respectively. The stereochemistry of the pentadienyl group in pyrethrins I and II is *Z* (cis) from the vicinal coupling constant (10 Hz) of the protons on C-2′ and C-3′, consistent with the relatively low extinction coefficient at 225 nm ($\Delta\epsilon_{diene}$ 22,000) for this system (Elliott, 1964b).

The NMR spectra also give the only direct proof of the stereochemistry of the olefinic side chain in pyrethrate esters by comparing chemical shifts (Crombie *et al.*, 1963) (Fig. 1) with those in related compounds of known stereochemistry.

Nuclear magnetic resonance spectra are also useful for determining the compositions of mixtures of pyrethroids. The intensities of the peaks at τ6.3 (methoxy) and τ3.6 (acid C=C*H*) are directly proportional to the quantity of esters of pyrethric acid (pyrethrins II) present. The size of the triplet at τ9.1 (CH_3CH_2) indicates the proportion of jasmolin I and jasmolin II. Similarly, pyrethrin I and pyrethrin II can be estimated from the almost unobscured olefinic peaks at about τ3.2 and τ3.9. Unfortunately, however, the characteristic doublet (CH_3CH=) at τ8.3 from cinerin I and cinerin II is obscured, so these esters can only be estimated

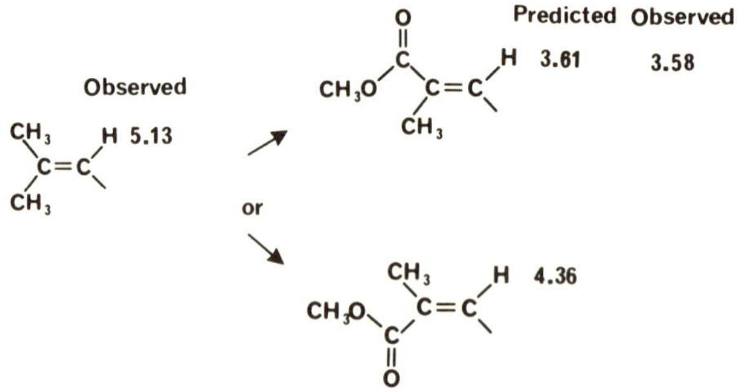

Fig. 1. Chemical shifts in the olefin side chain of pyrethrate esters.

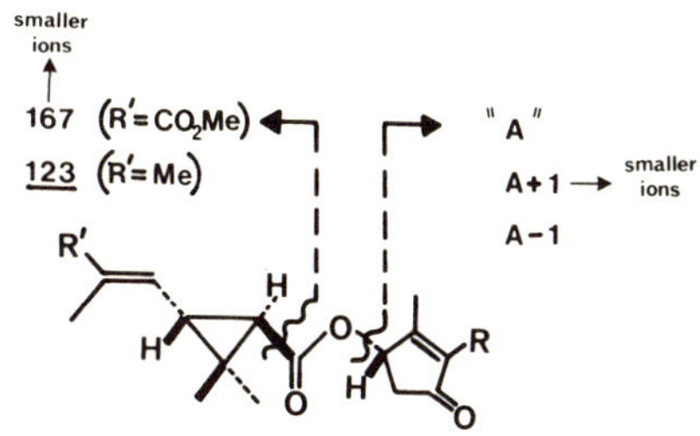

Fig. 2. Patterns of fragmentation of the six esters.

by difference or by examining the mixture of rethrolones obtained after ester cleavage when the interfering absorptions from the acids are removed. Thus NMR gives a rapid, but approximate, estimation of mixtures of pyrethroids to complement determinations by gas–liquid chromatography.

In the mass spectra of the six esters, the molecular ion peaks are small, but adequate to define their molecular weights. The patterns of subsequent fragmentations (King and Paisley, 1969) are indicated in Fig. 2. The central ester link is easily cleaved to give ions characteristic of the acid or alcoholic parts of the molecule. In chrysanthemates, the most intense peak (m/e 123) is produced when the bond to C-1 breaks; the analogous peak (m/e 167) from pyrethrates is relatively weaker because the methoxycarbonyl group induces subsequent fragmentations. The alcoholic moieties of the esters can give the fragment A (=161 for pyrethrins, 163 for jasmolins, and 149 for cinerins) (Fig. 2) directly by C-4–O cleavage or the fragments A + 1 or A — 1 by rearrangements involving hydrogen transfer. Further fragmentations give many peaks at smaller m/e values. A more detailed discussion of these and alternative fragmentations has recently been published (Pattenden *et al.*, 1973).

Under controlled conditions the intensities of the molecular ion peaks from the six esters could give an estimate of their relative abundances.

III. Isolation of Natural Acids and Alcohols

A. Chrysanthemic and Pyrethric Acids

The acid fraction from hydrolysis of the ester concentrate with excess alkali contains a mixture of chrysanthemic and chrysanthemum dicarboxylic acids (Staudinger and Ruzicka, 1924) which are easily separated, as in the early analytical method (Tattersfield *et al.*, 1929), for only chrysanthemic acid is steam volatile. With a little less than the theoretical quantity of alkali, chrysanthemic and pyrethric acids and some chrysanthemum dicarboxylic acid are obtained. Fractional distillation of the acid chlorides derived from this mixture (LaForge *et al.*, 1952a) gives chrysanthemic acid chloride followed by pyrethric acid chloride (olefinic proton at $\tau 3.6$) contaminated with a little chrysanthemum dicarboxylic acid dichloride (olefinic proton $\tau 3.4$); the pyrethric acid chloride can be purified by recrystallization (Elliott and Janes, 1969). This route is a convenient source of the acid components for optically and isomerically pure esters (see Section VI,E).

B. Pyrethrolone, Cinerolone, and Jasmolone

There is still no alternative to the method of Staudinger and Ruzicka (1924) for isolating the rethrolones from pyrethrum extract, because attempted saponification of the natural esters gives the acid by elimination, leaving a cyclopentadienone which dimerizes very readily (Fig. 3). The cyclopentadienones are too labile for much of their chemistry to have been established, but some aspects can be deduced by analogy with the reactions of other cyclopentadienones (LaForge *et al.*, 1952b; Elliott *et al.*, 1967; Nakada *et al.*, 1971). Elimination also occurs on attempted saponification of rethronyl acid phthalates (La Forge *et al.*, 1952b), but alcohols are obtained on very mild saponification of some alkylrethronyl acetates (Crombie *et al.*, 1950).

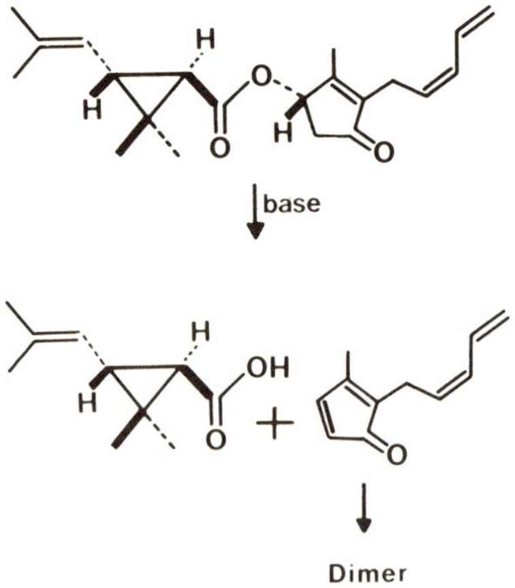

Fig. 3. Action of aqueous base on the natural esters.

To overcome these difficulties, Staudinger and Ruzicka (1924) suppressed the activating influence of the carbonyl group by forming the semicarbazones of the esters and cleaved this mixture with alcoholic sodium methoxide to the rethrolone semicarbazones. Even under these conditions some dimeric cyclopentadienones were obtained as "altered" semicarbazones. The rethrolone semicarbazones, with either potassium hydrogen sulfate, or dilute sulfuric acid (Elliott, 1964c), or pyruvic acid (Maciver, 1968), give a mixture of the natural ketols from which (+)-pyrethrolone can be separated and purified as its crystalline hydrate. This procedure is sufficiently mild to preserve the integrity of the optically active center at C-4, despite the ease with which this center is racemized in other reactions (see Section IV,B). It remains the most convenient route to substantial quantities of pure (+)-pyrethrolone, for synthetic (±)-pyrethrolone (Crombie *et al.*, 1969a) has not been resolved and preparative gas–liquid chromatography (King and Paisley, 1969) gives only limited quantities of the three alcohols.

(+)-Cinerolone contaminated with some (+)-jasmolone can be isolated by a tedious procedure (Elliott, 1964c) from residues after removing (+)-pyrethrolone.

IV. Absolute Stereochemistry

A. Chrysanthemic and Pyrethric Acids

(+)-*trans*-Chrysanthemic acid gives (—)-pyrocin and thence (+)-terebic acid (Crombie and Harper, 1954), the absolute stereochemistry of which is known (Fig. 4). This sequence destroys the stereochemical asymmetry at C-1 and establishes the absolute configuration at C-3 as (*R*).

In a complementary sequence (Crombie *et al.*, 1963) the stereochemistry of chrysanthemic acid at C-1 is stabilized by Arndt-Eistert homologation (without inversion) and the enantiomeric terebic acid is derived by lactonization and ozonolysis (Fig. 4) destroying asymmetry at C-3. The absolute configuration at C-1 is thus also (*R*).

The absolute configuration of (+)-pyrethric acid is shown to be also (1*R*, 3*R*) because it can be derived from (+)-*trans*-chrysanthemic acid (Fig. 5) (Matsui and Yamada, 1963, 1965; Sumitomo Chemical Co. Ltd., 1963). As in related oxidations of isopropylidene groups (Bhalerao and Rapoport, 1971) selenium dioxide attacks *t*-butyl (+)-*trans*-chrysanthemate almost exclusively at the *trans*-methyl group of the side

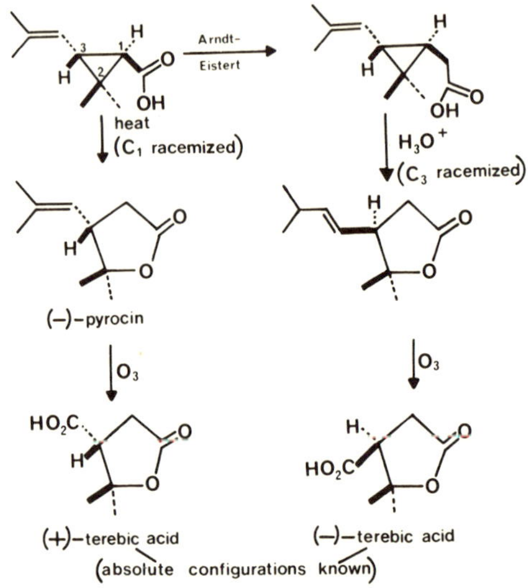

Fig. 4. Absolute stereochemistry of chrysanthemic acid.

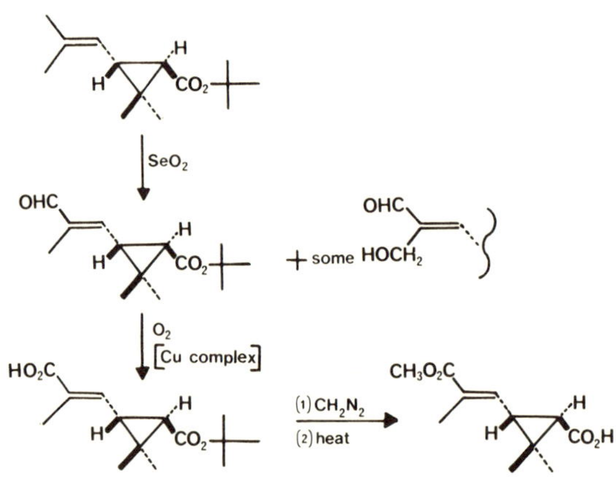

Fig. 5. Preparation of (+)-pyrethric acid from (+)-*trans*-chrysanthemic acid.

chain. Oxidation, esterification, and loss of isobutylene gives (+)-pyrethric acid, identical with the natural acid.

B. Rethrolones

(+)-Pyrethrolone methyl ether gave (−)-α-methoxysuccinic acid on exhaustive ozonolysis (Fig. 6) leading to the conclusion that the absolute configuration is (4S) (Inouye and Ohno, 1958; Katsuda *et al.,* 1958, 1959). However the pyrethrolone methyl ether used had

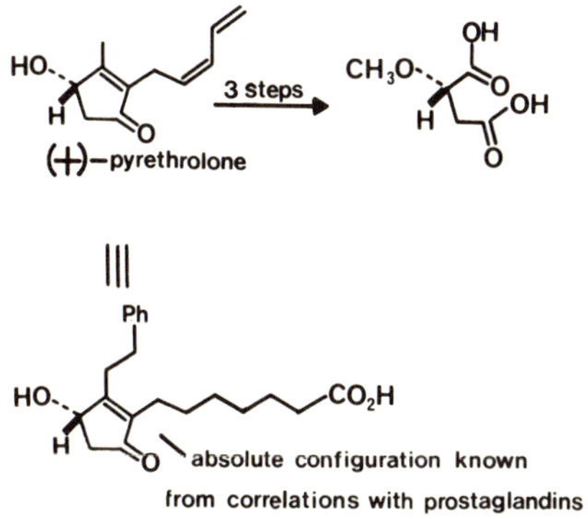

Fig. 6. Absolute configuration of (+)-pyrethrolone.

$[\alpha]_D = +10.5°$ and was made by a method (methanol-sulfuric acid on semicarbazone) shown by West (1944) to give a racemized product, whereas (+)-pyrethrolone methyl ether has $[\alpha]_D = +99°$ (Elliott, 1964a) when prepared by a method (dimethyl sulfate–potassium hydroxide in ether) known to avoid racemization (West, 1944). The conclusion as to the absolute configuration shown (Fig. 6) is supported by recent work (Miyano and Dorn, 1973). The ORD curves of (+)-pyrethrolone and the prostaglandin analog (of known absolute stereochemistry) shown were indistinguishable, with identical positive and negative Cotton effects. Recently, Begley *et al.* (1972) confirmed this result by x-ray determination of the crystal structure and absolute configuration of the related compound (+)-allethronyl (+)-*trans*-chrysanthemate 6-bromo-2,4-dinitrophenylhydrazone.

V. Preferred Conformations

Much evidence (reviewed by Elliott, 1969, 1971) indicates that the biological activity of the natural esters is associated with the intact molecules and depends on certain essential features in a specific spatial disposition to one another. It is therefore valuable to attempt to examine the most likely conformation for the molecules. However, free rotation is possible about all the acyclic single bonds, permitting a wide range of molecular shapes, because the energy barriers restricting rotation are frequently exceeded at room temperature. Only preferred conformations can therefore be deduced and spectral methods, such as NMR, can reflect only average populations of all the possible rotameric forms.

The information available about the seven single bonds (A–G, Fig. 7) will be considered in turn.

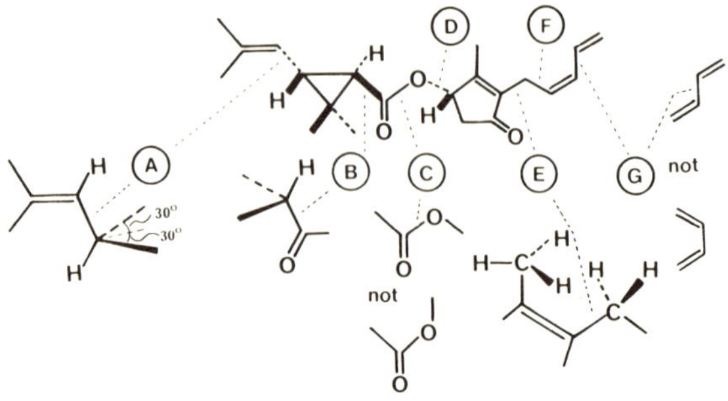

Fig. 7. Conformations of pyrethrin I.

Bond A. Models indicate that there is a minimum steric interaction when the bond is S-trans; the plane containing the atoms of the side chain then bisects the cyclopropyl angle at C-3. Pierre *et al.* (1970) found support for this as a preferred conformation from the magnitude of the coupling constant J_{31}, between the protons flanking bond A.

Bond B. The chemical shifts of the cyclopropyl substituents in ethyl *trans*-chrysanthemate relative to those in the cis isomer led Pierre *et al.* (1970) to suggest that the preferred conformation about this bond in *trans*-chrysanthemic esters approaches S-cis, the plane of the ester atoms nearly bisecting the cyclopropane angle at C-1 but rotated slightly to minimize steric hindrance.

Bond C. Sutton (1955) summarized evidence that in most esters (R·CO·OR′) the five atoms involved are coplanar and that the S-trans form (R and R′ opposed) is strongly preferred at room temperature. In ethyl chrysanthemate this S-trans conformation is indicated by NMR studies (Pierre *et al.*, 1970). Planarity of the ester link in rethronyl chrysanthemates would restrict interaction between substituents on the cyclopropane ring and those on the cyclopentenolone ring and limit the range of possible molecular shapes.

Bond D. Whereas in ethyl chrysanthemate Pierre *et al.* (1970) were able to suggest a conformation in which interaction between the terminal methyl group and the ester carbonyl was minimized, in rethronyl chrysanthemates the interactions are too numerous and too weak for a preferred conformation to be inferred.

Bond E. Models indicate that there is steric hindrance between the methyl group at C-3 and the α-methylene group of the side chain and probably therefore some preference for the rotamer with C-2′ in the plane of the ring.

Bond F. There should be no greater hindrance to rotation about this bond than the very small barrier in a normal aliphatic chain, and therefore no great preference for any one conformer. However, there is some indirect evidence, discussed in Section XI,C, about the conformation adopted at the site of action.

Bond G. There is no steric hindrance to the conformations in which all four olefinic carbon atoms are coplanar, giving maximum conjugation. Of the two possible forms, the S-trans is probably favored as in the directly analogous case of butadiene itself (Ingold, 1953).

VI. Synthesis

A. Chrysanthemic Acid

Chrysanthemic acid was the first of the five components of the natural esters to be synthesized. The numerous routes to it since devised greatly exceed requirements for structural proof, for it has been established as the most accessible and generally effective acid component of synthetic pyrethroids.

The units from which the cyclopropane ring has been constructed are shown (Fig. 8) and discussed in this section.

When the product is a mixture of cis and trans forms, the relative proportions are conveniently estimated from the intensities of the doublets from the olefinic protons at $\tau 5.1$ (trans) and $\tau 4.6$ (cis) (Bramwell *et al.*, 1969).

The cis component is conveniently removed from the acid mixture obtained on saponification as the neutral lactone (Formula in Fig. 26) formed with boron trifluoride (Matsui and Ueda, 1970). Alternatively, the cis isomers may be converted to the trans forms, as discussed in Section VII.

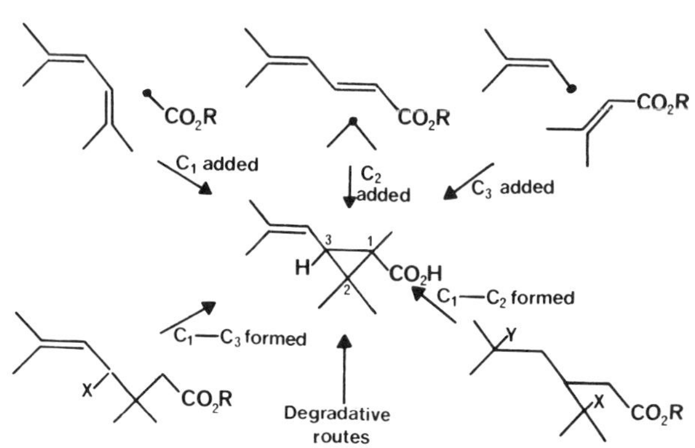

Fig. 8. Summary of route to chrysanthemic acid.

1. Addition of C-1

The original route (Staudinger *et al.*, 1924) from ethyl diazoacetate and 2,5-dimethylhexa-2,4-diene was greatly improved (Campbell and Harper, 1945; Harper *et al.*, 1951) and recently the addition has been adapted for continuous operation (Shim and Martin, 1971) (Fig. 9). The diene was originally synthesized from methallyl chloride but is now available from acetylene and acetone (Tedeschi, 1962; Hammond, 1955) or from isobutyraldehyde by the processes indicated. The commercial procedure has been described (Sanders and Taft, 1954). *t*-Butyl diazoacetate gives the (±)-trans isomer exclusively (Matsumoto *et al.*, 1963).

The C-1 addend can also be generated in lower yield from diiodoacetic ester and zinc–copper couple in tetrahydrofuran (Büchel and Korte, 1962).

2. Addition of C-2

The unit CMe_2 is generated (*in situ*) from diphenylsulfonium isopropylide (obtained from diphenyl sulfide, triethyloxonium fluoroborate, and lithium diisopropylamide) (Corey and Jautelat, 1967) and added

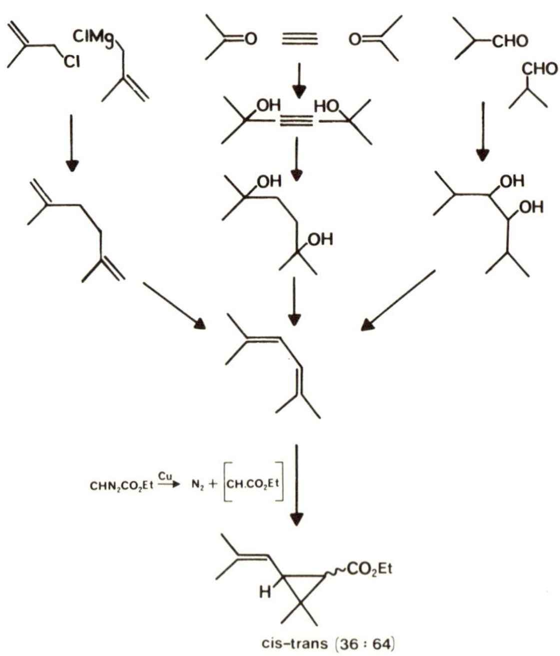

Fig. 9. Chrysanthemic acid from 2,5-dimethylhexa-2,4-diene.

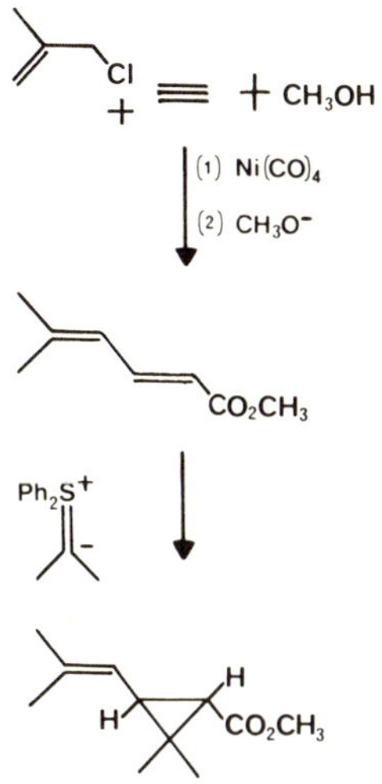

Fig. 10. Chrysanthemic acid by addition of the CMe_2 unit.

to the methyl methylsorbate, prepared as shown (Fig. 10). Despite the stereochemical purity of the product, the complexity of the reagents and stringent conditions make this route impractical commercially.

3. *Addition of C-3*

An elegant and practical alternative to the ethyl diazoacetate route, using readily available starting materials, was developed by Martel and Huynh (1967) (Fig. 11). A sulfone, derived as shown from an allylic

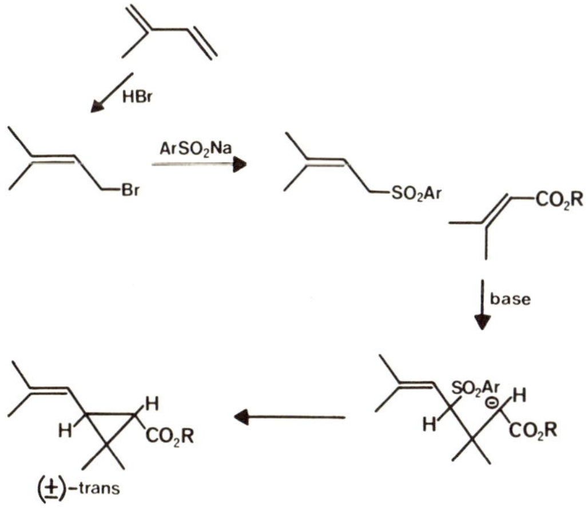

Fig. 11. Chrysanthemic acid via a sulphone.

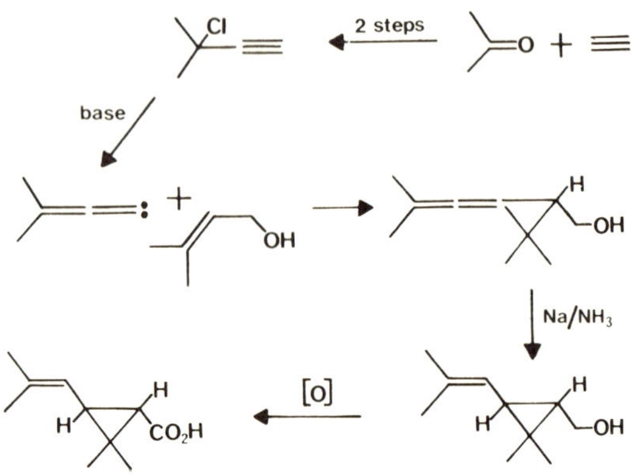

Fig. 12. Chrysanthemic acid from dimethylallyl carbene.

bromide, adds to ethyl senecioate in the presence of base. The adduct is either isolated and cyclized or, with sodium methoxide in dimethyl formamide (Martel and Goffinet, 1968), gives the chrysanthemate, almost exclusively (±)-trans, directly.

Alternatively (Julia and Guy-Rouault, 1967) the sulfone is added to isopropylidene malonate in the presence of a Grignard reagent and cuprous chloride. The 1,1-diethoxycarbonylcyclopropane is monosaponified and decarboxylates to a mixture of *cis*- and *trans*-chrysanthemates.

The most recent synthesis of chrysanthemic acid (Mills *et al.*, 1973) also adds C-3 but under novel conditions. Dimethylallyl carbene is generated as shown (Fig. 12) from acetone and acetylene and adds smoothly to dimethylallyl alcohol. The product is reduced regioselectively to give almost exclusively the cyclopropyl carbinol, oxidized in stages to the aldehyde, and then to the acid. The aldehyde was remarkably difficult to oxidize to the acid, in marked contrast to methyl 2,2-dimethyl-3-formylcyclopropane carboxylate, which gave the acid even by aerial oxidation (Crombie *et al.*, 1970).

4. *C-1 to C-2 Bond Formation*

Pyrocin, or its bond isomer (Fig. 13), has been synthesized by several methods (Julia, 1964; Julia *et al.*, 1960, 1962, 1963, 1965b) and converted to the chrysanthemate as shown (Matsui and Uchiyama, 1962; Julia *et al.*, 1964b, 1965b). The dichloro ester, produced with thionyl chloride in benzene, then ethanolic hydrogen chloride, is cyclized with sodium *t*-amylate in benzene or sodium hydride in dimethyl formamide. Saponification, followed by toluene 4-sulfonic acid in refluxing xylene to convert

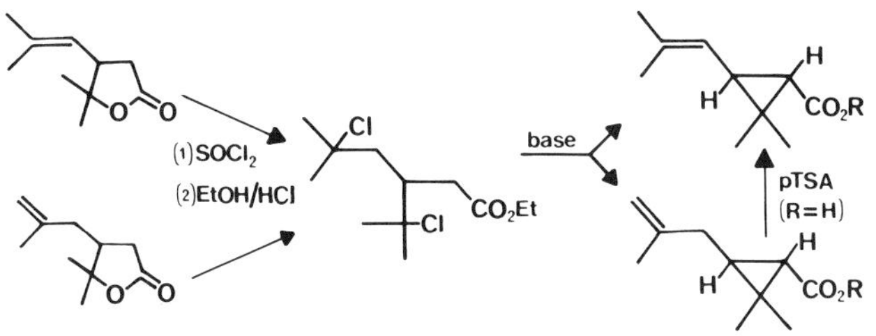

Fig. 13. Chrysanthemic acid from pyrocin or its bond isomer.

the 2-methylprop-2-enyl isomer, yields pure (±)-*trans*-chrysanthemic acid.

5. *C-1 to C-3 Bond Formation*

This alternative ring closure has also been investigated by Julia *et al.* (1964a, 1965a). The two routes, both complex, lead to 4-oxy derivatives of 3,3,6-trimethylhex-5-enoic acid, which are cyclized smoothly with strong base to (±)-*trans*-chrysanthemate as shown (Fig. 14).

6. *Degradative Routes*

Δ^3-Carene, a natural product containing a *gem*-dimethylcyclopropane ring, was degraded by Matsui *et al.* (1965, 1967) as outlined (Fig. 15) or by an alternative route (Cocker *et al.*, 1972) to *cis*-homocaronic anhydride. With methylmagnesium iodide this gave a mixture of lactones, one of which opened to (−)-*cis*-(1S,3*R*)-chrysanthemic acid. This acid was epimerized with base to (+)-*trans*-(1*R*,3*R*)-chrysanthemic acid.

In a related route S. Julia *et al.* (1964) (Fig. 16) synthesized a bicyclohexanone. The oxime gave the required chrysanthemate system directly with phosphorus pentachloride.

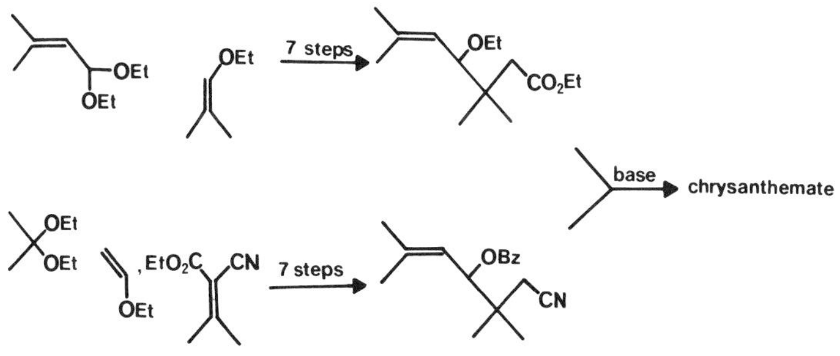

Fig. 14. Chrysanthemic acid by 1–3 ring closure.

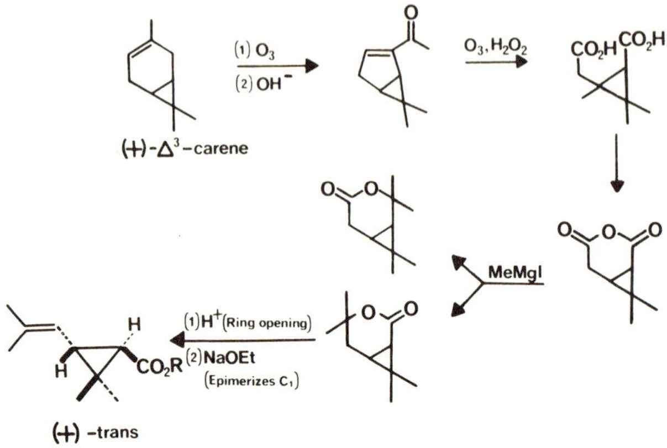

Fig. 15. Chrysanthemic acid from Δ^3-carene.

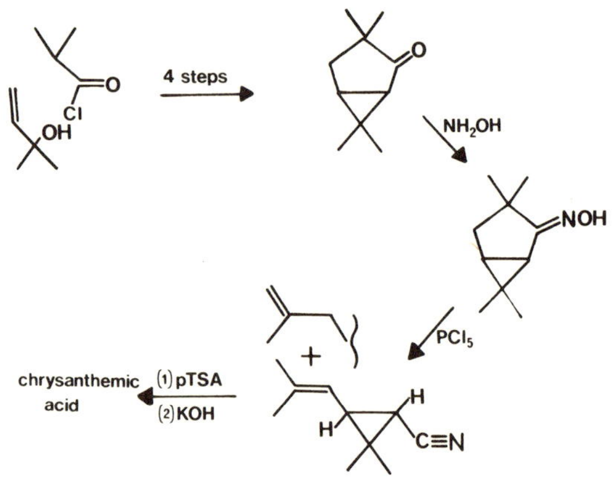

Fig. 16. Chrysanthemic acid from a bicyclohexanone.

7. *Other Routes*

Trans-chrysanthemic acid is also available from *cis*-chrysanthemic acid, as described in Section VII.

B. Pyrethric Acid

Various syntheses of chrysanthemum dicarboxylic acid and of its dimethyl ester have been reviewed (Crombie and Elliott, 1961) and the diacid is readily available from pyrethrum extract. Half hydrolysis of the dimethyl ester (Crombie *et al.*, 1957) gave a mixture of monomethyl

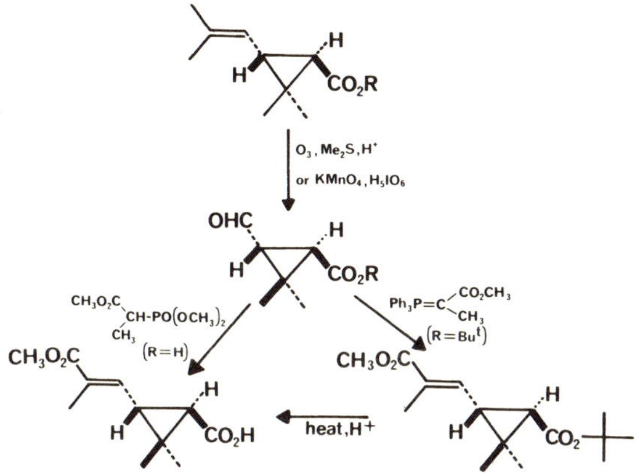

Fig. 17. Routes to (+)-pyrethric acid.

esters (Matsui and Meguro, 1963) in which the required isomer, pyrethric acid, predominates. The mixture is separated using the different solubilities of the two quinine salts in acetone.

Other syntheses, including that from chrysanthemic acid already described (Section IV,A), avoid the necessity to separate isomers. In the routes shown (Fig. 17) aldehydes are obtained by degradation of the side chain of *trans*-chrysanthemic acid or its *t*-butyl ester. The *t*-butyl aldehyde is condensed with the conventional Wittig reagent; then the *t*-butyl protecting group is removed by pyrolysis, forming the required monomethyl ester (Ueda and Matsui, 1970). Alternatively, in a convenient modification, the methoxycarbonylethylphosphonate ester is used directly on the aldehyde-acid (Martel, 1969; Martel and Buendia, 1970b). Both reactions form the pyrethric acid side chain with the required trans (*E*) stereochemistry.

C. Rethrolones

Despite numerous syntheses of cyclopentenones, there is no convenient route to the natural rethrolones by introduction of an hydroxyl group at C-4. N-Bromosuccinimide gave 4-bromocyclopentenones from alkylrethrones but the reaction failed with alkenylrethrones (Crombie *et al.*, 1950). Alkenylrethrones are converted to alkenylrethrolones and isomeric compounds in low yield by microbial action (Tabenkin *et al.*, 1969; LeMahieu *et al.*, 1970). Thus, allethrone and cinerone (Fig. 18) give the corresponding partially racemic rethrolones, together with products of hydroxylation at C-1′ and at the C-3 methyl. Rethrones are converted

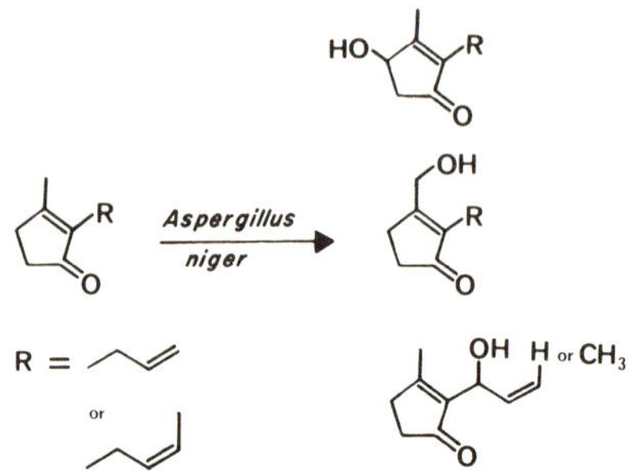

Fig. 18. Rethrolones from rethrones (biological).

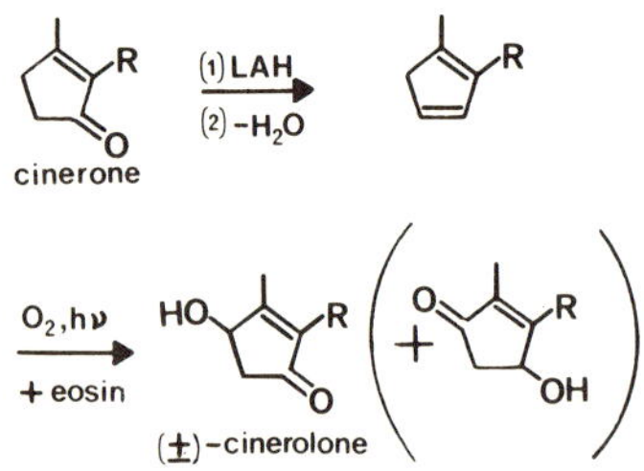

Fig. 19. Rethrolones from rethrones (chemical).

to rethrolones by an indirect route (Fig. 19) which involves a photochemical oxidation of the cyclopentadiene derived from the rethrone (LeMahieu *et al.*, 1968). The product is a mixture of the (±)-rethrolone and the isomer, as shown.

In all other syntheses of rethrolones the oxygen function is present at C-4 before cyclization. The original route of Schechter *et al.* (1949) using pyruvaldehyde, recently refined and improved by Crombie *et al.* (1969a) (Fig. 20), is still the only one versatile and mild enough to give (±)-pyrethrolone. The Wittig synthesis, under salt-free conditions, produces the olefinic side chains in the required cis (Z) configuration, a great improvement over earlier methods involving semihydrogenation of acetylenic intermediates (Crombie *et al.*, 1956). A further improvement is the use of methoxymagnesium carbonate to introduce, in one step, the carboxyl group needed to activate the methylene group for aldol condensation. Cinerolone and jasmolone were prepared by the same method as was the rethrolone with the allenic $CH_2CH{=}C{=}CHCH_3$, side chain once considered to be a constituent of the natural alcohols.

A modification of the pyruvaldehyde route (Fig. 21) has only been disclosed as a patent (Martel *et al.*, 1966). In place of the β-keto acid

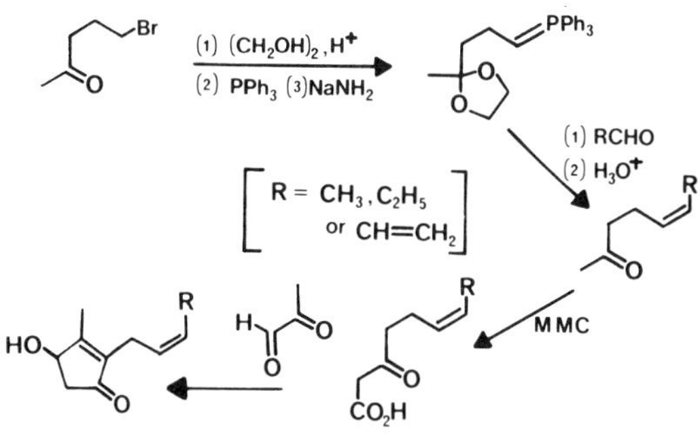

Fig. 20. Wittig synthesis of rethrolones.

and pyruvaldehyde, the β-keto ester is condensed with pyruvaldehyde acetal, more stable reagents that tolerate a strong base for the condensation.

Other recent syntheses are not applicable to or have not been extended to give pyrethrolone. In that of Vandewalle and Madeleyn (1970) (Fig. 22) the ring is constructed first; then the side chain, as a Grignard reagent, is introduced, and finally the keto group is reduced. Zinc in acetic acid is remarkably selective in attacking the less hindered carbonyl function, as required.

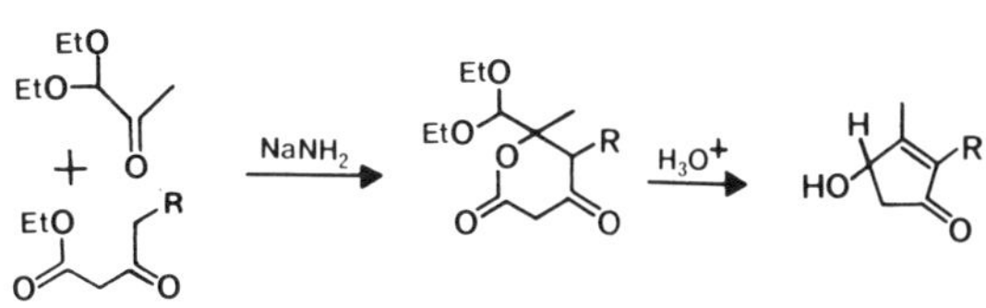

Fig. 21. Modified pyruvaldehyde route to allethrolone.

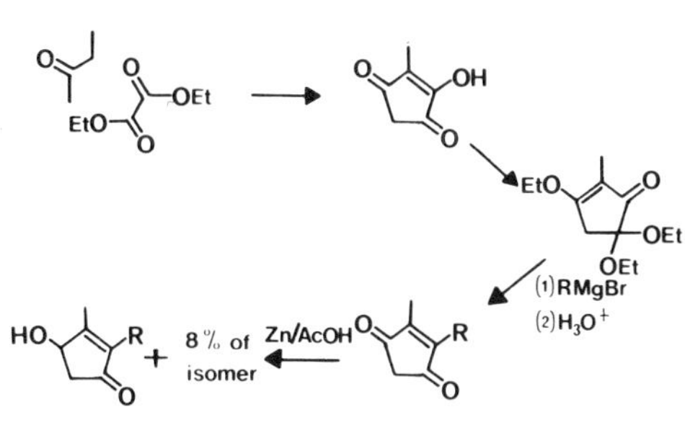

Fig. 22. Allethrolone from a cyclopentanetrione.

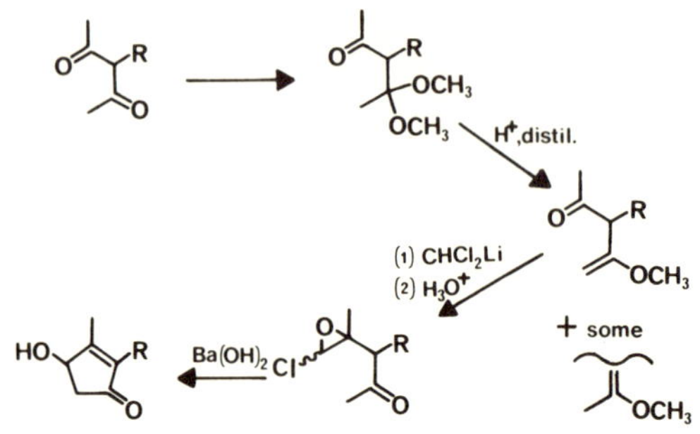

Fig. 23. Allethrolone from a chloro-epoxide.

Büchi *et al.* (1971) add a CHCl unit (eventually to be C-4) to a ketone to give a chloroepoxide containing C-1 as a masked ketone (Fig. 23). The chloroepoxide cyclizes smoothly under very mild conditions to the alkenylrethrolone.

D. Resolution

1. *Chrysanthemic Acid*

The reagents used to resolve $(\pm)$-*trans*-chrysanthemic acid (Fig. 24) are all amines that combine with the acid to form crystalline salts. In favorable cases the progress of the resolution can be monitored by NMR because the spectra of diastereoisomeric salts separated by fractional crystallization differ sufficiently to permit quantitative assessment of each isomer (cf. Barlow *et al.*, 1971).

The reagent of Goffinet and Locatelli (1968) is available in both $(+)$ and $(-)$ forms so direct crystallization of the two less-soluble salts is claimed to extract the $(+)$- and $(-)$-*trans*–acids, even from cis–trans mixtures.

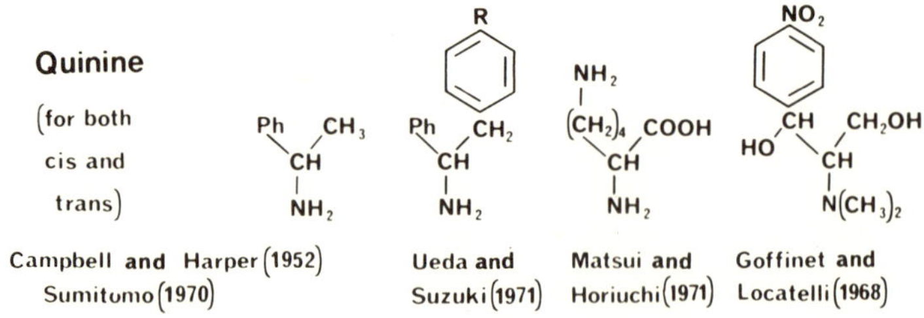

Fig. 24. Resolving agents for $(\pm)$-*trans*-chrysanthemic acid.

The four components of partially resolved (±)-cis–trans mixtures are conveniently estimated by GLC of their esters with (+)- or (—)-2-octanol (Murano, 1972).

(+)-*trans*-Chrysanthemic acid is readily available. Recent syntheses of pyrethric acid modify the side chain of chrysanthemic acid (Section VI,B) and therefore can produce (+)-pyrethric acid without a separate resolution.

2. Rethrolones

Surprisingly little attention has been given to resolving rethrolones. As noted (see Section III,B), (±)-pyrethrolone has not been resolved. (±)-Cinerolone and the related (±)-allethrolone have been resolved by esterification with (+)-*trans*-chrysanthemic acid, followed by fractional crystallization of the rethronyl chrysanthemate semicarbazone (LaForge and Green, 1952; LaForge *et al.*, 1954). (+)-Allethronyl (+)-*trans*-chrysanthemate is available commercially (S-bioallethron).

E. The Natural Esters

Isolation of (+)-pyrethrolone as the crystalline monohydrate (Elliott, 1964c) from pyrethrum extract removes all traces of cinerolone and jasmolone, without thermal isomerization or racemization. This product, esterified with naturally derived chrysanthemic and pyrethric acid chlorides, yields quantities of pure pyrethrins I and II adequate for most entomological and chemical studies (Elliott, 1961; Elliott and Janes, 1969; cf. LaForge and Barthel, 1947; Matsui and Meguro, 1964). Cinerins I and II have also been prepared by reconstitution (Elliott, 1961, 1964c; LaForge and Barthel, 1947; Matsui and Meguro, 1964). The esters are stable for long periods (more than 5 years) if stored in the dark in solution at —20°C (Godin, 1968).

Chrysanthemic anhydride is sometimes encountered as an impurity in esterifications with chrysanthemoyl chloride. It is best detected by thin-layer chromatography or IR (anhydride band at 1780 cm^{-1}) (Elliott *et al.*, 1970), because the NMR peaks of the anhydride are obscured by those of the ester. The anhydride is removed from the ester by chromatography on florisil (Elliott and Casida, 1972) but is better avoided by using absolutely dry conditions, and a slight excess of alcohol, which is more easily removed by chromatography on alumina (Elliott and Janes, 1969).

Before physical methods were developed to confirm purity, such reconstituted materials were compared with natural esters isolated by chromatography and found identical within close limits (Sawicki *et al.*,

1962). Reconstitution was thus established as a reliable method for preparing pure esters, even for the least stable (pyrethrin II) (Sawicki and Elliott, 1965).

VII. Interconversion of Isomers of Chrysanthemic Acid

Conversion of cis isomers to trans is usually achieved by epimerization at the more labile center, C-1, [i.e. (—)-*cis*–(1S,3*R*) gives (+)-*trans*–(1*R*,3*R*)], by heating the acid chloride at 150°C (Suzuki *et al.*, 1970; cf. Elliott *et al.*, 1970), the ester at 240°–260°C (Hanafusa *et al.*, 1970), or the ester with sodium ethoxide at 190°C (Julia *et al.*, 1959; Matsui and Yoshioka, 1965).

Both centers are epimerized by photochemical irradiation; all isomers of chrysanthemic acid then give the same (±)-*cis-trans* mixture (see Section X,A). More deviously, the two centers can also be epimerized chemically. The side chain is first oxidized with permanganate to a ketol, as shown later in Fig. 28, and, because C-1 and C-3 are now both adjacent to electrophilic centers, potassium *t*-butoxide produces a (±)-*trans*–ketol from either (+)- or (—)-*trans*–acid (Matsui *et al.*, 1963c).

Elegant routes to *cis*-chrysanthemic acid from the trans form have been developed. Ozonolysis of the trans side chain (Fig. 25) gives an aldehyde which epimerizes, forming a hemiacetal. As a masked aldehyde this reacts with the Wittig reagent to regenerate the isobutenyl side chain (Martel, 1970a,b). Alternatively, the side chain is hydrated and the cis form trapped as a lactone as epimerization occurs at C-1 (Fig.

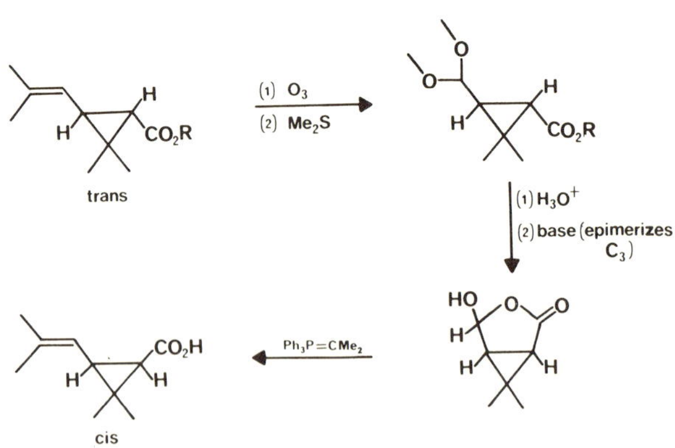

Fig. 25. (+)–*cis*- from (+)-*trans*-chrysanthemic acid.

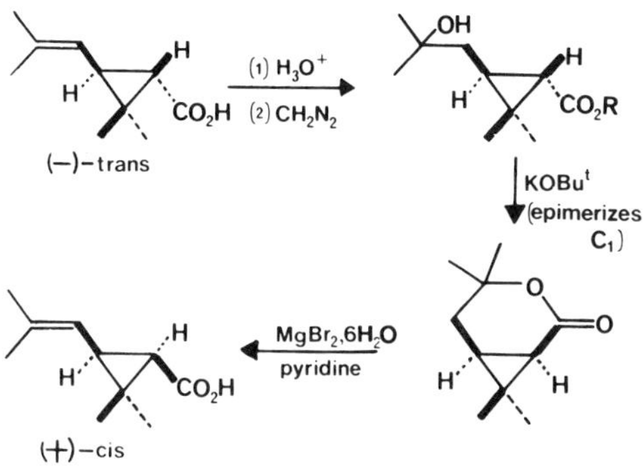

Fig. 26. (+)-*cis*- from (−)-*trans*-chrysanthemic acid.

26). The epimerized acid is obtained directly from the lactone with magnesium bromide hydrate in pyridine, so (−)-*trans*-chrysanthemic acid gives (+)-*cis*-chrysanthemic acid (Martel and Buendia, 1970a).

VIII. Radiolabeling

Table II summarizes methods for incorporating radiolabels into the pyrethrins.

A. Biological Methods

Pyrethrum plants grown in $^{14}CO_2$ incorporate the activity randomly and the resulting mixture has been partially separated by chromatography. However the specific activity was too low for many applications. Similarly, plants fed [^{14}C]mevalonate, [^{14}C]acetate, and [^{14}C]methionine (Godin *et al.*, 1963) give products of low activity, principally useful for biosynthetic studies.

B. Of Acids

Nishizawa and Casida (1965) synthesized (±)-*cis*, *trans*- and thence (+)-*trans*-chrysanthemic acid labeled with ^{14}C in the carboxyl carbon, from [^{14}C]ethyl diazoacetate. The acid was used to make [^{14}C]pyrethrin I (see Table II) (cf. Fine *et al.*, 1967). Crombie *et al.* (1970) reacted the Wittig reagent from [^{14}C]isopropyl bromide with the aldehyde from ozonolysis of methyl (+)-*trans*-chrysanthemate and thus obtained

TABLE II

Radiolabeled Preparations of the Natural Esters

Method	Product and label position	mCi/m*M*	Reference
Plants grown in $^{14}CO_2$	Random [^{14}C]pyrethrum	low	Pellegrini *et al.* (1952)
Chromatography of above mixture	Cinerin I	0.05	Chang and Kearns (1964)
	Pyrethrin I	0.05	Chang and Kearns (1964)
	Pyrethrins I	0.03	Winteringham *et al.* (1955)
Plants fed [^{14}C]mevalonate, etc., then chromatography	Various [^{14}C]pyrethrins I	low	Crowley *et al.* (1962)
	Various [^{14}C]pyrethrins II	low	Crowley *et al.* (1962)
(+)-Pyrethrolone + -*trans*-[^{14}C]acid	Ring-1-[^{14}C]pyrethrins I	0.017	Fine *et al.* (1967)
(+)-Pyrethrolone + (+)-*trans*-[^{14}C]acid	[^{14}C=O]pyrethrin I	1.3	Yamamoto and Casida (1968)
(+)-Pyrethrolone + [^{14}C]pyrethric acid	Methoxy-[^{14}C]pyrethrin II	~0.1	Elliott *et al.* (1972)
[^{3}H]Pyrethrolone + (+)-*trans*-chrysanthemic acid	[^{3}H]pyrethrin I (see Fig. 27)	675	Elliott and Casida (1972)
[^{3}H]Pyrethrolone + pyrethric acid	[^{3}H]pyrethrin II (see Fig. 27)	626	Elliott and Casida (1972)

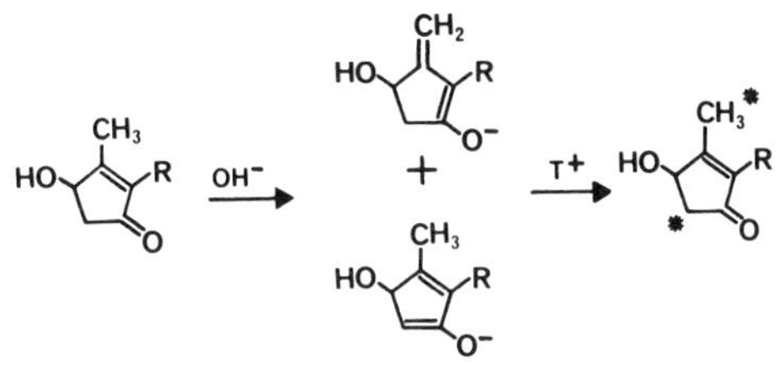

Fig. 27. Radiolabeling of rethrolones.

optically and geometrically pure labeled acid. [^{14}C]Chrysanthemum dicarboxylic acid was made similarly. Elliott *et al.* (1972) prepared [^{14}C]pyrethrin II labeled in the methoxyl position by reacting chrysanthemum dicarboxylic acid dichloride, first with (+)-pyrethrolone (1 equivalent), then with [^{14}C]methanol. The isomeric esters produced were separated by preparative thin-layer chromatography.

C. Of Rethrolones

(+)-Pyrethrolone was radiolabeled by exchanging the base labile protons in the methylene and methyl groups of the optically active alcohol with tritium (Elliott and Casida, 1972) (Fig. 27) under conditions that did not affect the optically active center. Esterification gave [^{3}H]pyrethrins I and II. The chemical reactions involved were monitored by NMR spectroscopy of the corresponding system of allethrolone with deuterium oxide.

Neither cinerolone nor jasmolone has been radiolabeled but the [^{14}C]pyruvaldehyde used to make (±)-[^{14}C]allethrolone labeled at C-4 and C-3-methyl (Winteringham *et al.*, 1955; Yamamoto and Casida, 1968) should yield [^{14}C](±)-pyrethrolone, [^{14}C](±)-cinerolone and [^{14}C](±)-jasmolone by the efficient routes recently developed (Section VI,C).

IX. Reactions

A. Acids

The olefinic side chain of chrysanthemic acid is more reactive than the cyclopropane ring. Thus it is hydrogenated without ring cleavage (Smejkal and Farkas, 1963) and gives a diol and a ketol with permanganate (Matsui *et al.*, 1963b) (Fig. 28); osmium tetroxide similarly gives a diol with chrysanthemate esters (Crombie *et al.*, 1970). The diol derived from the ester gives a triene acid by ring opening when treated with acid followed by saponification (Fig. 29) and the same

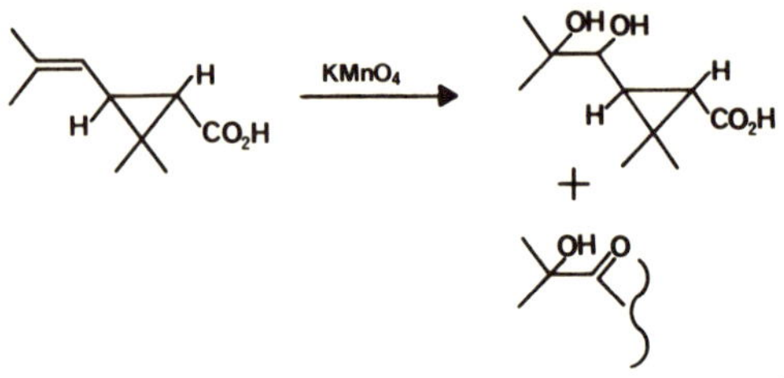

Fig. 28. Oxidation of the olefinic side chain of chrysanthemic acid.

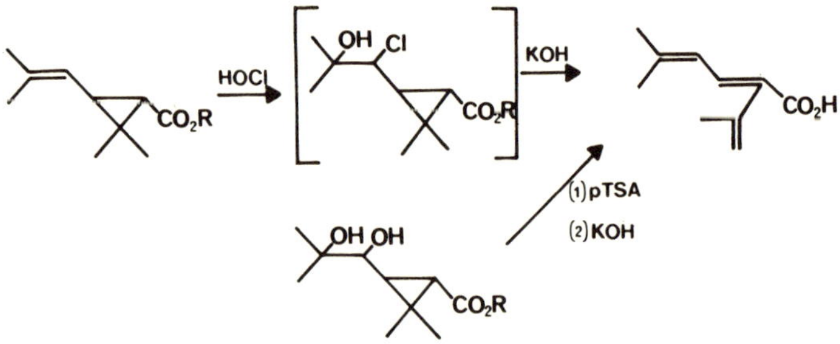

Fig. 29. Decomposition of oxidation products from chrysanthemic acid.

product is obtained from the chlorohydrin (not isolated) after saponification (Matsui *et al.*, 1963a).

The double bond in both *cis*- and *trans*-chrysanthemic esters reacts smoothly with *m*-chloroperbenzoic acid to give the epoxides (M. Elliott, D. H. S. Horn, and N. F. Janes, unpublished work, 1970) but it is reported that only the trans acid reacts with perphthalic acid (Sasaki *et al.*, 1968).

Selective attack at one of the isobutenyl methyl groups has been discussed in connection with the synthesis of pyrethric acid (Section VI,B); this reaction is also involved in biosynthesis and metabolism (Casida, Chapter 5 in this volume).

Thermal cleavage of *trans*-chrysanthemic esters requires temperatures of 500°C (Ohloff, 1965). The product (Fig. 30) is formed from C-1 to C-2 bond cleavage with concomitant proton transfer.

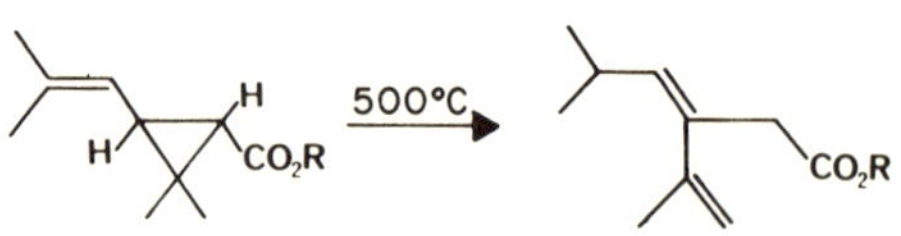

Fig. 30. Pyrolysis of chrysanthemates.

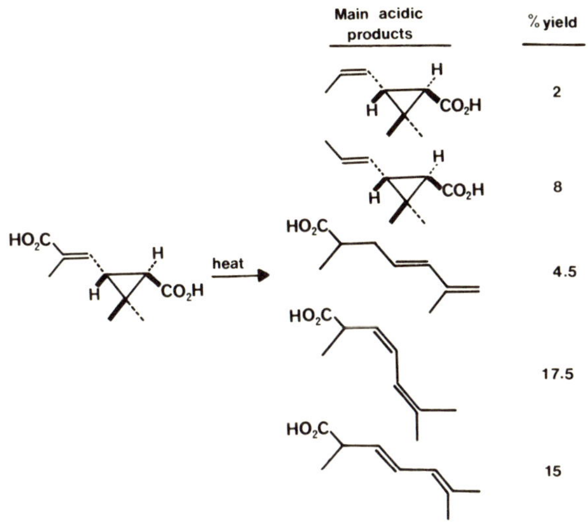

Fig. 31. Pyrolysis of chrysanthemum dicarboxylic acid (acidic products).

Thermal decomposition of chrysanthemum dicarboxylic acid gives a neutral and an acidic fraction (Crombie *et al.*, 1971b). Figure 31 shows percentage yields of the main acidic products. Contrary to earlier reports, summarized by Crombie and Elliott (1961), the carbon dioxide evolved arises from both carboxyl groups but predominantly from that at C-1.

B. Rethrolones

Both the nucleus and the side chain of pyrethrolone contain reactive and sensitive centers. The greater reactivity of the *cis*-pentadienyl side chain over the simpler olefinic systems in cinerolone and jasmolone is reflected in the much higher activity toward some insects of esters derived from pyrethrolone (see Section XI) and in some of the chemical reactions now considered.

Oxidation of pyrethrolone with manganese dioxide first converts C-4 to a carbonyl function; then the side chain hydrogen at C-1′ is abstracted and the radical formed dimerizes as shown (Fig. 32). Cinerolone and jasmolone have not been studied but allethrolone gives a similar dimer (Crombie *et al.*, 1971a).

The C-4 hydroxyl in (+)-pyrethrolone is methylated without race-

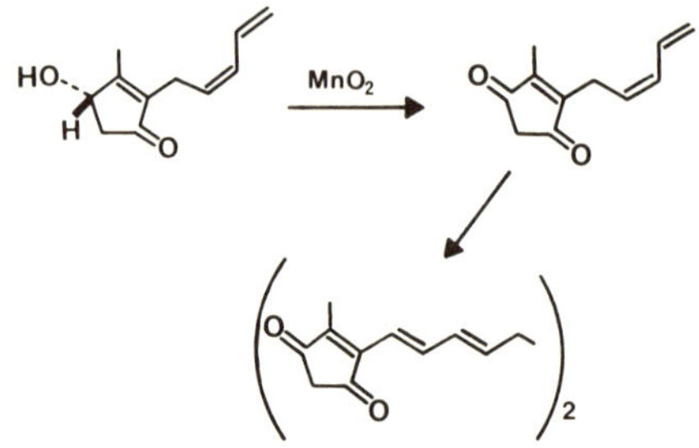

Fig. 32. Oxidation of pyrethrolone.

mization by using dimethyl sulfate and potassium hydroxide in ether (West, 1944).

Both ring and side chain in the pyrethrolone molecule are involved in reaction with sodium methoxide in hot methanol (Elliott, 1965). In a complex series of reactions the side chain is reduced and the ring oxidized (Fig. 33). A cyclopentanedione also formed arises from reaction within the ring. Under comparable conditions, allethrolone gives only a cyclopentanedione, and no products in which the side chain is reduced. The greater reactivity of the diene side chain in this reaction is comparable to that observed in mammalian metabolic systems, in which all metabolites of pyrethrin I and pyrethrin II have oxidized side chains but the simple allyl side chain in allethrin remains intact in some products (Elliott *et al.*, 1972).

The diene side chain in pyrethrolone and its derivatives is thermally isomerized at about 200°C to products containing an interesting cross-conjugated system, recognized by its characteristic UV spectrum (Fig. 34) (Elliott, 1964a; cf. Crombie *et al.*, 1969b). Presumably, at the high temperature the conformation of the side chain changes from the more

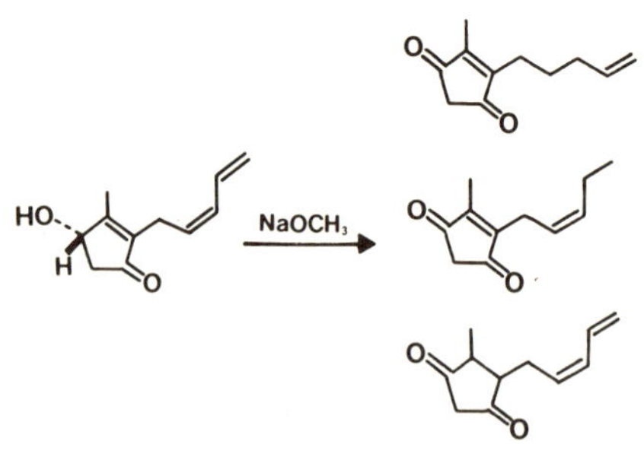

Fig. 33. Enols from pyrethrolone.

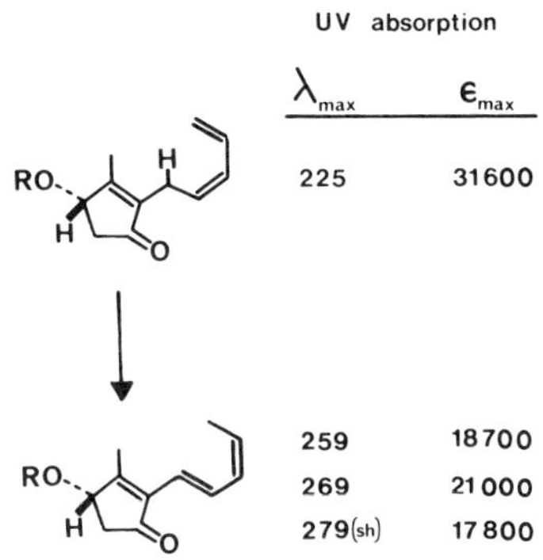

Fig. 34. Thermal isomerization of pyrethrolone and derivatives.

stable S-trans to the S-cis form appropriate for intramolecular hydrogen transfer.

C. Esters

Many reactions of the esters of the natural pyrethrins are predictable from those considered above for their acids and alcohols. In some instances, however, the relative rates at which groups in the alcoholic and acidic moieties react are important. For example, the three double bonds in pyrethrin I would be expected to compete fairly equally in epoxidation reactions, but in pyrethrin II one double bond is relatively deactivated by conjugation and epoxidation occurs preferentially on the pentadienyl side chain (Fig. 35) (M. Elliott, D. H. S. Horn, and N. F.

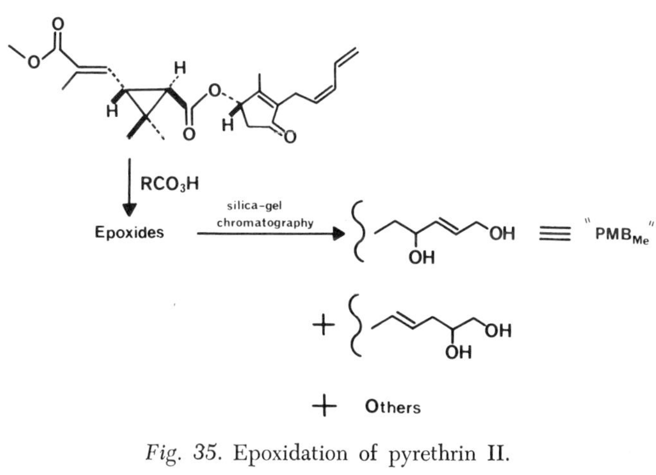

Fig. 35. Epoxidation of pyrethrin II.

Janes, unpublished work, 1970). From this product a series of diols is obtained on chromatography, one identical with a metabolic product of pyrethrin I and pyrethrin II in the rat (Elliott *et al.*, 1972).

X. Photochemistry

A. Of Acid Components

1. In the Absence of Chemically Reactive Agents

Sasaki *et al.* (1970), Ueda and Matsui (1971), and Bullivant and Pattenden (1971) independently examined the photochemical reactions of chrysanthemic acid and reached similar conclusions. A diradical (Fig. 36), formed by cleavage of the C-1 to C-3 bond, decomposes in one of four ways: (*a*) reformation of the 1–3 bond without rotation to regenerate starting material; (*b*) formation of the 1–3 bond after rotation; (*c*) fragmentation to a senecioate; or (*d*) rearrangement to the lactone shown. Epimerization to a cis–trans mixture [routes (*a*) and (*b*)] is faster in the presence of sensitizers, such as acetophenone, or better, isobutyrophenone (Ueda and Matsui, 1971). Whichever geometrical isomer is the starting material, the final cis-trans ratio is 36:64, the same as that in ethyl chrysanthemate from addition of ethyl diazoacetate to dimethylhexadiene (Fig. 9). This suggests that the same diradical intermediate may be involved in both processes.

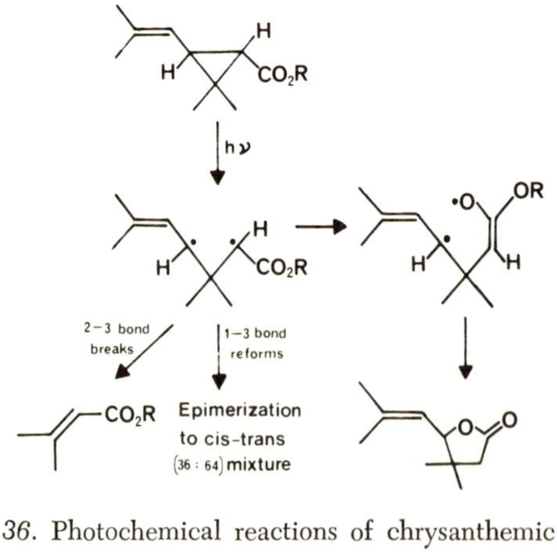

Fig. 36. Photochemical reactions of chrysanthemic acid.

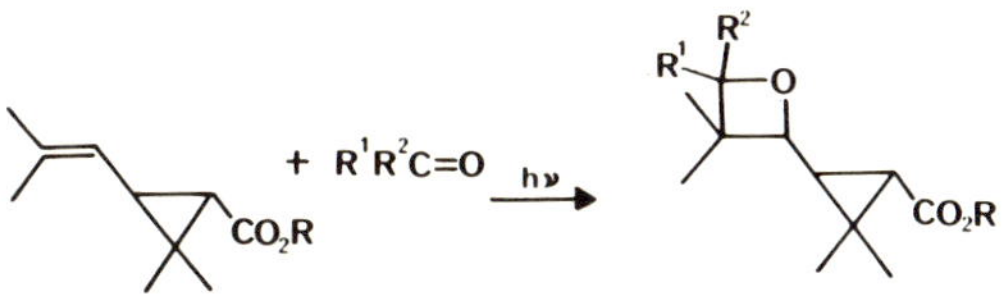

Fig. 37. Photo-sensitized reaction of chrysanthemate.

2. *In the Presence of Ketones*

With UV irradiation benzophenone adds to the olefinic bond of chrysanthemates to generate an oxetane (Fig. 37) (Sasaki *et al.*, 1968).

3. *In the Presence of Oxygen*

In chrysanthemates derived from photochemically stable alcohols, the main reaction is oxidative attack at the olefinic side chain. Thus, 2,4-dimethylbenzyl chrysanthemate (dimethrin) gives acidic oxidation products (isolated by saponification) (Fig. 38) (Chen and Casida, 1969) which are analogous to those formed by selenium dioxide (attack at the *trans*-methyl group—cf. Fig. 5), by epoxidation, or by potassium permanganate (attack at the double bond—cf. Fig. 28).

B. Of Alcohol Components

There is very little information on photochemical reactions of the natural alcohols or their simple derivatives. Bullivant and Pattenden (1972) recently showed a significant difference between the behaviour of derivatives of allethrolone and of jasmolone on irradiation under nitrogen (Fig. 39). Whereas the allyl side chain was converted to cyclo-

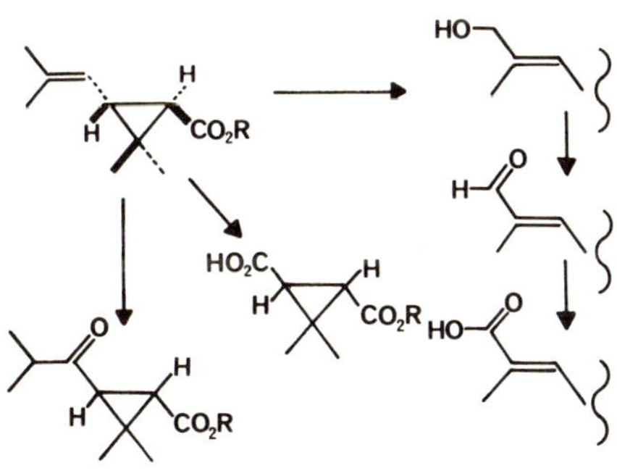

Fig. 38. Photo-oxidation of chrysanthemates.

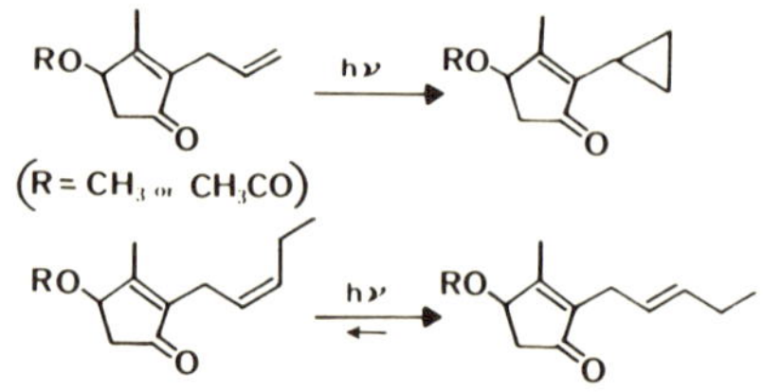

Fig. 39. Photochemical rearrangement of allethrolone and jasmolone.

propyl, the Z (*cis*)-pent-2-enyl group gave a steady-state mixture of 90% *E* (trans) and 10% Z (cis).

C. Of Esters

The photochemical oxidation of pyrethrin I was studied by Chen and Casida (1969), who summarize earlier work. As expected, pyrethrin I with a pentadienyl system which absorbs UV light strongly is significantly less stable than cinerin I and allethrin and than such esters as dimethrin. Some oxidation of the acid component in irradiated pyrethrin I was proved by saponification, but the alcoholic moiety was the more susceptible to photodecomposition although the products, probably polymeric, were not identified. The intensity of absorption at 225 nm by samples of natural pyrethrins kept in the dark in either nitrogen or oxygen for 12 hr remained constant, but on irradiation only the oxygenated sample showed a rapid decrease in absorption, proving the importance of both oxygen and light in photodecomposition (M. Elliott, unpublished work, 1956).

Miskus and Andrews (1972) recently reviewed methods and additives for stabilizing the pyrethrins and showed that the range of UV radiation that induces pyrethroid destruction is from 290 to 320 nm: this range of radiation would be excluded by soft window glass, explaining the persistence of pyrethrins in glass houses. They showed that the decomposition of the natural pyrethrins was slower in the presence of selective light absorbers and antioxidants, usually phenolic compounds.

XI. Insecticidal Activity of the Natural Esters

A. Introduction

The natural pyrethrins are valued for their remarkable lethal action against a wide range of insect species (Gnadinger, 1936, 1936–1945;

TABLE III

Relative Toxicities of the Natural Esters to Houseflies (*Musca domestica* L.) and Mustard Beetles (*Phaedon cochleariae* Fab.)

Ester	Houseflies[a]	Mustard beetles[a]
Pyrethrin I	100[b]	100[c]
Pyrethrin II	180	30
Cinerin I	40	20
Cinerin II	70	13
Jasmolin II	60	—

[a] Topical application of 1-μl drops in acetone to adult females (houseflies) or males and females (mustard beetles).

[b] Taken as standard. LD_{50} = 0.59 μg per insect.

[c] Taken as standard. LD_{50} = 0.005 μg per insect.

Negherbon, 1959); they also show very low toxicity to mammals and very fast "knockdown" action, and few insects have become resistant to them under practical conditions [although resistant strains can be developed in the laboratory (Farnham, 1971)]. The insecticidal activity of the pyrethrins is enhanced by relatively cheap compounds (synergists) (Casida, 1970; Yamamoto, Chapter 10 in this volume) that suppress detoxification within the insect; this is important commercially.

The most detailed examination of the relative activities of pyrethrin I, pyrethrin II, cinerin I, cinerin II, and jasmolin II against houseflies was made by Sawicki (1962a,b,c; Sawicki and Thain, 1961, 1962; Sawicki *et al.*, 1962). His results are shown in Table III, together with comparable data for mustard beetles (Ward, 1953; P. H. Needham, unpublished work, 1955–1962). Sawicki found that pyrethrin I and cinerin I are synergized more than pyrethrin II and cinerin II, and the cinerins are synergized more than the corresponding pyrethrins. Synergists had relatively little effect on knockdown.

The relative toxicities of the natural constituents vary with the insect examined (as illustrated for two species in Table III). This is because the observed toxicity depends on many factors, such as rate of penetration, rate of detoxication, and mode of application, which ideally should be studied separately.

B. Recognition of Pyrethrin-like Activity

Conclusions about the structural features essential for insecticidal activity of the pyrethrin type are limited if only the six natural esters are considered. However, before results with synthetic compounds are introduced, it is necessary to decide how to recognize true pyrethrin-like activity. Caution is necessary because with compounds of great physiological activity, such as the pyrethrins, systems isolated from the insect, but not necessarily related to the site of fatal lesion, may give apparently significant responses. It is therefore important to choose a preparation in which the measured phenomenon parallels the action of the compound on the intact insect.

Burt and Goodchild (1970) concluded that toxicities of pyrethrin I and a range of synthetic pyrethroids to whole cockroaches were reflected by their activities on the central nervous system. In further work to locate the site of action more precisely, the concentration of pyrethrin I in the hemolymph following topical application of a just lethal dose was estimated to be less than 2×10^{-8} *M* (Burt and Goodchild, 1971a). This was less than the minimum concentration of pyrethrin I (10^{-7} *M*) that blocked the action in the giant fiber axons of cockroaches (cf. Narahashi, 1962, 1971), but spontaneous activity in the sixth abdominal ganglion was increased by concentrations as low as 10^{-10} *M*. Consequently, results with isolated preparations, such as the giant fiber axons of crayfish (Berteau *et al.*, 1968; Berteau and Casida, 1969), must be used with caution in considering structure–toxicity relationships. Special care is necessary for compounds that are much less toxic to whole insects than, for example, pyrethrin I.

C. Essential Features for Insecticidal Activity

Firm conclusions as to the structural features necessary for pyrethrin-like activity rely on results from insects reared and tested under standardized conditions. Such data for houseflies (A. W. Farnham) and mustard beetles (P. H. Needham) have been discussed (Elliott, 1969, 1971).

The observed insecticidal activity of any compound depends on its ability to penetrate and on its susceptibility to detoxication as well as on its intrinsic activity, all influenced by changes in the chemical structure. Separate studies to correlate each factor with chemical constitution are therefore desirable. Yamamoto (Chapter 10 in this volume) emphasizes inhibition of detoxication by synergists. In this chapter, the relationship between structure and effectiveness at the site of action is considered.

Especially in the housefly, an important route for detoxication involves the mixed function oxidase system. Sawicki and Farnham (1967, 1968) suppressed the effect of this by pretreatment of each insect with a large dose (2 μg) of a synergist (sesamex) several hours before the insecticide.

TABLE IV

Relative Toxicities of Various Pyrethroids to Houseflies (*Musca domestica* L.) (plus Synergist) and Mustard Beetles (*Phaedon cochleariae* Fab.)[a]

Compound	Structure[b]	Relative toxicities: Houseflies[c]	Relative toxicities: Mustard beetles[d]
Pyrethrin I	A (R = Me)	100[e]	100[f]
Pyrethrin II	A (R = CO_2Me)	26	38
Benzylnorthrin	B	180	11
Bioresmethrin	C	190	62
5-Benzyl-3-furylmethyl (+)-*trans*-2,2-dimethyl-3-cyclopentylidene-methylcyclopropane carboxylate	D	200	110
(+)-Pyrethronyl 2,2,3,3-tetramethyl-cyclopropane carboxylate	E	130	6
3-Benzylbenzyl (+)-*trans*-chrysan-themate	F (X = CH_2)	84	32
3-Phenoxybenzyl (+)-*trans*-chrysan-themate	F (X = O)	44	14

[a] Data from Elliott (1971), Barlow *et al.* (1971).

[b]

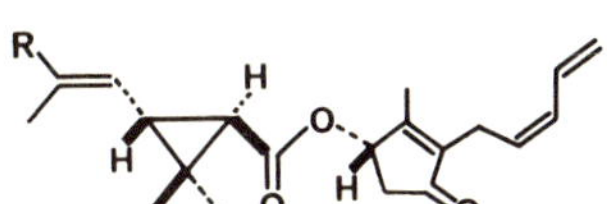
(A)

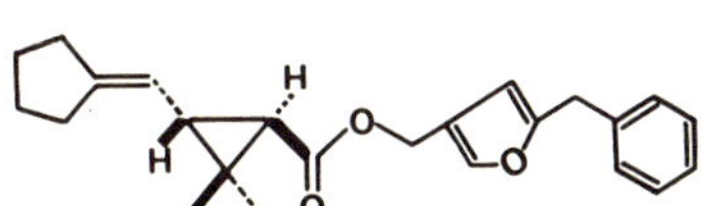
(D)

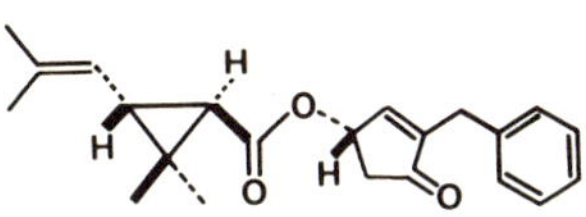
(B)

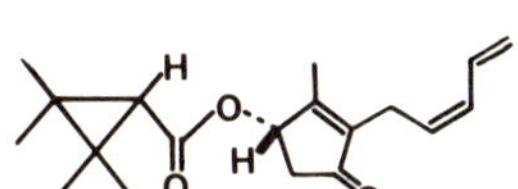
(E)

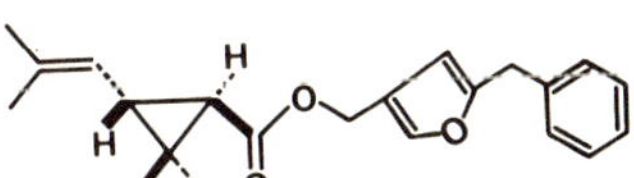
(C)

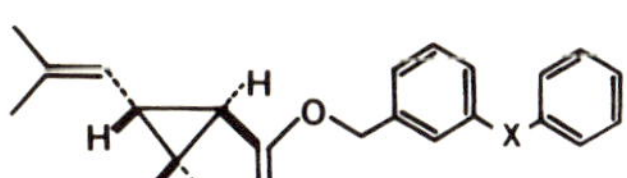
(F)

[c] Topical application of 1-μl drops of acetone solution to adult females, 3 hr after applying sesamex (2 μg).

[d] Topical application of 1-μl drops of acetone solution to adult males and females.

[e] LD_{50} 0.0011 μg per insect (LD_{50} unsynergized 0.33 μg per insect: synergistic factor = 300).

[f] LD_{50} 0.0024 μg per insect.

Table IV shows toxicities obtained by this technique of various pyrethroids relative to pyrethrin I. Also in Table IV are relative toxicities without synergist of the same compounds to mustard beetles; in this insect the mixed function oxidase system is apparently unimportant, because only very little synergism is observed. Data for the two species are therefore directly comparable and indicate more closely the relative potencies at the site of action.

From these and other data (e.g., Barlow *et al.*, 1971) the following conclusions may be drawn about the features in the structure of the natural pyrethrins required for high toxicity.

(*a*) The *gem*-dimethyl group on the cyclopropane ring is essential since compounds without it are almost completely inactive.

(*b*) A side chain on the alcoholic component is also necessary.

(*c*) Unsaturation in the side chain of the alcohol is important but can be alkenyl, cycloalkenyl, or aromatic (furyl, benzyl, or phenoxy).

(*d*) The methyl on C-3 of the cyclopentenolone ring can be replaced by hydrogen.

(*e*) The whole cyclopentenolone ring can be replaced by structures which maintain the stereochemistry between the ester link and the unsaturation in the alcohol side chain. The 3-furylmethyl system is the only substitute known so far that does not lower the toxicity.

(*f*) The substituent on C-3 of the cyclopropane ring is not necessarily isobutenyl (as in pyrethrin I) or substituted propenyl (pyrethrin II). Other effective substituents there are dimethyl (houseflies only) and cyclopentylidenemethyl.

These conclusions indicate that in compounds of the highest activity, the essential groups (the *gem*-dimethyl group on the cyclopropane ring and the unsaturated center in the alcohol side chain) are disposed in the correct steric relationship with one another. This concept is consistent with the effects on toxicity when changes are made only at the three optical centers in the molecule, as follows: (+)-*cis*-(1*R*,3*S*)- and (+)-*trans*-(1*R*,3*R*)-chrysanthemates, in which this relationship is maintained, show high toxicity, but the (—)-*cis*-(1*S*,3*R*)- and (—)-*trans*-(1*S*,3*S*)-esters are almost nontoxic. The relationship is also modified, but less drastically, on epimerization ($S \rightarrow R$) of the asymmetric center at C-4 of the alcohol, which causes a smaller change in toxicity. Changes, apart from inversions at optical centers, necessarily entail alteration in physical properties, which govern penetration into and partition within the insect. Consequently, compounds in which the steric relationship discussed above is present show a range of activities. The physical properties of pyrethrin I clearly approach the optimum for no other compound shows a much superior synergized toxicity.

The precise steric relationship between the *gem*-dimethyl group and the unsaturated center in pyrethrin I at its site of action is difficult to predict, because the molecule is so flexible, as discussed in Section V. However, the high toxicity (Table IV) of 3-benzyl and 3-phenoxybenzyl (+)-*trans*-chrysanthemates may indicate the conformation adopted by pyrethrin I. In these compounds the two benzene rings are constrained by steric hindrance to be nonplanar, so the corresponding groups in pyrethrin I (butadienyl and cyclopentenonyl) may adopt a similarly nonplanar conformation by rotation about the C-1′ to C-2′ bond.

D. Knockdown Action

Recently, Burt and Goodchild (1971b) showed, in a comparison based on relative EC_{50}'s at 3 min, that pyrethrin I is at least ten times more effective in knocking down houseflies when it is injected than when it is applied topically. They concluded that knockdown and kill may be stages in the same process occurring at the same site in the central nervous system, rather than separate phenomena, one occurring very rapidly at a peripheral nervous site. Differences in speed of knockdown by different compounds were associated with differences between rates of penetration and detoxication. The structural requirements for knockdown action differ from those for kill (Barlow *et al.*, 1971). Rapid knockdown appears to be associated with compounds containing the groups for kill (see Section XI,C) correctly oriented but modified at less important parts of the molecule to give different physical properties and hence rates of penetration (Elliott, 1969). This is well illustrated by the examples of pyrethrin I (good kill) and pyrethrin II (good knockdown), where the change of methyl to methoxycarbonyl increases polarity.

References

Barlow, F., Elliott, M., Farnham, A. W., Hadaway, A. B., Janes, N. F., Needham, P. H., and Wickham, J. C. (1971). Insecticidal activity of the pyrethrins and related compounds. IV. Essential features for insecticidal activity in chrysanthemates and related cyclopropane esters. *Pestic. Sci.* **2**, 115–118.

Barthel, W. F., Haller, H. L., and LaForge, F. B. (1944). The preparation of 98% pure pyrethrins for use in freon aerosol bombs. *Soap Sanit. Chem.* **20**(7), 121–135.

Beevor, P. S., Godin, P. J., and Snarey, M. (1965). Jasmolin I, cinerin I, and a new method for isolating research quantities of the pyrethroids. *Chem. Ind. (London)*, pp. 1342–1343.

Begley, M. J., Crombie, L., Simmonds, D. J., and Whiting, D. A. (1972). Absolute configuration of the pyrethrins. Configuration and structure of (+)-allethronyl (+)-*trans*-chrysanthemate 6-bromo-2,4-dinitrophenylhydrazone by X-ray methods. *J. Chem. Soc., D* pp. 1276–1277.

Berteau, P. E., and Casida, J. E. (1969). Synthesis and insecticidal activity of some pyrethroid-like compounds including ones lacking cyclopropane or ester groupings. *J. Agr. Food Chem.* **17**, 931–938.

Berteau, P. E., Casida, J. E., and Narahashi, T. (1968). Pyrethroid-like biological activity of compounds lacking cyclopropane and ester groupings. *Science* **161**, 1151–1153.

Bhalerao, U. T., and Rapoport, H. (1971). Stereochemistry of allylic oxidation with selenium dioxide. Stereospecific oxidation of *gem*-dimethyl olefins. *J. Amer. Chem. Soc.* **93**, 4835–4840.

Bramwell, A. F., Crombie, L., Hemesley, P., Pattenden, G., Elliott, M., and Janes, N. F. (1969). Nuclear magnetic resonance spectra of the natural pyrethrins and related compounds. *Tetrahedron* **25**, 1727–1741.

Büchel, K. H., and Korte, F. (1962). Eine synthese der DL-chrysanthemumsäure. *Z. Naturforsch. B* **17**, 349–350.

Büchi, G., Minster, D., and Young, J. C. F. (1971). A new synthesis of rethrolones. *J. Amer. Chem. Soc.* **93**, 4319–4320.

Bullivant, M. J., and Pattenden, G. (1971). Photochemical decomposition of chrysanthemic acid and its alkyl esters. *Pyrethrum Post* **11**(2), 72–76.

Bullivant, M. J., and Pattenden, G. (1972). Triplet di-π-methane rearrangement of a "free-rotor" 2-prop-2-enylcyclo-pent-2-enone. *J. Chem. Soc., D* pp. 864–865.

Burt, P. E., and Goodchild, R. E. (1970). Mode of action of pyrethroids. *Rothamsted Exp. Sta., Rep.* pp. 166–167.

Burt, P. E., and Goodchild, R. E. (1971a). The site of action of pyrethrin I in the nervous system of the cockroach, *Periplaneta americana. Entomol. Exp. Appl.* **14**, 179–189.

Burt, P. E., and Goodchild, R. E. (1971b). Mode of action of pyrethroids. *Rothamsted Exp. Sta., Rep.*, pp. 185–187.

Campbell, I. G. M., and Harper, S. H. (1945). Experiments on the synthesis of the pyrethrins. I. Synthesis of chrysanthemum monocarboxylic acid. *J. Chem., Soc., London* pp. 283–286.

Campbell, I. G. M., and Harper, S. H. (1952). The chrysanthemumcarboxylic acids. IV. Optical resolution of the chrysanthemic acids. *J. Sci. Food Agr.* **3**, 189–192.

Casida, J. E. (1970). Mixed function oxidase involvement in the biochemistry of insecticide synergists. *J. Agr. Food Chem.* **18**, 753–772.

Chang, S. C., and Kearns, C. W. (1964). Metabolism *in vivo* of C^{14}-labelled pyrethrin I and cinerin I by house flies with special reference to the synergistic mechanism. *J. Econ. Entomol.* **57**, 397–404.

Chen, Y-L., and Casida, J. E. (1969). Photodecomposition of pyrethrin I, allethrin, phthalthrin, and dimethrin. *J. Agr. Food Chem.* **17**, 208–215.

Cocker, W., Lauder, H. St. J., and Shannon, P. V. R. (1972). A two-stage synthesis of (+)-*cis*-homocaronic acid from (+)-car-3-ene. *J. Chem. Soc., D* pp. 684–685.

Copius-Peereboom, J. W. (1969). *In* "Thin Layer Chromatography" (E. Stahl, ed.), pp. 645–647. Springer-Verlag, Berlin and New York.

Corey, E. J., and Jautelat, M. (1967). Construction of ring systems containing the gem-dimethylcyclopropane unit using diphenylsulfonium isopropylide. *J. Amer. Chem. Soc.* **89**, 3912–3913.

Crombie, L., and Elliott, M. (1961). Chemistry of the natural pyrethrins. *Fortsch. Chem. Org. Naturst.* **19**, 120–164.

Crombie, L., and Harper, S. H. (1954). The chrysanthemumcarboxylic acids. IV. The configurations of the chrysanthemic acids. *J. Chem. Soc., London* p. 470.

Crombie, L., Elliott, M., and Harper, S. H. (1950). Experiments on the synthesis of the pyrethrins. Part III. Synthesis of dihydrocinerin-I and tetrahydropyrethrin-I; A study of the action of N-bromosuccinimide on 3-methyl-2-*n*-alkyl (and alkenyl) cyclopent-2-en-1-ones. *J. Chem. Soc., London* pp. 971–978.

Crombie, L., Harper, S. H., and Newman, F. C. (1956). Experiments on the synthesis of the pyrethrins. XI. Synthesis of *cis*-pyrethrolone and pyrethrin I. Introduction of the *cis*-penta-2,4-dienyl system by selective hydrogenation. *J. Chem. Soc., London* pp. 3963–3971.

Crombie, L., Harper, S. H., and Sleep, K. C. (1957). Experiments on the synthesis of the pyrethrins. XIII. Total synthesis of (±)-*cis*- and *trans*-chrysanthemumdicarboxylic acid, (±)-*cis* and *trans*-pyrethric acid, and rethrins II. *J. Chem. Soc., London* pp. 2743–2754.

Crombie, L., Crossley, J., and Mitchard, D. A. (1963). Synthesis, absolute configuration and ring fission of *cis*- and *trans*-homocaronic acid. Their configurative relation to natural terpenes. *J. Chem. Soc., London* pp. 4957–4969.

Crombie, L., Hemesley, P., and Pattenden, G. (1969a). Synthesis of ketols of the natural pyrethrins. *J. Chem. Soc., C* pp. 1016–1024.

Crombie, L., Hemesley, P., and Pattenden, G. (1969b). Synthesis of *cis*-jasmone and other *cis*-rethrones. *J. Chem. Soc., C* pp. 1024–1027.

Crombie, L., Doherty, C. F., and Pattenden, G. (1970). Syntheses of ^{14}C-labelled (+)-*trans*-chrysanthemum mono-, and dicarboxylic acids, and of related compounds. *J. Chem. Soc., C* pp. 1076–1080.

Crombie, L., Ellis, J. A., Gould, R., Pattenden, G., Elliott, M., Janes, N. F., and Jeffs, K. A. (1971a). Oxidative dimerisations of natural rethrolones and related compounds with manganese dioxide. *J. Chem. Soc., C* pp. 9–13.

Crombie, L., Doherty, C. F., Pattenden, G., and Woods, D. K. (1971b). The acid thermal decomposition products of natural chrysanthemum dicarboxylic acid. *J. Chem. Soc., C* pp. 2739–2743.

Crowley, M. P., Godin, P. J., Inglis, H. S., Snarey, M., and Thain, E. M. (1962). The incorporation of ^{14}C-labelled compounds into the flowers of *Chrysanthemum cinerariaefolium* and the biosynthesis of chrysanthemum monocarboxylic acid. *Biochim. Biophys. Acta* **60**, 312–319.

Donegan, L., Godin, P. J., and Thain, E. M. (1962). The separation and estimation of the insecticidal constituents of pyrethrum extract by gas-liquid chromatographic analysis. *Chem. Ind.* (*London*) p. 1420.

Elliott, M. (1961). The pyrethrins and related compounds. II. Infra-red spectra of the pyrethrins and of other constituents of pyrethrum extract. *J. Appl. Chem.* **11**, 19–23.

Elliott, M. (1964a). The pyrethrins and related compounds. III. Thermal isomerization of *cis*-pyrethrolone and its derivatives. *J. Chem. Soc., London* pp. 888–892.

Elliott, M. (1964b). The pyrethrins and related compounds. IV. The ultraviolet absorption of the conjugated *cis*-pentadiene in pyrethrolone. *J. Chem. Soc., London* pp. 1854–1855.

Elliott, M. (1964c). The pyrethrins and related compounds. V. Purification of (+)-pyrethrolone as the monohydrate, and the nature of "pyrethrolone-C." *J. Chem. Soc., London* pp. 5225–5228.

Elliott, M. (1965). The pyrethrins and related compounds. VI. The structures of the "enols" of pyrethrolone. *J. Chem. Soc., London* pp. 3097–3101.

Elliott, M. (1969). Structural requirements for pyrethrin-like activity. *Chem. Ind. (London)* pp. 776–781.

Elliott, M. (1971). The relationship between the structure and the activity of pyrethroids. *Bull. WHO* **44**, 315–324.

Elliott, M., and Casida, J. E. (1972). Optically pure pyrethroids labeled with deuterium and tritium in the methylcyclopentenonyl ring. *J. Agr. Food Chem.* **20**, 295–299.

Elliott, M., and Janes, N. F. (1969). Pyrethrin II and related esters obtained by reconstitution. *Chem. Ind. (London)* pp. 270–271.

Elliott, M., Harper, S. H., and Kazi, M. A. (1967). Experiments on the synthesis of the pyrethrins. XIV. Rethrins and the cyclopentadienone related to 3-methylcyclopent-2-enone. *J. Sci. Food Agr.* **18**, 167–170.

Elliott, M., Janes, N. F., and Jeffs, K. A. (1970). The pyrethrins and related compounds. X. The methylbenzyl chrysanthemates. *Pestic. Sci.* **1**, 49–52.

Elliott, M., Janes, N. F., Kimmel, E. C., and Casida, J. E. (1972). Metabolic fate of pyrethrin I, pyrethrin II, and allethrin administered orally to rats. *J. Agr. Food Chem.* **20**, 300–312.

Farnham, A. W. (1971). Changes in cross-resistance patterns of houseflies selected with natural pyrethrins or resmethrin (5-benzyl-3-furylmethyl (±)-*cis-trans*-chrysanthemate). *Pestic. Sci.* **2**, 138–143.

Fine, B. C., Godin, P. J., Thain, E. M., and Marks, T. B. (1967). Resistance to pyrethrins and DDT in a strain of houseflies, *Musca domestica* L. *J. Sci. Food Agr.* **18**, 220–224.

Fujitani, J. (1909). Chemistry and pharmacology of insect powder. *Arch. Exp. Pathol. Pharmakol.* **61**, 47.

Gnadinger, C. B. (1936). "Pyrethrum Flowers," 2nd ed. McLaughlin, Gormley, King, Minneapolis, Minnesota.

Gnadinger, C. B. (1936–1945). "Pyrethrum Flowers," Supplement to 2nd ed. McLaughlin, Gormley, King, Minneapolis, Minnesota.

Godin, P. J. (1968). The stability of pyrethroid concentrates and pure esters. *Pyrethrum Post* **9**(4), 17 and 21.

Godin, P. J., Inglis, H. S., Snarey, M., and Thain, E. M. (1963). Biosynthesis of the pyrethrins. Part II. Pyrethric acid and the origin of ester methyl groups. *J. Chem. Soc., London* pp. 5878–5880.

Godin, P. J., Inglis, H. S., and Stevenson, J. H. (1965). Insecticidal activity of fresh and dry pyrethrum flowers. *J. Sci. Food Agr.* **16**, 186–191.

Goffinet, B., and Locatelli, A. (1968). Separation of d-*trans*-chrysanthemic acid from its optical and geometric isomers. French Patent 1,536,458; *Chem. Abstr.* **71**, 90923w (1969).

Hammond, J. A. S. (1955). 2,5-dimethyl-2,4-hexadiene. U.S. Patent 2,715,649; *Chem. Abstr.* **50**, 7840 (1956).

Hanafusa, T., Ohnishi, M., and Yakawa, Y. (1970). Thermal *cis-trans* isomerisation of methyl chrysanthemate. *Chem. Ind. (London)* pp. 1050–1052.

Harper, S. H., Reed, H. W. B., and Thompson, R. A. (1951). The chrysanthemum-carboxylic acids. I. Preparation of the chrysanthemic acids. *J. Sci. Food Agr.* **2**, 94–100.

Head, S. W. (1964). The identification of the active constituents of pyrethrum extract when separated by gas-liquid chromatography. *Pyrethrum Post* **7**(4), 12–14.

Head, S. W. (1966). The quantitative determination of pyrethrins by gas-liquid chromatography. Part I. Detection by electron capture. *Pyrethrum Post* **8**(4), 3–7.

Head, S. W. (1967). The quantitative determination of pyrethrins by gas-liquid chromatography. Part II. Detection by hydrogen flame ionization. *Pyrethrum Post* **9**(1), 12–17.

Ingold, C. K. (1953). "Structure and Mechanisms in Organic Chemistry," p. 56. Bell, London.

Inouye, Y., and Ohno, M. (1958). Absolute configuration of natural pyrethrins. *Kagaku (Tokyo)* **28**, 636.

Julia, M. (1964). Preparation of DL-*trans*-chrysanthemic acid. French Patent 1,356,954; *Chem. Abstr.* **61**, 5533 (1964).

Julia, M., and Guy-Rouault, A. (1967). Synthèse de cyclopropanes à partir de sulfones. Application à l'acide chrysanthémique. *Bull. Soc. Chim. Fr.* p. 1411.

Julia, M., Julia, S., Bémont, B., and Tchernoff, G. (1959). Sur l'isomerisation des esters cyclopropaniques *cis* en *trans*. *C. R. Acad. Sci.* **248**, 242–244.

Julia, M., Julia, S., and Jeanmart, C. (1960). Synthèses de l'acide *trans*-dihydrochrysanthémique. *C. R. Acad. Sci.* **251**, 249–251.

Julia, M., Julia, S., Jeanmart, C., and Langlois, M. (1962). Synthèses de la pyrocine et de l'acide *trans*-chrysanthémique. *Bull. Soc. Chim. Fr.* pp. 2243–2246.

Julia, M., Julia, S., and Langlois, M. (1963). Synthèse de l'acide chrysanthémique à partir de l'acide lévulique. *C. R. Acad. Sci.* **256**, 436–438.

Julia, M., Julia, S., and Cochet, B. (1964a). Synthèses de l'acide (±)-*trans*-chrysanthémique à partir du β-ethoxy-isovaleraldehyde et de l'acide (±)-nor-*trans*-chrysanthémique à partir du β-ethoxy-butyraldehyde. *Bull. Soc. Chim. Fr.* pp. 1476–1486.

Julia, M., Julia, S., and Cochet, B. (1964b). Synthèses de la β-β-dimethyl-γ-isobutény1-γ-butyrolactone et transformation en acide chrysanthémique. *Bull. Soc. Chim. Fr.* pp. 1487–1492.

Julia, M., Julia, S., and Langlois, M. (1965a). Synthèses de l'acide *trans* chrysanthémique à partir de la diméthylacroléine et de l'acide *trans nor*-chrysanthémique à partir du crotonaldehyde. *Bull. Soc. Chim. Fr.* pp. 1007–1014.

Julia, M., Julia, S., and Langlois, M. (1965b). Synthèse de l'acide *trans* chrysanthémique à partir de l'acide lévulique. *Bull. Soc. Chim. Fr.* pp. 1014–1019.

Julia, S., Julia, M., and Linstrumelle, G. (1964). Synthèse des acides (±)-*cis*-homocaronique et (±)-*trans*-chrysanthémique par l'intermediaire de bicyclo (3,1,0) héxanones-2-substituées. *Bull. Soc. Chim. Fr.* pp. 2693–2694.

Katsuda, Y., Chikamoto, T., and Inouye, Y. (1958). The absolute configuration of naturally derived pyrethrolone and cinerolone. *Bull. Agr. Chem. Soc. Jap.* **22**, 427–428.

Katsuda, Y., Chikamoto, T., and Inouye, Y. (1959). Relationship between stereoisomerism and biological activity of pyrethroids. V. The absolute configuration of (+)-pyrethrolone and (+)-cinerolone. *Bull. Agr. Chem. Soc. Jap.* **23**, 174–178.

King, T. A., and Paisley, H. M. (1969). Mass spectra and isolation of pyrethroids and related compounds. *J. Chem. Soc., C* pp. 870–874.

LaForge, F. B., and Barthel, W. F. (1947). Constituents of pyrethrum flowers. XX. The partial synthesis of pyrethrins and cinerins and their relative toxicities. *J. Org. Chem.* **12**, 199.

LaForge, F. B., and Green, N. (1952). Constituents of pyrethrum flowers. XXV. The synthesis of d-cinerolone, cinerin I, and its optical isomers. *J. Org. Chem.* **17**, 1635–1640.

LaForge, F. B., Gersdorff, W. A., Green, N., and Schechter, M. S. (1952a). Allethrin-type esters of cyclopropanecarboxylic acids and their relative toxicities to house flies. *J. Org. Chem.* **17**, 381–389.

LaForge, F. B., Green, N., and Schechter, M. S. (1952b). Dimerized cyclopentadienones from esters of allethrolone. *J. Amer. Chem. Soc.* **74**, 5392–5394.

LaForge, F. B., Green, N., and Schechter, M. S. (1954). Allethrin, resolution of dl-allethrolone and synthesis of four optical isomers of *trans*-allethrin. *J. Org. Chem.* **19**, 457–462.

LeMahieu, R. A., Carson, M., and Kierstead, R. W. (1968). The conversion of cinerone into cinerolone. *J. Org. Chem.* **33**, 3660–3662.

LeMahieu, R. A., Tabenkin, B., Berger, J., and Kierstead, R. W. (1970). Microbiological hydroxylation of allethrone. *J. Org. Chem.* **35**, 1687–1688.

Maciver, D. (1968). A modified laboratory scheme for the preparation of (+)-pyrethrolone from crude pyrethrum extract. *Pyrethrum Post* **9**(4), 41.

Martel, J. (1969). d-*trans*-Pyrethric acid. German Patent 1,807,091; *Chem. Abstr.* **72**, 89876x (1970).

Martel, J. (1970a). Cyclopropanecarboxylic acids. German Patent 1,935,986; *Chem. Abstr.* **72**, 100136d (1970).

Martel, J. (1970b). Cyclopropanecarboxylic acids. German Patent 1,935,320; *Chem. Abstr.* **72**, 121078b (1970).

Martel, J., and Buendia, J. (1970a). Racemic and optically active *cis*-chrysanthemumic acids. German Patent 2,010,182; *Chem. Abstr.* **73**, 109362c (1970).

Martel, J., and Buendia, J. (1970b). New syntheses of the natural pyrethric acid. *Proc. Riga Symp. Natur. Prod.*

Martel, J., and Goffinet, B. (1968). (±)-*trans*-Chrysanthemic acid and its esters. French Addition 90,564; *Chem. Abstr.* **70**, 37280n (1969).

Martel, J., and Huynh, C. (1967). Synthèse de l'acide chrysanthémique. II. Accès stéréoséléctif au (±)-*trans*-chrysanthémate d'éthyle. *Bull. Soc. Chim. Fr.* pp. 985–986.

Martel, J., Huynh, C., and Nominé, G. (1966). Unsaturated cyclopentanes. French Patent 1,434,224; *Chem. Abstr.* **65**, 16878 (1966).

Matsui, M., and Horiuchi, F. (1971). Resolution of the chrysanthemic acids with L-lysine. *Agr. Biol. Chem.* **35**, 1984–1985.

Matsui, M., and Meguro, H. (1963). Studies on chrysanthemic acid. IX. Alternate preparation of (+)-*trans*-pyrethric acid and methyl (+)-*trans*-2,2-dimethyl-3-(2-carboxy-1-propyl)cyclopropanecarboxylate. *Agr. Biol. Chem.* **27**, 379–380.

Matsui, M., and Meguro, H. (1964). Studies on chrysanthemic acid. Part XIII. Preparation of rethrins II and rethrins II isomers and their relative toxicities to rethrins I. *Agr. Biol. Chem.* **28**, 27–31.

Matsui, M., and Uchiyama, M. (1962). Studies on chrysanthemic acid. VII. Selective synthesis of (±)-*trans* chrysanthemic acid from (±) pyrocin. *Agr. Biol. Chem.* **26**, 532–534.

Matsui, M., and Ueda, K. (1970). Racemisation of optically active *trans*-cyclopropane carboxylic acids and their derivatives. German Patent 2,013,924; *Chem. Abstr.* **74**, 12697y (1971).

Matsui, M., and Yamada, Y. (1963). Studies on chrysanthemic acid. VIII. Synthesis of pyrethric acids. *Agr. Biol. Chem.* **27**, 373–378.

Matsui, M., and Yamada, Y. (1965). Studies on chrysanthemic acid. Part XV. Selenium dioxide oxidation of tert-butyl (±)-*trans*-chrysanthemate. The isolation of tert-butyl (±)-*trans*-2,2-dimethyl-3-(2′-formyl-3′-hydroxyl-1′-propenyl) cyclopropanecarboxylate. *Agr. Biol. Chem.* **29**, 956–958.

Matsui, M., and Yamamoto, I. (1971). *In* "Naturally Occurring Insecticides" (M. Jacobson and D. G. Crosby, eds.), Chapter 1, p. 3. Dekker, New York.

Matsui, M., and Yoshioka, H. (1965). *trans*-Chrysanthemummonocarboxylic acid. Japanse Patent 6457 ('65); *Chem. Abstr.* **63**, 1822 (1965).

Matsui, M., Uchiyama, M., and Yoshioka, H. (1963a). Studies on chrysanthemic acid. Part X. Oxidation and rearrangement products from chrysanthemic acid, 2-6-dimethylhepta-2,4,6-triene-5-carboxylic acid. *Agr. Biol. Chem.* **27**, 549–553.

Matsui, M., Uchiyama, M., and Yoshioka, H. (1963b). Studies on chrysanthemic acid. Part XI. Oxidation products from chrysanthemic acid. 2,2,-dimethyl-3-(1′-oxo-2′-hydroxy-2′-methyl)propyl-1,3-*trans*-cyclopropane-1-carboxylic acid. *Agr. Biol. Chem.* **27**, 554–557.

Matsui, M., Uchiyama, M., and Yoshioka, H. (1963c). Studies on chrysanthemic acid. Part XII. Racemization of optically active 2,2-dimethyl-3-(1′-oxo-2′-hydroxy-2′-methyl) propyl-1,3-*trans*-cyclopropane-1-carboxylic acid and reduction to give (±)-*trans*-chrysanthemic acid. *Agr. Biol. Chem.* **27**, 558–561.

Matsui, M., Yoshioka, H., Sakamoto, H., Yamada, Y., and Kitahara, T. (1965). Selective synthesis of optically pure (+)- and (−)-*trans*-chrysanthemic acids from (+)–Δ^3-carene. *Agr. Biol. Chem.* **29**, 784–786.

Matsui, M., Yoshioka, H., Sakamoto, H., Yamada, Y., and Kitahara, T. (1967). Studies on chrysanthemic acid. Part XVII. Stereospecific synthesis of optically pure (+)-*trans*-chrysanthemic acid from (+)-Δ^3-carene. *Agr. Biol. Chem.* **31**, 33–37; *Ind. Chim. Belge* **32**, 890.

Matsumoto, T., Nagai, A., and Takahashi, Y. (1963). The stereoselective synthesis of *trans*-chrysanthemum monocarboxylic acid. *Bull. Chem. Soc. Jap.* **36**, 481–482.

Mills, R. W., Murray, R. D. H., and Raphael, R. A. (1973). A new stereoselective synthesis of *trans*-Chrysanthemic Acid [2,2-di-methyl-3-(2-methylprop-1-enyl)cyclopropanecarboxylic acid]. *J. Chem. Soc., Perkin Trans.* **1**, 133–137.

Miskus, R. P., and Andrews, T. L. (1972). Stabilization of thin films of pyrethrins and allethrin. *J. Agr. Food Chem.* **20**, 313–315.

Miyano, M., and Dorn, C. R. (1973). Prostaglandins. VI. Correlation of the absolute configuration of pyrethrolone with that of prostaglandins. *J. Amer. Chem. Soc.* **95**, 2664–2669.

Murano, A. (1972). Gas-liquid chromatographic separation and determination of optical isomers of insecticidal pyrethroids. *Agr. Biol. Chem.* **36**, 917.

Nakada, Y., Yura, Y., and Murayama, K. (1971). Pyrolysis of allethrin. *Bull. Chem. Soc. Jap.* **44**, 1724.

Narahashi, T. (1962). Effect of the insecticide allethrin on membrane potentials of cockroach giant axons and nature of the negative after-potential increased by the insecticide allethrin in cockroach giant axons. *J. Cell. Comp. Physiol.* **59**, 61–66 and 67–76.

Narahashi, T. (1971). Mode of action of pyrethroids. *Bull.* WHO **44**, 337–345.

Negherbon, W. O. (1959). "Handbook of Toxicology," Vol. III. Saunders, Philadelphia, Pennsylvania.

Nishizawa, Y., and Casida, J. E. (1965). Synthesis of d-*trans*-chrysanthemumic acid-1-C^{14}, and its antipode on a semi micro scale. *J. Agr. Food Chem.* **13**, 525–527.

Ohloff, G. (1965). Thermische umlagerung von derivaten des (+)-Δ^4-carens und der (±)-*cis-trans*-chrysanthemumsäure. *Tetrahedron Lett.* pp. 3795–3800.

Pattenden, G., Crombie, L., and Hemesley, P. (1973). Mass spectra of the pyrethrins and related compounds. *Org. Mass Spectrom.* **7**, 719–735.

Pellegrini, J. P., Jr., Miller, A. C., and Sharpless, R. V. (1952). Biosynthesis of radioactive pyrethrins using $^{14}CO_2$. *J. Econ. Entomol.* **45**, 532–536.

Pierre, J. L., Perraud, R., and Arnaud, P. (1970). Résonance magnétique nucléaire des petits cycles. XII. Etude conformationelle des esters et alcools (primaires) *cis* et *trans* chrysanthémiques. *Bull. Soc. Chim. Fr.* pp. 1539–1550.

Rickett, F. E. (1972). Preparative scale separation of pyrethrins by liquid-liquid partition chromatography. *J. Chromatogr.* **66**, 356–360.

Sanders, H. J., and Taft, A. W. (1954). "Allethrin." *Ind. Eng. Chem.* **46**, 414–426.

Sasaki, T., Eguchi, S., and Ohno, M. (1968). Studies on chrysanthemic acid. I. Some reactions of the isobutenyl group in chrysanthemic acid. *J. Org. Chem.* **33**, 676–679.

Sasaki, T., Eguchi, S., and Ohno, M. (1970). Studies on chrysanthemic acid. IV. Photochemical behaviour of chrysanthemic acid and its derivatives. *J. Org. Chem.* **35**, 790–793.

Sawicki, R. M. (1962a). Insecticidal activity of pyrethrum extract and its four insecticidal constituents against house flies. II. Synergistic activity of piperonyl butoxide with the four constituents. *J. Sci. Food Agr.* **13**, 260–264.

Sawicki, R. M. (1962b). Insecticidal activity of pyrethrum extract and its four insecticidal constituents against house flies. III. Knockdown and recovery of flies treated with pyrethrum extract with and without piperonyl butoxide. *J. Sci. Food Agr.* **13**, 283–292.

Sawicki, R. M. (1962c). Insecticidal activity of pyrethrum extract and its four insecticidal constituents against house flies. V. Knockdown activity of the four constituents with piperonyl butoxide. *J. Sci. Food Agr.* **13**, 591–598.

Sawicki, R. M., and Elliott, M. (1965). Insecticidal activity of pyrethrum extract and its four insecticidal constituents against house flies. VI. Relative toxicity of pyrethrin I and pyrethrin II against four strain of house flies. *J. Sci. Food Agr.* **16**, 85–89.

Sawicki, R. M., and Farnham, A. W. (1967). Genetics of resistance to insecticides of the SKA strain of *Musca domestica*. I. Location of the main factors responsible for the maintenance of high DDT-resistance in diazinon selected SKA flies. *Entomol. Exp. Appl.* **10**, 253–262.

Sawicki, R. M., and Farnham, A. W. (1968). Examination of the isolated autosomes of the SKA strain of houseflies (*Musca domestica* L.) for resistance to several insecticides with and without pretreatment with sesamex and TBTP. *Bull. Entomol. Res.* **59**, 409–421.

Sawicki, R. M., and Thain, E. M. (1961). The chemical and biological examination of commercial pyrethrum extracts for insecticidal constituents. *J. Sci. Food Agr.* **12**, 137–145.

Sawicki, R. M., and Thain, E. M. (1962). Insecticidal activity of pyrethrum extract and its four insecticidal constituents against house flies. IV. Knock-down activities of the four constituents. *J. Sci. Food Agr.*, **13**, 292–297.

Sawicki, R. M., Elliott, M., Gower, J. C., Snarey, M., and Thain, E. M. (1962). Insecticidal activity of pyrethrum extract and its four insecticidal constituents against house flies. I. Preparation and relative toxicity of the pure constituents; statistical analysis of the action of mixtures of these components. *J. Sci. Food Agr.* **13**, 172–185.

Schechter, M. S., Green, N., and LaForge, F. B. (1949). Constituents of pyrethrum flowers. XXIII. Cinerolone and the synthesis of related cyclopentenolones. *J. Amer. Chem. Soc.* **71**, 3165–3173.

Shim, K. S., and Martin, D. J. (1971). Ethyl *cis*- and *trans*-chrysanthemumates. German Patent 2,123,989; *Chem. Abstr.* **76**, 45812b (1972) to Stauffer Chemical Co.

Smejkal, J., and Farkas, J. (1963). Hydrogenolytic ring cleavage in substituted cyclopropane carboxylic acids. *Collect. Czech. Chem. Commun.* **28**, 1557–1568.

Smith, H. J. (1960). Detection and estimation of the biologically active constituents of pyrethrum. *J. Sci. Food Agr.* **11**, 172–176.

Stahl, E., and Pfeifle, J. (1966). Thin layer and gas chromatography of pyrethrum-based insecticides. *Pyrethrum Post* **8**(4), 8.

Staudinger, H., and Ruzicka, L. (1924). Insektentotende stoffe, I–V and X. *Helv. Chim. Acta* **7**, 177–201, 201–211, 212–235, 236–244, 245–259, and 448–458.

Staudinger, H., Muntwyler, O., Ruzicka, L., and Seibt, S. (1924). Insektentotende stoffe. VII. *Helv. Chim. Acta* **7**, 390–406.

Stephenson, H. (1960). The separation and estimation of the four insecticidal constituents of *Chrysanthemum cinerariaefolium* by elution chromatography on a column of absorptive charcoal. *Pyrethrum Post* **5**(4), 22–32.

Sumitomo Chemical Co. Ltd. (1963). Preparation of pyrethric acid. Belgian Patent 632,851; *Chem. Abstr.* **61**, 591 (1964).

Sumitomo Chemical Co. Ltd. (1970). Resolution of (±)-*trans*-chrysanthemum monocarboxylic acid, as its quinine salt. British Patent 1,178,423; *Chem. Abstr.* **72**, 90671h (1970).

Sutton, L. E. (1955). *In* "Determination of Organic Structures by Physical Methods" (E. A. Braude and F. C. Nachod, eds.), Vol. 1, p. 405. Academic Press, New York.

Suzuki, Y., Hirai, H., Toyoura, A., and Magara, O. (1970). *trans*-Chrysanthemic acid from *cis*-chrysanthemoyl chloride. German Patent 2,003,065; *Chem. Abstr.* **73**, 76735w (1970).

Tabenkin, B., LeMahieu, R., Berger, J., and Kierstead, R. W. (1969). Microbiological hydroxylation of cinerone to cinerolone. *Appl. Microbiol.* **17**, 714–717.

Tattersfield, F., Hobson, R. P., and Gimingham, C. T. (1929). Pyrethrin I and II. Their insecticidal value and estimation in pyrethrum (*Chrysanthemum cinerariaefolium*). I. *J. Agr. Sci.* **19**, 266–296.

Tedeschi, R. J. (1962). Hydrogenation-hydrogenolysis studies of symmetrically substituted acetylenic glycols. *J. Org. Chem.* **27**, 2398–2402.

Ueda, K., and Matsui, M. (1970). Studies on chrysanthemic acid. Part XIX. Conversion of optically active *trans*-chrysanthemic acid to the racemic one via pyrocine. *Agr. Biol. Chem.* **34**, 1115–1118.

Ueda, K., and Matsui, M. (1971). Studies on chrysanthemic acid. XXI. Photochemical isomerization of chrysanthemic acid and its derivatives. *Tetrahedron* **27**, 2771–2774.

Ueda, K., and Suzuki, Y. (1971). (+)-*trans*-Chrysanthemumic acid by resolution of (±)-*trans*-chrysanthemumic acid. German Patent 2,032,097; *Chem. Abstr.* **74**, 87484k (1971).

Vandewalle, M., and Madeleyn, E. (1970). Cyclopentanones. III. A new synthesis of (±)-allethrolone. *Tetrahedron* **26**, 3551–3554.

Ward, J. (1953). Separation of the pyrethrins by displacement chromatography. *Chem. Ind.* (*London*) p. 586.

West, T. F. (1944). The structure of pyrethrolone and related compounds. Part IV. *J. Chem. Soc., London* pp. 239–242.

Winteringham, F. P., Harrison, A., and Bridges, P. M. (1955). Absorption and metabolism of ^{14}C-pyrethroids by the adult housefly, *Musca domestica, in vivo. Biochem. J.* **61**, 359–367.

Yamamoto, I., and Casida, J. E. (1968). Syntheses of ^{14}C-labeled pyrethrin I, allethrin, phthalthrin and dimethrin on a submillimole scale. *Agr. Biol. Chem.* **32**, 1382–1391.

Yamamoto, R. (1919). The insecticidal principle in *Chrysanthemum cinerariaefolium*. Part I. *J. Tokyo Chem. Soc.* **40**, 126.

Yamamoto, R. (1923). The insecticidal principle in *Chrysanthemum cinerariaefolium*. Parts II and III. On the constitution of pyrethronic acid. *Nippon Kagaku Kaishi* **44**, 311.

Chapter 5

Biochemistry of the Pyrethrins

JOHN E. CASIDA

Introduction

The safe and efficient use of any pesticide chemical, whether it is new or old, persistent or nonpersistent, depends on a thorough knowledge of its biochemical effects and its biodegradation. With some pesticide chemicals it is possible to explain most of their effects at acute lethal and chronic sublethal doses in pest organisms and in mammals by assaying a particular biochemical lesion; with other compounds no such lesion occurs or is known. As natural products, it is of interest to understand the biosynthesis of the pyrethrins because this information may lay the background for improved yields. It is becoming increasingly evident that biodegradable or photolabile pesticides have much less chance of producing permanent detrimental effects on the ecosystem. The specificity of many of the most selectively toxic insecticides results from their relative rates of metabolism in sensitive and nonsensitive organisms. Pyrethrins are almost always used in combination with a synergist (usually piperonyl butoxide); therefore, the biochemical interactions involved between the pyrethrins and synergists must be evaluated.

This review considers the current knowledge of pyrethrins biochemistry and is based on studies that have been conducted, in the most part, during the past dozen years.

II. Action on Enzyme Systems and Intermediary Metabolism

The insecticidal action of pyrethrins has not been definitely associated, to date, with the inhibition of any specific enzyme system or the disruption of a particular biochemical pathway. This is true of both knockdown and lethal effects, whether or not these are independent phenomena. The relatively few biochemical disruptions that have been noted do not appear to be specific for poisoning or blocking with a pyrethroid and may sometimes be secondary affects resulting from the hyperactivity and paralysis. Neuropharmacological studies indicate that the primary mode of action of the pyrethrins probably occurs at a biophysical level and involves the disruption of ion transport at nerve membranes (Camougis, this volume, Chapter 11; Narahashi, 1971).

The inhibitory effect of pyrethrins or allethrin has been tested on several *in vitro* enzyme systems. Cholinesterase activity of the heads from American cockroaches, *Periplaneta americana,* is not inhibited by pyrethrins at 1×10^{-3} *M* (Hartley and Brown, 1955); honey bee, *Apis mellifica,* brain cholinesterase is also insensitive to pyrethrins (Richards and Cutkomp, 1945b). The oxygen uptake of the intact femoral muscle of *P. americana* is not altered by pyrethrins at 6.6×10^{-5} *M* (Fukami, 1956). Three dehydrogenases (citric, malic, and succinic) of the adzuki bean weevil, *Callosobruchus chinensis,* are not affected by pyrethrins at 2×10^{-5} *M* (Suzuki *et al.*, 1951). Cytochrome c oxidase activity of coxal muscle from *P. americana* is not affected by pyrethrins or allethrin at 10^{-5} *M* but complete inhibition occurs at 10^{-3} M; however, many other insecticides produce the same effect on this system and so the results do not establish a mode of action (Morrison and Brown, 1954). The Na^+-K^+-adenosinetriphosphatases (ATPase's) in swine erythrocyte membrane fragments and rat brain microsomes are inhibited by relatively high levels of allethrin, the interaction being competitive with Na^+ in the case of the enzyme from erythrocytes. The apparent blocking of Na^+ activation and the common mode of inhibition of the ATPase's by allethrin and DDT suggest that inhibition of this enzyme may be analogous to effects on nerve function (Schneider, 1972). Na^+-K^+-ATPase of the *P. americana* nerve cord is inhibited by pyrethrum, but the relatively high level required and the shallow slope of the regression line indicate a lack of highly specific action. Other ATPase's inhibited by pyrethrum, requiring even higher levels, include the mitochondrial Mg^{2+}-ATPase of cockroach nerve cord and coxal muscle and the Na^+-K^+- and mitochondrial Mg^{2+}-ATPases of fish brain (Desaiah and Cutkomp, 1972).

Insects paralyzed with pyrethrins are similar to unpoisoned insects with respect to most of the biochemical parameters investigated. With *P. americana,* the staining of the nerve cord, muscle, and digestive organs with 2,3,5-triphenyltetrazolium chloride is not decreased in poisoned roaches as compared with normal roaches and pyrethrins added *in vitro* at 6.6×10^{-5} *M* to the nerve cord do not affect the staining reaction; these observations suggest that the pyrethrins do not inhibit the respiratory metabolism of these tissues (Fukami, 1956). Poisoning of *P. americana* with pyrethrins does not produce detectable changes in the tissue glutathione levels (Forgash, 1951) or in the distribution of cholinesterase activity and sodium ion in either brain or ganglionic tissue (Campbell *et al.,* 1968). Although the acetylcholine content of the whole thorax is increased on pyrethrins poisoning of *P. americana,* this increase is less than that produced by lindane, aldrin, or DDT (Waller and Lewis, 1961). Ingestion of pyrethrins by *P. americana* is associated with alterations of the midgut and caeca involving loss of acid and alkaline phosphatase activity, reduction in phospholipids, and increased glycogen content (Saxena and Srivastava, 1968, 1969a,b); however, these changes are probably not related to the primary mode of insecticidal action of the pyrethrins. The effect of poisoning with pyrethrins is much less than that of poisoning with DDT and related chlorinated hydrocarbon insecticides on the metabolism of formate in houseflies, *Musca domestica* (Cline and Pearce, 1963). Pyrethrins are more toxic to last instar mosquito larvae than they are to cells of mosquito larval tissues cultured *in vitro,* suggesting that the functions of highly organized tissues are more readily interfered with than those of individual cell types comprising them (Mitsuhashi *et al.,* 1970).

Neurotoxins, hormones, and other substances are sometimes released from the nervous system into the hemolymph of insects poisoned with pyrethrins or allethrin. A blood-carried toxin that does not contain fragments of the pyrethrins as a part of its structure is released in paralyzed *P. americana* (Blum and Kearns, 1956). The hemolymph of poisoned *P. americana* causes symptoms resembling those of pyrethroid poisoning when this hemolymph is applied to nerve preparations from normal cockroaches even though the pyrethrins content of the hemolymph is less than 2×10^{-8} *M* (Burt *et al.,* 1971). Pyrethrins poisoning of this cockroach may result in the release of a specific substance from the nerve cord that interferes in the action of another insecticide, malathion, on heart beat (Sudershan and Naidu, 1967). A diuretic hormone is liberated in the blood-sucking bug, *Rhodnius prolixus* (Casida and Maddrell, 1971; Maddrell and Casida, 1971). Although these phenomena may be important in determining the ultimate death or survival of the poisoned insect, they are probably not related to the primary action of the pyrethrins because many other types of insecticide chemicals at paralytic doses initiate the same or a similar sequence of events. The diuretic hormone released from the nervous system in insecticide-induced paralysis of *Rhodnius* acts on the Malpighian tubules to stimulate their secretion, possibly by inducing adenyl cyclase synthesis necessary for increased production of cyclic adenosine 3′,5′-monophosphate (Casida and Maddrell, 1971). The enhanced rate of water loss in *P. americana* and *M. domestica* poisoned with pyrethrins may be caused by a related regulatory mechanism involving nervously controlled secretory activity on the part of the epidermal cells (Ingram, 1955). Allatectomy of *M. domestica* does not alter its susceptibility to pyrethrins poisoning (Sawicki, 1965).

The heart action and hemolymph composition are also of interest in other respects in relation to the pyrethrins, as shown in studies with *P. americana.* The heart beat continues long after the appendages are paralyzed and, in severe poisoning, the blood circulation ceases before the heart beat stops (Coon, 1944). Blocking the hemocytes with carbon particles or Trypan Blue does not alter the susceptibility of the insect to pyrethrins poisoning (McGovran *et al.,* 1944). The effect of pyrethrins on the beat of isolated heart preparations appears to result both from an action on the ganglia and at the neuromuscular junction by release of acetylcholine and not by an anticholinesterase action (Naidu, 1955).

The negative temperature coefficient of pyrethrins action on cockroaches and many other insects has not been explained in biochemical terms; it is not related to the penetration rate of the pyrethrins through the cuticle (Blum and Kearns, 1956) but may be associated, in part, with temperature-dependent changes in the sodium conductance mechanism of the nerve itself (Narahashi, 1971). The finding of alterations in oxygen uptake and carbon dioxide production in insects poisoned with pyrethrins (Harvey and Brown, 1951; Lord, 1949, 1950; Nel and Dürr, 1946) does not suggest a primary mode of action for the pyrethrins since several other insecticidal nerve poisons of different types produce similar effects. The histological changes in insect nerve tissue resulting from pyrethrins poisoning are sometimes of quite specific types (Hartzell, 1945; Lhoste, 1959; Moussa and Banhawy, 1959; Richards and Cutkomp, 1945a; Wigglesworth, 1941) but they are also not defined in biochemical terms. However, it should be noted in this respect that electron microscopic examination of the ultrastructure of *P. americana* brain and ganglia shows no change on pyrethrins poisoning (Campbell *et al.*, 1968). Finally, the action of pyrethrins in reducing dietary uptake by *Corcyra cephalonica* larvae (Gupta and Radhakrishnamurty, 1971), producing latent external injury on *Thermobia domestica* (Sweetman and Gyrisko, 1944), and reducing egg deposition of *Lasioderma serricorne* (Tenhet, 1947) have not been examined as to their biochemical basis, if any.

III. Biosynthesis

Biosynthetic studies are of fundamental interest because of the unique class of compounds involved with a variety of unusual groupings in a precise stereochemical configuration. Also, knowledge of the biosynthesis may provide the basis for increased yields of pyrethrins by the addition of suitable stimulants.

It is important to produce flowers with high amounts of the insecticidal esters, and particularly with high amounts of pyrethrin I when they will be used in the production of mosquito coils. The pyrethrum plant is amazingly efficient in this biosynthesis, producing 2–4 mg of pyrethrins per flower head or 1–2% of pyrethrins relative to the flower dry weight (Head, 1966, 1967). The six insecticidal esters of pyrethrum flowers are localized, for the most part, in the developing seeds or achenes. Although each ester is detectable in the roots, leaves and petioles, and stems of the pyrethrum plant, the highest content is in the flower heads, and 92–94% of the amount in the flower heads is localized in the achenes (Gnadinger and Corl, 1930; Head, 1966); the pyrethrins content is about two-fold greater, on a percent dry weight basis, in the achenes than in the flower as a whole (Head, 1966). The distribution of pyrethrins in the plant might be related to the sites of biosynthesis and/or to the sites of localization after translocation, if any. This aspect warrants further study.

The relative proportions of the six insecticidal constituents in the mature flower heads vary widely in different pyrethrum clones (Head, 1966, 1967). The ratios of pyrethrins–cinerins–jasmolins and of chrysanthemates–pyrethrates are characteristic of a particular clone (Head, 1967) suggesting that large variations occur in different clones in the activity of enzymes or availability of precursors for the various biosynthetic steps. The proportion of pyrethrins–cinerins–jasmolins within the chrysanthemate components is remarkably similar to that within the pyrethrate components (Head, 1967); this supports other evidence, discussed below, that the chrysanthemates serve as precursors for the pyrethrates or at least that chrysanthemic acid is the precursor for the pyrethrates.

Early biosynthesis studies using whole plants or detached flowers of *C. cinerariaefolium* exposed to $^{14}CO_2$ provided samples of labeled pyrethrins but no information on the biosynthetic pathways involved in their formation (Levy and Usubillaga, 1957; Levy *et al.*, 1960; Pelligrini *et al.*, 1952). Later studies by Crowley, Godin, and co-workers and by Pattenden and co-workers, described below, involving the feeding of labeled precursors have led to a partial definition of the biosynthetic pathway for the acid moieties of the pyrethrins (Fig. 1) but, for lack of direct information, only a speculative route can be given for the alcohol moieties (Fig. 2). The following labeled organic acids were fed as the free acids or potassium salts to isolated ovules by vacuum impregnation followed by 15–48 hr metabolism time: [2-^{14}C]acetate, [1-^{14}C]mevalonic acid, [2-^{14}C]mevalonic acid, and [^{14}C]methyl-methio-

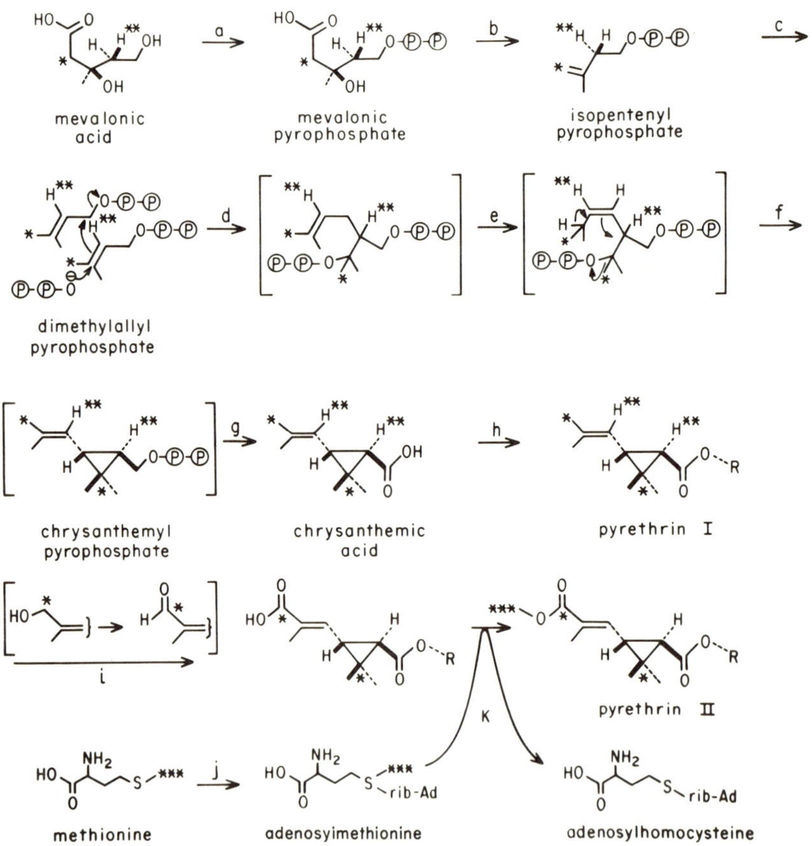

Fig. 1. Biosynthetic pathway for the acid moieties of pyrethrins I and II. The positions of labeling following the feeding of labeled precursors are indicated: *[2-^{14}C]mevalonic acid; **[2-^{14}C]-4*R*[4-^{3}H]mevalonic acid; ***[^{14}C]methyl-methionine.

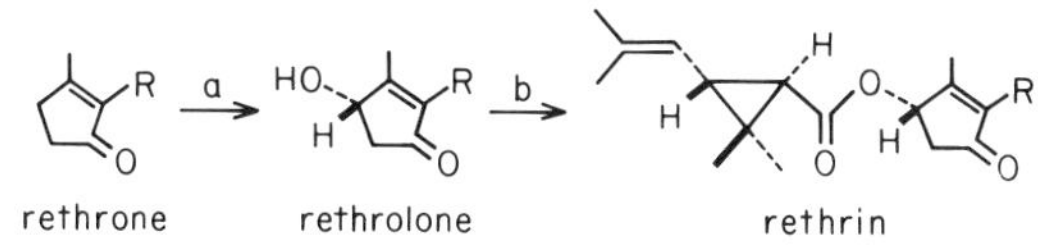

Fig. 2. Possible partial biosynthetic pathway for the alcohol moiety of rethrins.

nine (Crowley *et al.*, 1961, 1962; Godin *et al.*, 1963); [2-^{14}C]-4*R*[4-^{3}H]- and [2-^{14}C]-4S[4-^{3}H]mevalonic acid (G. Pattenden and R. Storer, private communication, unpublished results, 1972); (+)-*trans*-[^{14}C]vinylmethyl chrysanthemic acid (G. Pattenden, private communication, 1972). The sodium salt of [^{14}C]chrysanthemic acid was also administered to isolated flowers followed by 2–3 days metabolism time (Pattenden, 1970). There is reason to believe that both of these feeding procedures yield results indicative of the actual biosynthetic pathways in normal growing plants. Some of the degradation procedures used to establish the positions of ^{14}C labeling in chrysanthemum dicarboxylic acid (Godin *et al.*, 1963) were subsequently found to involve reactions other than those considered at the time of the biosynthesis studies (Crombie *et al.*, 1971); accordingly, some of this early biosynthetic work ought to be reexamined.

The chrysanthemate portion arises from mevalonic acid, losing a ^{14}C label in the 1 position (Crowley *et al.*, 1962) but incorporating a 2-^{14}C label in the *trans*-methyl group of the isobutenyl moiety and in the *gem*-dimethyl groups (Crowley *et al.*, 1961, 1962; Godin *et al.*, 1963). Tritium at the 4*S* position of mevalonic acid is lost, whereas tritium at the 4*R* position is retained in the chrysanthemate (G. Pattenden and R. Storer, private communication, unpublished results, 1972). [^{14}C]Chrysanthemic acid is converted, after uptake by the flowers, to pyrethrins I and II in about a 2:1 ratio, maintaining the original labeling position in the esters ultimately formed (Pattenden, 1970). In another experiment with impregnated ovules obtained early in the flowering season, [^{14}C]chrysanthemic acid gave a high yield (19%) of pyrethrin I but a low yield (0.1%) of pyrethrin II; it appears possible that the unexpected low yield of pyrethrin II might be caused by seasonal differences in the activity of oxidative or other enzymes involved in converting the chrysanthemate to the pyrethrate (G. Pattenden, private communication, 1972). Methionine is the donor of the methyl group for the methoxycarbonyl moiety of pyrethrin II (Godin *et al.*, 1963).

These findings can be combined with the available information on enzyme-mediated isoprene pathways in yeast and mammalian tissue (Cornforth, 1969; Dagley and Nicholson, 1970; Epstein and Rilling, 1970; van Tamelen and Schwartz, 1971) to propose the overall pathway given in Fig. 1. Mevalonic acid is converted to mevalonic pyrophosphate (step *a*) on reaction with two equivalents of adenosine triphosphate (ATP). Decarboxylation and dehydration of mevalonic acid pyrophosphate on reaction with 1 equivalent of ATP yields isopentenyl pyrophosphate (step *b*). Dimethylallyl pyrophosphate is formed (step *c*) by the action of an isomerase on isopentenyl pyrophosphate, the stereochemistry of this isomerization in *C. cinerariaefolium* (G. Pattenden and R. Storer, private communication, unpublished results, 1972) being the same as that found in yeast and mammalian tissue (Cornforth, 1969). Mechanistically, the unusual "middle-to-tail" combination of two dimethylallyl pyrophosphates is probably initiated (step *d*) by the attack of an external or internal pyrophosphate and interaction of the π bond of one dimethylallyl pyrophosphate with the allylic methylene of the other dimethylallyl pyrophosphate. The action of an isomerase (step *e*), followed by proton loss and ring closure (step *f*), then leads to chrysanthemyl pyrophosphate, which undergoes hydrolysis and oxidation (step *g*) to chrysanthemic acid. Esterification to form pyrethrin I (step *h*, R = pyrethronyl) probably involves coenzyme A esterified with chrysanthemic acid as an intermediate. The pyrethronyl esters of palmitic and linoleic acids are also formed (Acree and LaForge, 1937); therefore, the acyltransferase(s) is not specific, utilizing some common fatty acids as well as chrysanthemic acid. Oxidation of pyrethrin I (step *i*) via the alcohol and aldehyde derivatives leads to the carboxylic acid (Pattenden, 1970). It has been claimed that chrysanthemum dicarboxylic acid is present in both the free and combined state in pyrethrum flowers (Ripert, 1934). The final conversion to the pyrethrate requires methylation, undoubtedly by conversion of methionine to adenosylmethionine (step *j*) and transmethylation by an appropriate transferase (step *k*) to form pyrethrin II and adenosylhomocysteine (Dagley and Nicholson, 1970).

The alcohol moiety does not incorporate label from either [1-^{14}C] or [2-^{14}C]mevalonic acid but it is labeled on feeding [2-^{14}C]acetate (Crowley *et al.*, 1961, 1962). Therefore, the rethrolone portion does not arise from isoprene units, as does the acid portion, but [2-^{14}C]acetate is incorporated into both portions, with about twice the incorporation in the alcohol as in the acid moiety (Crowley *et al.*, 1961, 1962). The acetate-labeling pattern in the rethrolone moiety has not been established. The methyl group of L-methionine is also incorporated to a small degree into the alcohol moiety, again without definition as to the labeling sites (Godin *et al.*, 1963). The further biosynthetic steps involve speculation based on the occurrence of other natural compounds, on the known pathways of biosynthesis of related materials and on certain microbial metabolism studies. Jasmone (Fig. 2, rethrone, R = $-CH_2CH\overset{c}{=}CHCH_2CH_3$)

occurs in the oil of *Jasminium grandiflorum* and certain *Mentha* species (Ruzicka and Pfeiffer, 1933; Schmidt, 1947). The related allethrone (Fig. 2, rethrone, $R = -CH_2CH{=}CH_2$) and cinerone (Fig. 2, rethrone, $R = -CH_2CH\overset{c}{=}CHCH_3$) are oxidized by *Aspergillus niger*, producing high yields of the corresponding alcohols, allethrolone and cinerolone (LeMahieu *et al.*, 1970; Tabenkin *et al.*, 1969). Jasmone occurs together with *cis*-hex-3-en-1-ol in *Mentha* and so this alcohol can be considered as a possible precursor of the five carbons of the side chain and one of the ring of jasmone or jasmololone (Crombie and Elliott, 1961). Oxidation of jasmone to jasmololone and esterification via chrysanthemyl coenzyme A and an acyltransferase is then a possible source of jasmolin I. Dehydrogenation or ω-oxidation followed by dehydration of the jasmone, jasmololone, or jasmolins would yield pyrethrone, pyrethrolone, or pyrethrins on the same basis. The biosynthetic origin of the cyclopentenolone ring, or, in fact, of any specific portions of the rethrolone molecule, is unknown but susceptible to study with the available methodology. For example, it is not known whether the pyrethrins, cinerins, and jasmolins arise by independent or by sequential biosynthetic pathways or whether the alcohol side chain synthesis is completed before or after any one of the steps of cyclization, cyclopentenone hydroxylation, and esterification. The simple labeling procedure used to prepare [^{3}H]allethrolone and [^{3}H]pyrethrolone (Elliott and Casida, 1972) should be useful in preparing labeled precursors, such as rethrones and rethrolones, to aid in establishing certain of these points.

It would be of interest to examine the biosynthesis of the alcohol or acid components of the pyrethrins in microbial systems using suitable precursors or in higher plants other than the one examined but, to date, no appropriate system for such studies is known. The claim that pyrethrins occur in the petals of the peony, *Paeonia albiflora* (Pall.) (Chmielewska and Kasprzyk, 1962), has proved to be unfounded (Godin *et al.*, 1967).

IV. Metabolism

It is important to understand the persistence and mechanisms of degradation of insecticide chemicals in mammals and in the environment. Studies on detoxication of the pyrethrins became possible with increasing knowledge of the chemistry and, particularly, the stereochemical features of the pyrethrum constituents. These biodegradation studies have yielded meaningful results only during recent years because of the increased need for this information, the development of procedures for radiosynthesis of the pyrethrins and derivatization of their metabolites, and the application of newer chromatographic and spectroscopic techniques. A technical difficulty that plagued earlier studies and that must still be carefully considered is that of instability; the pyrethrins and related esters including metabolites may decompose during cleanup and analysis unless special precautions are taken to insure against this problem.

The first steps in metabolism of pyrethrins in insects and mammals are now understood and, by considering the metabolism of related compounds, some additional reactions to seek in further studies can be anticipated. The metabolic systems that have been examined include esterase and oxidase enzyme systems, living houseflies, mice and rats, and various microorganisms, applied either to the pyrethrins or compounds with many of the same functional groups. This information has been useful in preparing related synthetic compounds for evaluation as insecticide chemicals. Studies on the biochemistry of these related compounds, particularly allethrin, have, in turn, been helpful in gaining a better understanding of the action and degradation of the pyrethrins. The detoxication reactions involved are given in Fig. 3.

Much of the present knowledge of the metabolism of pyrethrins and allethrin in mammals comes from the studies by Elliott, Casida, and co-workers (Casida *et al.*, 1971; Elliott *et al.*, 1969, 1972a,b; Yamamoto *et al.*, 1971). The cyclopropane carboxylicester group is cleaved to a small extent (Fig. 3, a) in living rats in the case of allethrin but almost no cleavage occurs with the pyrethrins. This differentiates these cyclopentenonyl esters from synthetic pyrethroids that are chrysanthemates of primary alcohols; the primary alcohol esters of the *trans*-chrysanthemates (resmethrin, tetramethrin, and others) are extensively hydrolyzed in rats (Ambrose, 1963, 1964; Masri *et al.*, 1964; Miyamoto *et al.*, 1968, 1971) and an esterase in mouse liver microsomes readily

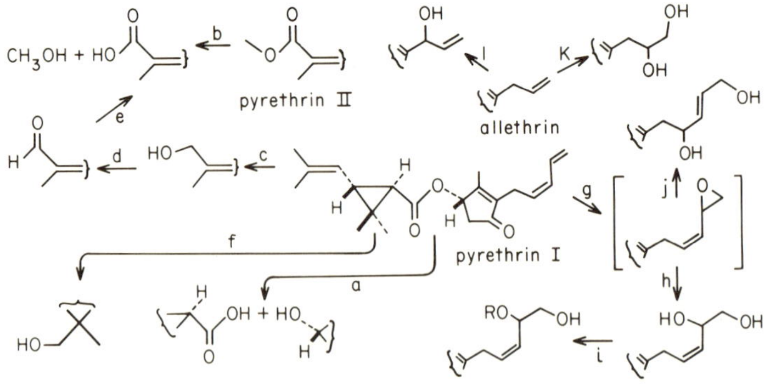

Fig. 3. Detoxication reactions of pyrethrin I, pyrethrin II, and allethrin based on studies with living rats, mice, and houseflies, and various oxidase and esterase enzyme systems from these organisms.

cleaves the *trans*- but not the *cis*-esters (Abernathy and Casida, 1973). The pyrethrate methoxycarbonyl group is readily cleaved (Fig. 3, b) by esterases present in rat or mouse liver resulting in rapid detoxication of the pyrethrates. The majority of the metabolites of pyrethrin I and allethrin arise by various sites of oxidative attack (Fig. 3, c–3, l). The microsomal mixed function oxidase system of houseflies (Yamamoto and Casida, 1966) or rat liver (Elliott *et al.*, 1972b), as well as living houseflies (Yamamoto *et al.*, 1969), quickly oxidize the *trans*-methyl group of the isobutenyl moiety to the alcohol (Fig. 3, c), which is then oxidized by other enzymes to the aldehyde (Fig. 3, d) and the acid (Fig. 3, e) ultimately producing the same acid moiety resulting from pyrethrin II by simple esteratic hydrolysis. Rats fed chrysanthemic acid excrete chrysanthemum dicarboxylic acid, among other products, so a portion of any chrysanthemic acid liberated undergoes oxidative steps c, d, and e of Fig. 3. The acid moiety also may be oxidatively modified, as noted with allethrin, by converting one of the geminal dimethyl groups to the alcohol (Fig. 3, f). The final metabolic modification on the acid moiety, liberating $^{14}CO_2$ from chrysanthemates, including allethrin and pyrethrin I labeled with ^{14}C in the 1 position, must involve cleavage of both the cyclopropane ring and the ester grouping, but the products of the reaction other than $^{14}CO_2$ are not defined.

The alcohol side chain of the pyrethrins and allethrin is rapidly modified in mammals resulting in a series of mono- and dihydroxylated derivatives (Fig. 3, g–l). With pyrethrins administered orally to rats, the initial attack probably involves epoxidation of the terminal double bond of the pentadienyl group (Fig. 3, g) and the epoxide rearranges to the 4′,5′-diol (Fig. 3, h) which undergoes conjugation with an unidentified moiety (R, Fig. 3, i) which possibly is a phenylacetic acid derivative. Alternatively, the epoxide can open and rearrange to the 2′,5′-diol (Fig. 3, j). With allethrin, the double bond of the allyl group is dihydroxylated (Fig. 3, k) probably via the epoxide intermediate and the allylic methylene group is hydroxylated (Fig. 3, l). There are many unidentified metabolites that possibly arise from metabolic attack at two or more of the sites already identified and from conjugation of the metabolites. For example, it is likely that houseflies conjugate the alcohol derivative of allethrin (Fig. 3, c) to form a glucoside (Yamamoto *et al.*, 1969).

The sites of metabolic alteration on the acid moieties of cinerins I and II and jasmolins I and II undoubtedly are the same as with pyrethrins I and II, with a small degree of cyclopropanecarboxylic ester hydrolysis. The variable portion of these compounds is the alcohol side chain and, with this, the terminal methyl group of the cinerins and the 4′-methylene of the jasmolins are expected to undergo hydroxylation with further conjugation or oxidation of the initially formed hydroxy derivatives. Also, the *cis* double bond may be converted to an epoxide that would cleave to the diol and the α-methylene position may be hydroxylated. A final possibility, suggested from fungal metabolism studies indicated below, is that the double bond may be reduced to yield the alkane side chain and its further metabolites, which are expected to undergo subsequent oxidation of the terminal methyl group.

Several streptomycetes, bacteria, and fungi are known to hydroxylate one or both of cinerone (Fig. 4, butenyl side chain) (Tabenkin *et al.*, 1969) and allethrone (Fig. 4, allyl side chain) (LeMahieu *et al.*, 1970). With cinerone, the best organisms for hydroxylation were *Aspergillus niger* and *Streptomyces aureofaciens*. Several of the metabolites of the *A. niger* cultures are identified and suggest sites of oxidation or reduction that might be expected to occur with pyrethrins in other organisms or in mammals. The allyl side chain of allethrone is hydroxylated at the methylene group (Fig. 4, a), in this case the product being analogous to that formed on mammalian metabolism of allethrin (Fig. 3, l). The butenyl side chain of cinerone is hydroxylated on the terminal methyl group (Fig. 4, b) or is reduced to the butyl derivative (Fig. 4, c). Another reaction with allethrone is oxidation of the cyclopentenone methyl group to the alcohol (Fig. 4, d); this reaction should be sought among the metabolic products of the pyrethrins. A final, and a major, reaction of cinerone and allethrone is oxidation to cinerolone and allethrolone (Fig. 4, e). The cinerolone derived from cinerone is sometimes optically active, depending on the organism used, but the desired (+) isomer does not predominate; partial racemization of (+)-cinerolone occurs when this material is incubated in the culture. As pointed out before, this observation of ring hydroxylation suggests that the cyclopentenones may serve as precursors for the cyclopentenolones in biosynthesis of the pyrethrins in pyrethrum plants.

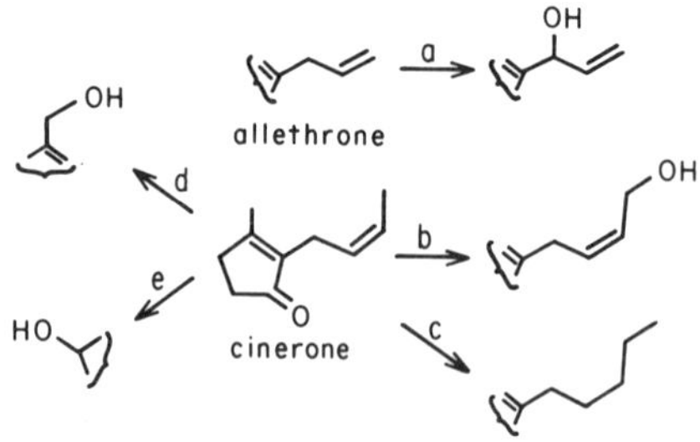

Fig. 4. Metabolites of allethrone and cinerone formed on incubation with *Aspergillis niger* cultures.

Products excreted	Amount, %, following oral dosing with indicated compound	
	pyrethrin I	pyrethrin II
unmetabolized	18.0	4.0
metabolites, R group	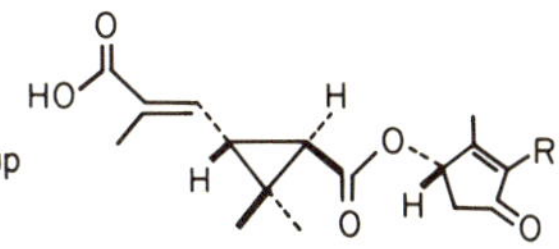	
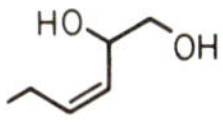	14.3	21.1
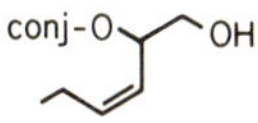	3.9	6.2
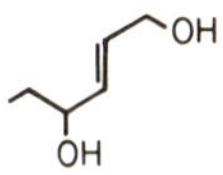	4.4	3.3

Fig. 5. Metabolites of pyrethrins I and II excreted by rats administered these compounds at 3 mg/kg orally.

Some quantitative data is available on the metabolites of pyrethrins I and II excreted by rats administered the labeled compounds at 3 mg/kg orally (Fig. 5) (Elliott *et al.*, 1972b). In addition to the identified compounds, 30% of the administered dose is excreted as unidentified metabolites in each case, the majority of these being very polar compounds. The unmetabolized compound that is excreted appears entirely in the feces; this failure to absorb a portion of the oral dose in rats is also found with tetramethrin (Miyamoto *et al.*, 1968). The ease of *in vivo* hydrolysis of the methoxycarbonyl group probably contributes to the low mammalian toxicity of the pyrethrates, whereas the rapid oxidation of the *trans*-methyl group of the isobutenyl moiety is likely to be a significant factor in reducing the toxicity of the chrysanthemates. The ease of oxidation of the pentadienyl side chain is also important in minimizing the toxicity of the pyrethrins (Elliott *et al.*, 1972a,b).

A portion of the human exposure to pyrethrum involves intake by inhalation. The metabolite fate of inhaled pyrethrum constituents is not known but it is known (C. O. Abernathy and J. E. Casida, unpublished results, 1972) that the mixed function oxidase system of lung microsomes of mice metabolizes allethrin to give the same products, in the most part, as liver microsomes; therefore, it appears likely that inhaled pyrethrins will be metabolized to give similar products to the ingested compounds.

The biochemical basis for the action of synergists in increasing the insecticidal activity of the pyrethrins is relatively well understood. Piperonyl butoxide combines with a site(s) in the microsomal mixed function oxidase system at or near the cytochrome P-450 component, causing slow metabolism of the synergist and inhibition of the oxidative metabolism of the pyrethrins. Certain *O*-propynyl phenylphosphonates inhibit not only oxidative but also esteratic attack on pyrethroids (Casida, 1970; Yamamoto, this volume, Chapter 10). The degree of interaction of pyrethrins and various synergists in toxic actions on rats is considered by Bond, Mauger, and De Feo (this volume, Chapter 9). The level of piperonyl butoxide present as residues in food is unlikely to cause inhibition of microsomal mixed function oxidases in man; therefore, there is no reason to believe that this synergist alters the susceptibility of humans to various drugs, carcinogens, or other agents metabolized by the hepatic mixed function oxidase system (N. C. Brown, 1971; Casida, 1970; Conney *et al.*, 1972).

Insect detoxication of the pyrethrins is partially understood, particularly with houseflies (Yamamoto, this volume, Chapter 10). Pyrethrin I is converted *in vivo* in houseflies and by housefly abdomen homogenates, when these homogenates are fortified with reduced nicotinamide-adenine dinucleotide phosphate, to the alcohol, aldehyde, and carboxylic acid derivatives (Fig. 3, c, d, e) (Yamamoto and Casida, 1966; Yamamoto *et al.*, 1969). There are species and strain differences in the ease of oxidation of pyrethroids by insects. Some housefly strains readily oxidize allethrin, whereas others do not (Tsukamoto and Casida, 1967; Plapp and Casida, 1969; Yamamoto *et al.*, 1969). It appears likely that milkweed bugs do not readily detoxify pyrethrins (Jao and Gordon, 1969) by either oxidative or hydrolytic processes.

Pyrethrins resistance appears to result, in some cases, from an increased ability of the resistant strain to detoxify the insecticide (Fine, 1963; Snyder and Chadwick, 1964). Although this type of resistance is, for now, largely a matter of academic interest rather than practical importance, it can be expected to become of greater importance with the continuing or increased use of pyrethrins. One type of pyrethroid (allethrin) resistance in houseflies appears to be the result of the action of an oxidase(s) controlled by a gene(s) on autosome 2, as shown with the R-Baygon strain (Plapp and Casida, 1969). There is reason to believe that in another housefly strain, the R-Fc, the oxidase(s) are conferred by autosome 5 (Plapp and Casida, 1969). Differences in the cross-resistance patterns of housefly strains selected with pyrethrins or resmethrin were interpreted as indicating that two detoxication factors are involved, one conferring resistance resulting from metabolic attack on the acid moiety and the other from attack on the alcohol moiety, and that the balance of these factors depends on the compound used for selection (Farnham, 1971). Resistance factors other than those associated with pyrethrins detoxication, such as decreased absorption or penetration and reduced nerve sensitivity, also probably contribute to the pyrethrins resistance in houseflies (Farnham, 1971; Fine, 1963; Fine *et al.*, 1963; Plapp and Hoyer, 1968; Tsukamoto, 1969). Selection of blue ticks and several insect species with DDT results in strains that are resistant not only to DDT but to pyrethrins as well; also, several pyrethrins-resistant pests are cross-resistant to DDT (A. W. A. Brown, 1971; Cole and Clark, 1961; Farnham, 1971; Fine, 1963; Whitehead, 1959). Therefore, the previous history of selection of the strain for reistance to DDT or other insecticides influences the susceptibility of this strain to pyrethrins and this may have implications in the continued effectiveness of pyrethrins for use in the field (Fine, 1963). It is also known that DDT and dieldrin fed to housefly strains resistant to these insecticides induce an increased level of allethrin-detoxifying oxidases (Plapp and Casida, 1970). Therefore, pyrethrins resistance can result from induction of detoxifying enzymes within a given population as well as from selection of the population for those individuals with the highest pyrethrins-detoxifying activity.

V. Summary

The mechanism of insecticidal action of the pyrethrins is not definitely defined at the biochemical level, despite many studies in this area. Possibly a biophysical rather than a biochemical action is involved. The chrysanthemate and pyrethrate moieties of the pyrethrins are formed from mevalonic acid in pyrethrum flowers. The pathways of biosynthesis of the alcohol moieties are not established. Houseflies detoxify pyrethrin I by oxidation of the *trans*-methyl group of the isobutenyl moiety but there are also many unidentified metabolites with the ester group intact. Oxidative metabolism of the pyrethrins in insects may constitute one mechanism of resistance to these insecticides. The utility of synergists in reducing the dose of pyrethrins necessary for insect control appears to result from inhibition of the oxidative metabolism of the pyrethrins so that they persist longer in the insect. In living rats, pyrethrin II undergoes carbomethoxy hydrolysis, whereas pyrethrin I is oxidized at the *trans*-methyl group of the isobutenyl moiety; the *cis*-pentadienyl side chain is also oxidized, yielding two diols, a portion of one of them being conjugated prior to excretion. The enzymes primarily involved in limiting the persistence and toxicity of pyrethrin II to rats appear to be esterases, whereas those involved with pyrethrin I are microsomal mixed function oxidases. The biodegradability of the pyrethrins probably accounts for their low toxicity to mammals and the failure of these compounds to yield persisting residues in environmental organisms.

Acknowledgment

Preparation of this review was supported in part by Public Health Service grant 2 P01 ES00049 and The Rockefeller Foundation. The author thanks G. Pattenden of the University of Nottingham and S. W. Head of the Pyrethrum Marketing Board for useful suggestions.

References

Abernathy, C. O., and Casida, J. E. (1973). Pyrethroid insecticides: Esterase cleavage in relation to selective toxicity. *Science* **179**, 1235–1236.

Acree, F., Jr., and LaForge, F. B. (1937). Constituents of pyrethrum flowers. X. Identification of the fatty acids combined with pyrethrolone. *J. Org. Chem.* **2**, 308–313.

Ambrose, A. M. (1963). Toxicologic studies on pyrethrin-type esters of chrysanthemumic acid. I. Chrysanthemumic acid, 6-chloropiperonyl ester (Barthrin). *Toxicol. Appl. Pharmacol.* **5**, 414–426.

Ambrose, A. M. (1964). Toxicologic studies on pyrethrin-type esters of chrysanthemumic acid. II. Chrysanthemumic acid, 2,4-dimethylbenzyl ester. *Toxicol. Appl. Pharmacol.* **6**, 112–120.

Blum, M. S., and Kearns, C. W. (1956). Temperature and the action of pyrethrum in the American cockroach. *J. Econ. Entomol.* **49**, 862–865.

Brown, A. W. A. (1971). Pest resistance to pesticides. *In* "Pesticides in the Environment" (R. White-Stevens, ed.), Vol. 1, Part II, pp. 457–552. Dekker, New York.

Brown, N. C. (1971). A review of the toxicology of piperonyl butoxide. *Pyrethrum Post* **11**(2), 66–68.

Burt, P. E., Lord, K. A., Forrest, J. M., and Goodchild, R. E. (1971). The spread of topically-applied pyrethrin I from the cuticle to the central nervous system of the cockroach *Periplaneta americana*. *Entomol. Exp. Appl.* **14**, 255–269.

Campbell, W. A., Patel, K., Becker, R. A., and Hartmann, J. F. (1968). Resistance of insect nerve ultrastructure to alteration by pyrethrins. *Pyrethrum Post* **9**(4), 13–16.

Casida, J. E. (1970). Mixed-function oxidase involvement in the biochemistry of insecticide synergists. *J. Agr. Food Chem.* **18**, 753–772.

Casida, J. E., and Maddrell, S. H. P. (1971). Diuretic hormone release on poisoning *Rhodnius* with insecticide chemicals. *Pestic. Biochem. Physiol.* **1**, 71–83.

Casida, J. E., Kimmel, E. C., Elliott, M., and Janes, N. F. (1971). Oxidative metabolism of pyrethrins in mammals. *Nature (London)* **230**, 326–327; *Pyrethrum Post* **11**(2), 58–59 and 71.

Chmielewska, I., and Kasprzyk, Z. (1962). Pyrethrin and gallotannin in the flowers of peony. *Nature (London)* **196**, 776.

Cline, R. E., and Pearce, G. W. (1963). Unique effects of DDT and other chlorinated hydrocarbons on the metabolism of formate and proline in the housefly. *Biochemistry* **2**, 657–662.

Cole, M. M., and Clark, P. H. (1961). Development of resistance to synergized pyrethrins in body lice, and cross resistance to DDT. *J. Econ. Entomol.* **54**, 649–651.

Conney, A. H., Chang, R., Levin, W. M., Garbut, A., Munro-Faure, A. D., Peck, A. W., and Bye, A. (1972). Effects of piperonyl butoxide on drug metabolism in rodents and man. *Arch. Environ. Health* **24**, 97–106.

Coon, B. F. (1944). Effects of paralytic insecticides on heart pulsations and blood circulation in the American cockroach as determined with a fluorescein indicator. *J. Econ. Entomol.* **37**, 785–789.

Cornforth, J. W. (1969). Exploration of enzyme mechanisms by asymmetric labelling. *Quart. Rev., Chem. Soc.* **23**, 125–140.

Crombie, L., and Elliott, M. (1961). Chemistry of the natural pyrethrins. *Fortschr. Chem. Org. Naturst.* **19**, 120–164.

Crombie, L., Doherty, C. F., Pattenden, G., and Woods, D. K. (1971). The acid thermal decomposition products of natural chrysanthemumdicarboxylic acid. *J. Chem. Soc., C* pp. 2739–2743.

Crowley, M. P., Inglis, H. S., Snarey, M., and Thain, E. M. (1961). Biosynthesis of the pyrethrins. *Nature (London)* **191**, 281–282.

Crowley, M. P., Godin, P. J., Inglis, H. S., Snarey, M., and Thain, E. M. (1962). The biosynthesis of the "pyrethrins." I. The incorporation of ^{14}C-labelled compounds into the flowers of *Chrysanthemum cinerariaefolium* and the biosynthesis of chrysanthemum monocarboxylic acid. *Biochim. Biophys. Acta* **60**, 312–319.

Dagley, S., and Nicholson, D. E. (1970). "An Introduction to Metabolic Pathways," pp. 186–187 and 205. Wiley, New York.

Desaiah, D., and Cutkomp, L. K. (1972). Sensitivity of the cockroach, *Periplaneta americana* ATPases to pyrethrum. *Annu. Meet., Entomol. Soc. Amer.* Paper 30, Section B.

Elliott, M., and Casida, J. E. (1972). Optically pure pyrethroids labeled with deuterium and tritium in the methylcyclopentenonyl ring. *J. Agr. Food Chem.* **20**, 295–299.

Elliott, M., Kimmel, E. C., and Casida, J. E. (1969). ^{3}H-Pyrethrin I and -pyrethrin II: Preparation and use in metabolism studies. *Pyrethrum Post* **10**(2), 3–8.

Elliott, M., Janes, N. F., Kimmel, E. C., and Casida, J. E. (1972a). Mammalian metabolites of pyrethroids. *In* "Insecticides" (A. S. Tahori, ed.), Vol. 1, pp. 141–162. Gordon & Breach, New York: *Pyrethrum Post* **11**(3), 94–103.

Elliott, M., Janes, N. F., Kimmel, E. C., and Casida, J. E. (1972b). Metabolic fate of pyrethrin I., pyrethrin II, and allethrin administered orally to rats. *J. Agr. Food Chem.* **20**, 300–313.

Epstein, W. W., and Rilling, H. C. (1970). Studies on the mechanism of squalene biosynthesis. The structure of presqualene pyrophosphate. *J. Biol. Chem.* **245**, 4597–4605.

Farnham, A. W. (1971). Changes in cross-resistance patterns of houseflies selected with natural pyrethrins or resmethrin (5-benzyl-3-furylmethyl (±)-*cis-trans*-chrysanthemate). *Pestic. Sci.* **2**, 138–143.

Fine, B. C. (1963). The present status of resistance to pyrethroid insecticides. *Pyrethrum Post* **7**(2), 18–21 and 27.

Fine, B. C., Godin, P. J., and Thain, E. M. (1963). Penetration of pyrethrin I labelled with carbon-14 into susceptible and pyrethroid-resistant houseflies. *Nature (London)* **199**, 927–928.

Forgash, A. J. (1951). The effect of insecticides and other toxic substances upon the reduced glutathione of *Periplaneta americana*. *J. Econ. Entomol.* **44**, 870–878.

Fukami, J. (1956). Effect of some insecticides on the respiration of insect organs, with special reference to the effects of rotenone. *Botyu-Kagaku* **21**, 122–128.

Gnadinger, C. B., and Corl, C. S. (1930). Studies on pyrethrum flowers. II. Relation between maturity and pyrethrins content. *J. Amer. Chem. Soc.* **52**, 680–684.

Godin, P. J., Inglis, H. S., Snarey, M., and Thain, E. M. (1963). Biosynthesis of the pyrethrins. Part II. Pyrethric acid and the origin of the ester-methyl groups. *J. Chem. Soc., London* pp. 5878–5880.

Godin, P. J., King, T. A., Stahl, E., and Pfeifle, J. (1967). "Pyrethrum" in peonies. *Nature (London)* **214**, 319.

Gupta, M. R., and Radhakrishnamurty, R. (1971). Estimation of diet intake by the larvae of *Corcyra cephalonica* and *Tribolium castaneum* using a ^{45}calcium-labelled compound. *J. Insect Physiol.* **17**, 2491–2497.

Hartley, J. B., and Brown, A. W. A. (1955). The effects of certain insecticides on the cholinesterase of the American cockroach. *J. Econ. Entomol.* **48**, 265–269.

Hartzell, A. (1945). Histological effects of certain sprays and activators on the nerves and muscles of the housefly. *Contrib. Boyce Thompson Inst.* **13**, 443–454.

Harvey, G. T., and Brown, A. W. A. (1951). The effect of insecticides on the rate of oxygen consumption in *Blattella*. *Can. J. Zool.* **29**, 42–53.

Head, S. W. (1966). A study of the insecticidal constituents in *Chrysanthemum cinerariaefolium*. (1) Their development in the flower head. (2) Their distribution in the plant. *Pyrethrum Post* **8**(4), 32–37.

Head, S. W. (1967). A study of the insecticidal constituents of *Chrysanthemum cinerariaefolium*. (3) Their composition in different pyrethrum clones. *Pyrethrum Post* **9**(2), 3–7.

Ingram, R. L. (1955). Water loss from insects treated with pyrethrum. *Ann. Entomol. Soc. Amer.* **48**, 481–485.

Jao, L. T., and Gordon, H. T. (1969). Toxicity of certain pyrethroids and carbamates to the CS strain of *Oncopeltus fasciatus*. *J. Econ. Entomol.* **62**, 612–616.

LeMahieu, R. A., Tabenkin, B., Berger, J., and Kierstead, R. W. (1970). Microbiological hydroxylation of allethrone. *J. Org. Chem.* **35**, 1687–1688.

Levy, L. W., and Usubillaga, A. (1957). Biosynthesis of radioactive pyrethrins. *Bol. Inform. Cient. Nac.* **9**, 42–49.

Levy, L. W., Muñoz, M. O., Muggia, F., and Jimenez, J. O. (1960). Formation of pyrethrins in detached pyrethrum flowers. A study with radio-isotopes. *Pyrethrum Post* **5**(4), 3–6.

Lhoste, J. (1959). Histopathologie des insectes intoxiques par les pyrethrines associees a des synergistes. *Pyrethrum Post* **5**(2), 19–21 and 26.

Lord, K. A. (1949). The effect of insecticides on the respiration of *Oryzaephilus surinamensis:* An attempt to compare the speeds of action of a number of D.D.T. analogues. *Ann. Appl. Biol.* **36**, 113–138.

Lord, K. A. (1950). The effect of insecticides on respiration. II. The effects of a number of insecticides on the oxygen uptake of adult *Tribolium castaneum* Hbst. at 25°C. *Ann. Appl. Biol.* **37**, 105–122.

McGovran, E. R., Yeager, J. F., Mayer, E. L., and Munson, S. C. (1944). Effect of KF and pyrethrum on *Periplaneta americana* (L.) after hemocytes are blocked with Chinese ink or nephrocytes are strained with trypan blue. *Ann. Entomol. Soc. Amer.* **37**, 414–419.

Maddrell, S. H. P., and Casida, J. E. (1971). Mechanism of insecticide-induced diuresis in *Rhodnius*. *Nature (London)* **231**, 55–56.

Masri, M. S., Jones, F. T., Lundin, R. E., Bailey, G. F., and DeEds, F. (1964). Metabolic fate of two chrysanthemumic acid esters: Barthrin and dimethrin. *Toxicol. Appl. Pharmacol.* **6**, 711–715.

Mitsuhashi, J., Grace, T. D. C., and Waterhouse, D. F. (1970). Effects of insecticides on cultures of insect cells. *Entomol. Exp. Appl.* **13**, 327–341.

Miyamoto, J., Sato, Y., Yamamoto, K., Endo, M., and Suzuki, S. (1968). Biochemical studies on the mode of action of pyrethroidal insecticides. I. Metabolic fate of phthalthrin in mammals. *Agr. Biol. Chem.* **32**, 628–640.

Miyamoto, J., Nishida, T., and Ueda, K. (1971). Metabolic fate of resmethrin, 5-benzyl-3-furylmethyl dl-*trans*-chrysanthemate in the rat. *Pestic. Biochem. Physiol.* **1**, 293–306.

Morrison, P. E., and Brown, A. W. A. (1954). The effects of insecticides on cytochrome oxidase obtained from the American cockroach. *J. Econ. Entomol.* **47**, 723–730.

Moussa, T. A., and Banhawy, M. (1959). Effect of insecticides on the bodies of Nissl and neurofibrillae of the locust, *Schistocerca gregaria*. *Nature (London)* **184**, 915–916.

Naidu, M. B. (1955). Physiological action of drugs and insecticides on insects. *Bull. Entomol. Res.* **46**, 205–220.

Narahashi, T. (1971). Mode of action of pyrethroids. *Bull. WHO* **44**, 337–345.

Nel, R. G., and Dürr, H. J. R. (1964). Rate of carbon dioxide production by cockroaches dusted with DDT and other insecticidal dusts. *J. Entomol. Soc. S. Afr.* **9**, 115–126.

Pattenden, G. (1970). Some studies on the biosynthesis of the pyrethrins. *Pyrethrum Post* **10**(4), 2–5.

Pellegrini, J. P., Jr., Miller, A. C., and Sharpless, R. V. (1952). Biosynthesis of radioactive pyrethrins using $C^{14}O_2$. *J. Econ. Entomol.* **45**, 532–536.

Plapp, F. W., Jr., and Casida, J. E. (1969). Genetic control of house fly NADPH-dependent oxidases: Relation to insecticide chemical metabolism and resistance. *J. Econ. Entomol.* **62**, 1174–1179.

Plapp, F. W., Jr., and Casida, J. (1970). Induction by DDT and dieldrin of insecticide metabolism by house fly enzymes. *J. Econ. Entomol.* **63**, 1091–1092.

Plapp, F. W., Jr., and Hoyer, R. F. (1968). Possible pleitropism of a gene conferring resistance to DDT, DDT analogs, and pyrethrins in the house fly and *Culex tarsalis*. *J. Econ. Entomol.* **61**, 761–765.

Richards, A. G., Jr., and Cutkomp, L. K. (1945a). Neuropathology in insects. *J. N.Y. Entomol. Soc.* **53**, 313–354.

Richards, A. G., Jr., and Cutkomp, L. K. (1945b). The cholinesterase of insect nerves. *J. Cell. Comp. Physiol.* **26**, 57–61.

Ripert, J. (1934). Sur un nouveau procédé d'analyse des produits contenant des extraits de pyrèthre. *Ann. Fals. Fraudes* **27**, 580–595.

Ruzicka, L., and Pfeiffer, M. (1933). Über Jasminriechstoffe. I. Die Konstitution des Jasmons. *Helv. Chim. Acta* **16**, 1208–1214.

Sawicki, R. M. (1965). Similarity in response to diazinon, dieldrin, and pyrethrum extract by allatectomized and normal houseflies (*Musca domestica*). *Bull. Entomol. Res.* **55**, 727–732.

Saxena, S. C., and Srivastava, J. P. (1968). On histopathology and histochemistry of insecticide-treated insects. I. Non-specific phosphatases in the midgut and caeca of pyrethrins treated *Periplaneta americana* L. *Pyrethrum Post* **9**(3), 9–13.

Saxena, S. C., and Srivastava, J. P. (1969a). The histopathology and histochemistry of insecticide-treated insects. II. Glycogen in the mid-gut of *Periplaneta americana* L. treated with pyrethrum. (Blattaria:Blattidae). *Pyrethrum Post* **10**(1), 12–13 and 23.

Saxena, S. C., and Srivastava, J. P. (1969b). The histopathology and histochemistry of insecticide-treated insects. III. The phospholipids in the mid-gut of *Periplaneta americana* L. treated with pyrethrum. (Blattaria:Blattidae). *Pyrethrum Post* **10**(1), 14–16.

Schmidt, H. (1947). Zur Kenntnis des Pfefferminzöls. Vorkommen von Jasmon im ätherischen Öl von *Mentha piperita* L. *Chem. Ber.* **80**, 538–546.

Schneider, R. P. (1972). Kinetics in inhibition of (Na + K)-ATPase by DDT and allethrin. *Fed. Proc., Fed. Amer. Soc. Exp. Biol.* **31**, 591.

Snyder, F. M., and Chadwick, L. E. (1964). The course of poisoning of normal and pyrethrins-resistant houseflies by pyrethrins-piperonyl butoxide residues: A kinetic analysis. *Entomol. Exp. Appl.* **7**, 229–240.

Sudershan, P., and Naidu, M. B. (1967). Effect of insecticides and insecticide treated cockroach blood on the heart beat of *Periplaneta americana* L. *Indian J. Exp. Biol.* **5**, 215–218.

Suzuki, R., Yamasaki, T., and Ishii, T. (1951). The effects of some insecticides on the dehydrogenase of the adzuki-bean weevil, *Callosobruchus chinensis*. *Oyo Kontyu* **7**, 58; *Chem. Abstr.* **48**, 3624e (1954).

Sweetman, H. L., and Gyrisko, G. G. (1944). Latent injury from pyrethrum and rotenone dusts. *J. Econ. Entomol.* **37**, 746–749.

Tabenkin, B., LeMahieu, R. A., Berger, J., and Kierstead, R. W. (1969). Microbiological hydroxylation of cinerone to cinerolone. *Appl. Microbiol.* **17**, 714–717.

Tenhet, J. N. (1947). Effect of sublethal dosages of pyrethrum on oviposition of the cigarette beetle. *J. Econ. Entomol.* **40**, 910–911.

Tsukamoto, M. (1969). Biochemical genetics of insecticide resistance in the housefly. *Residue Rev.* **25**, 289–314.

Tsukamoto, M., and Casida, J. E. (1967). Metabolism of methylcarbamate insecticides by the $NADPH_2$-requiring enzyme system from houseflies. *Nature (London)* **213**, 49–51.

van Tamelen, E. E., and Schwartz, M. A. (1971). Mechanism of presqualene pyrophosphate-squalene biosynthesis. *J. Amer. Chem. Soc.* **93**, 1780–1782.

Waller, J. B., and Lewis, S .E. (1961). The effect of γ-isomer of 1,2,3,4,5,6-hexachlorocyclohexane and other insecticides on acetylcholine levels in the cockroach. *J. Insect Physiol.* **7**, 315–323.

Whitehead, G. B. (1959). Pyrethrum resistance conferred by resistance to DDT in the blue tick. *Nature (London)* **184**, 378–379.

Wigglesworth, V. B. (1941). The effect of pyrethrum on the spiracular mechanism of insects. *Proc. Roy. Entomol. Soc. London, Ser. A* **16**, 11–14.

Yamamoto, I., and Casida, J. E. (1966). *O*-Demethyl pyrethrin II analogs from oxidation of pyrethrin I, allethrin, dimethrin, and phthalthrin by a house fly enzyme system. *J. Econ. Entomol.* **59**, 1542–1543.

Yamamoto, I., Kimmel, E. C., and Casida, J. E. (1969). Oxidative metabolism of pyrethroids in houseflies. *J. Agr. Food Chem.* **17**, 1227–1236.

Yamamoto, I., Elliott, M., and Casida, J. E. (1971). The metabolic fate of pyrethrin I, pyrethrin II, and allethrin. *Bull. WHO* **44**, 347–348.

Part III

TOXICOLOGY AND PHARMACOLOGY

Chapter 6

Toxicity of Pyrethrum and Its Constituents to Mammals

WILLIAM F. BARTHEL

Introduction

Pyrethrum is one of the oldest organic insecticides in use in the world today. There is some evidence that the crusaders brought the information back to Europe that the ground flower heads of a daisy (*Chrysanthemum roseum*) helped keep body lice under control. Certainly the epidemics of louse-borne diseases in Europe through the ages showed that this information went unheeded. O'Brien (1967) states that a Mr. Jumtikoff of Armenia discovered that certain Caucasian tribes used ground flowers as an insecticide in the early 1800's, with subsequent commercialization of the flowers by his son in 1828.

Despite this long history of use, O'Brien (1967) was still impelled to say, "The mechanism of pyrethrin toxicity and of its remarkable selectivity remain a mystery." Yet, in spite of this lack of knowledge of the mechanism of toxicity of the pyrethrins, we know that the rapid metabolism of the pyrethrins in mammals accounts for their selective toxicity. This was most fully established by Elliott *et al.* (1972) in their report on the metabolic fate of pyrethrin I, pyrethrin II, and allethrin.

Many reviews (Ambrose and Robbins, 1951; Anonymous, 1970; Gaudin, 1937; Lehman, 1952; Metcalf, 1955; Negherbon, 1959) on the toxicology of pyrethrins have appeared. What is disappointing about these reviews is that the same authors are cited and little new information is introduced. However, some definitive reports have come out since the studies of Shimkin and Anderson in 1936. These new reports include those of Bond and DeFeo (1969), Gaudin (1937), Kimbrough *et al.* (1968), Leonard (1942), Malone and Brown (1968), Camougis *et al.* (1971), Carpenter *et al.* (1950), Moore (1966, 1972), and most recently Verschoyle and Barnes (1972).

This review will be restricted to a discussion of the active principles of pyrethrum, *Chrysanthemum cinerariaefolium*. It considers synergists and synthetic pyrethroids only to the extent needed to illustrate the toxicity of the pyrethrins.

Staudinger and Ruzicka (1924a–d) first described the active principles of pyrethrum and designated them as pyrethrin I and pyrethrin II. Later, LaForge and Barthel (1944, 1945a,b) showed that two additional active compounds were involved and called them cinerin I and cinerin II. Further, LaForge and Barthel (1945c, 1947) elucidated the structure of both the pyrethrins and cinerins. Subsequently, LaForge and Soloway (1947a,b) corrected the proposed structural formula by shifting the location of the hydroxyl group of the alcoholic moiety to its correct and more logical position. More recently, Godin and co-workers (1964, 1965, 1966) showed that there are two additional active ingredients, jasmolin I and jasmolin II. This discussion then, by necessity, will cover the toxicology of the natural mixture of these six compounds.

Although the use of pyrethrum and pyrethrins dates back to antiquity, detailed accounts of well-conducted toxicological studies on this mixture of natural compounds are few.

II. Toxicity to Mammals

A. Acute Toxicity

O'Brien (1967), in his text on action and metabolism of insecticides, refers to the toxicity of pyrethrum (in this case the natural mixture of pyrethrins, cinerins, and jasmolins) but cites data from Negherbon

(1959) as do many of the reviewers of the toxicity of pyrethrins. Negherbon, however, was citing earlier work by other investigators so it is more appropriate and correct to consider the earlier work at this point rather than to continue to cite Negherbon.

Carpenter *et al.* (1950), at the Mellon Institute, made a rather exhaustive study of the toxicity of pyrethrins and allethrin to rats, rabbits, and dogs with inhalation studies, oral and skin exposure. They reported the oral LD_{50} of the pyrethrum oleoresin at 820 mg/kg on male and female Sherman strain white rats, whereas the LD_{50} of the purified 20% pyrethrins extract was 1870 mg/kg. Although the procedures used in the two tests varied a little, the comparison is still valid enough to show the greater toxicity of the unpurified oleoresin.

In an early study of the acute oral toxicity of pyrethrins, Lehman (1954) reported the LD_{50} to be 200 mg/kg for rats. He gave no details of the study, but the work was done in the Division of Pharmacology of the United States Food and Drug Administration (FDA). In the same report Lehman gave the intravenous LD_{50} at 10.5 mg/kg, which points out the high intrinsic toxicity of pyrethrins, which is not realized in practice because it is so readily broken down or metabolized to nontoxic metabolites.

These figures are somewhat at variance with data summarized by Negherbon (1959). Negherbon cites references giving the LD_{50} of pyrethrins for rats at 820 to about 1500 mg/kg. The variation among these figures is too great to be ignored. In subsequent citations perhaps we may be able to see some of the causes for this seeming discrepancy. LD_{50} values of pyrethrins against rats are listed in Table I.

Shimkin and Anderson (1936) conducted a number of studies with mice, rats, and guinea pigs using dilute pyrethrum solutions. The interpretation of their results is difficult because one cannot tell whether they used 0.1 to 2.4% dilutions or 14% oleoresin in any particular series. The distinction is important because the very dilute solutions may not act at all like the more concentrated oleoresin. They report that pyrethrins killed at 100 to 150 mg/kg intraperitoneally with rats, mice, and guinea pigs. Because they did not give data on untreated controls, one cannot tell whether the mortality at lower doses is significant. In oral tests they seem to show an LD_{50} of 1500 mg/kg with guinea pigs, although they used only 18 animals to get this figure. With rats they report no mortality at doses of 500, 1000, and 1500 mg/kg.

Bond and DeFeo (1969) conducted the most recent detailed and definitive study on the toxicity of pyrethrum to be reported. Using five groups of ten rats they found the LD_{50} to be 7200 mg/kg. As they were working with a pyrethrum concentrate containing 20% pyrethrins,

TABLE I

Reported LD_{50} of Pyrethrins Using White Rats as Test Animals

Reference	Date	Test material	Route	Reported LD_{50} (mg/kg)
Shimkin and Anderson	1936	Dilute solution and oleoresin	oral	>1500
Carpenter *et al.*	1950	oleoresin	oral	820
Carpenter *et al.*	1950	purified concentrate	oral	1870
Ambrose and Robbins	1951	86.2% concentrate	oral	>2600
Ambrose and Robbins	1951	86.2% concentrate	subcutaneous	>1600
Lehman	1954		oral	200
Lehman	1954		dermal	200–400
Hayes	1965	20% concentrate	oral	100–300
Malone and Brown	1968	partially dewaxed oleoresin	oral	794
Malone and Brown	1968	pale extract	oral	584
Malone and Brown	1968	oleoresin	oral	634
Malone and Brown	1968	77.8% concentrate	oral	715–900
Beck	1969	commercial concentrate	oral	480
Bond and DeFeo	1969	20% concentrate	oral	1440
Casida *et al.*	1971	pyrethrin I	oral	260–420
Casida *et al.*	1971	pyrethrin II	oral	>600
T. B. Gaines and R. D. Kimbrough, personal communication	1972	74% concentrate	oral	470
Verschoyle and Barnes	1972		oral	>1400

the 7200 must be divided by 5 if one wishes to have the LD_{50} for 100% material. Such a crude assumption is not desirable, but few people have used highly concentrated pyrethrins extracts for toxicological work. Using this simple arithmetic, the Bond and DeFeo LD_{50} for male and female white rats is calculated as 1440 mg/kg.

In 1951 Ambrose and Robbins reported that purified pyrethrins concentrate containing 86.2% total pyrethrins administered gastrically to rats in doses of 2600 mg/kg produced no toxic reactions. Even doses subcutaneously of 1600 mg/kg produced no toxic reaction. These studies were done to assess the toxicology of allethrin, and pyrethrins were included only as a standard. The authors state, "From these preliminary observations it is concluded that purified allethrin is no more toxic than purified pyrethrins." Ambrose and Robbins were working with a highly purified pyrethrins concentrate prepared by the method of Barthel *et al.* (1944).

All too little toxicological woık has been done with such concentrates, so we have little evidence of what the other constituents of pyrethrum contribute to the overall toxicity. The great variation in toxicities reported by various investigators may be caused more by variation in the purity of the starting material than any other single factor.

T. B. Gaines and R. D. Kimbrough (personal communication, 1972) did extensive work with pyrethrum, determining the LD_{50} and the effects of chronic ingestion on reproductive potential; but shortly after their work was completed a fire destroyed their laboratory and the records of the study. However, they repeated some of their work on 74% purified pyrethrins furnished by J. B. Moore of McLaughlin Gormley King Company.[1] The results of their study with Sherman strain white rats are presented in Table II. From four of the rats that died from single doses of 450 mg/kg given as 9% pyrethrins in peanut oil the following histopathological notes were made, "Brains normal; lungs, intraaveolar hemorrhage; heart, normal." Some evidence of liver vacuolation was found in two of these rats. In another study (T. B. Gaines and R. D. Kimbrough, personal communication, 1972) with ten rats dosed at 300 mg/kg of pyrethrins in a 20% concentrate, six out of ten female rats were killed.

In yet another study, Kimbrough *et al.* (1968) reported on the combined effects of DDT, pyrethrum, and piperonyl butoxide. At a dietary level of 1000 ppm pyrethrins, there was some enlargement of the liver but no mortality. Even when combined with 10,000 ppm piperonyl butoxide, there was still no mortality, but the liver weights were still higher as percentages of body weight. Only when DDT was added to the pyrethrins or to pyrethrins and piperonyl butoxide was there any mortality. With respect to pathology, these latter authors report that pyrethrins cause liver changes similar to these caused by DDT:

TABLE II

Oral Toxicity of 74% Pyrethrins Concentrate to Sherman Strain White Rats

Sex	LD_{50} (mg/kg)	Range in LD_{50} (mg/kg)
Male	470	431–512
Female	263	231–300

[1] Use of trade names is for identification only and does not constitute endorsement by the Health Services and Mental Health Administration or by the United States Department of Health, Education, and Welfare.

uniform enlargement of the liver, smoother cytoplasm, many cytoplasmic inclusions, and moderate vacuolization.

Weir (1966a) studied the acute oral toxicity of pyrethrins with rats. He administered by stomach tube pyrethrins as a 20% solution in corn oil. This first study gave an acute LD_{50} for pyrethrins for male albino rats of between 681 and 1000 mg/kg. When, in a second study, they repeated this to obtain more data for statistical analysis, all the animals given 562 and 826 mg/kg died. Therefore, a third study was done with freshly prepared 20% pyrethrins solution. In this third study the LD_{50} was found to be 710 mg/kg with confidence limits between 568 and 888 mg/kg. The animals that died showed depression, labored or rapid breathing, ataxia, sprawling of limbs, tremors, and weight loss. Those that recovered did so between the second and sixth day. At necropsy, the findings were congestion of the lungs, liver, kidneys, and adrenals and slight gastric inflammation.

Casida *et al.* (1971), using pure samples of pyrethrin I and pyrethrin II, found a low order of toxicity for pyrethrin II, although the LD_{50} for pyrethrin I fell in the range of 260–420 mg/kg. Some rats survived total doses of 450 to 2000 mg/kg of pyrethrin I administered in two to six portions over 12–54 hr. The LD_{50} of pyrethrin II could not be calculated, although it was higher than 600 mg/kg with rats surviving cumulative doses of 2900 mg/kg.

Beck (1969), in comparing pyrethrins, allethrin, and bioallethrin [(±)-allethronyl (+)-*trans*-chrysanthemate], found the pyrethrins to be of about the same toxicity as the other two compounds, as shown in Table III. The greater toxicity they found for males than females is the opposite of that found by Gaines and Kimbrough with highly concentrated pyrethrins.

Hayes (1965), without citing the source of the data, reported the LD_{50} of pyrethrum to rats as 100–300 mg/kg. Presumably this work

TABLE III

Comparison of Toxicity of Allethrin, Pyrethrins, and Bioallethrin to White Rats

Compound	LD_{50} (mg/kg)
Allethrin	603 ± 94 (males)
	774 ± 146 (females)
Pyrethrins	480 ± 97 (males)
	840 ± 14 (females)
Bioallethrin	500 ± 75 (males)
	860 ± 15 (females)

was done in his laboratory and reflects data lost in the fire that occurred at about the time his report was published.

Malone and Brown (1968) reported on results of studies with several grades of pyrethrum extract, including the nitromethane concentrate prepared according to Barthel *et al.* (1944) and containing 77.8% pyrethrins. Results of their studies are given in Tables IV and V as presented in their article. These data show the difference in susceptibility of the rat, mouse, and chick, and that the degree of purity of the extract or concentrate is important. The 77.8% concentrate is less toxic to all species studied by all authors.

Leonard (1942) obtained concentrates of pyrethrin I and pyrethrin II from LaForge for animal toxicity studies. The pyrethrin I concentrate assayed 65% pyrethrin I, 22% pyrethrin II, and 12% unidentified material. The pyrethrin II concentrate contained 95% pyrethrin II and 5% pyrethrin I. When the pyrethrin II concentrate was injected intraperitoneally in mice at the dose of 40 mg/kg reported by Gaudin (1937) to be toxic, only transient diarrhea developed. It was not until the dose was raised to 240 mg/kg that the LD_{50} was approached. Four of six mice died from 240 mg/kg. The survivors were normal. Another mouse given 480 mg/kg suffered convulsions for 6 hr but survived. Convulsions and tremors at these higher doses were similar to those reported by Gaudin for lower doses. Leonard reported that 240 mg/kg pyrethrin II in rats only caused some diarrhea with no convulsions or locomotor effects. In guinea pigs, at 240 mg/kg, again diarrhea was the only symptom outside of some deepened respiratory amplitude.

Verschoyle and Barnes (1972), using pyrethrin I and pyrethrin II prepared by reconstituting the component alcohols and acids obtained from the natural material, determined LD_{50}'s against Porton strain female white rats using glycerol formal solutions. The oral LD_{50} of natural pyrethrins was found to be greater than 1400 mg/kg. On intravenous treatment, the LD_{50} values were 5 mg/kg for pyrethrin I, 1 mg/kg for pyrethrin II, and 5 mg/kg for the natural pyrethrins. They interpret the intravenous toxicity results as indicating that the nervous systems of insects and mammals are almost equally sensitive to the pyrethrins and respond in a similar manner. The data on the dose differential for oral and intravenous toxicity dramatically point out the ease of detoxication of the pyrethrins. Animals surviving an intravenous dose bordering on the lethal quickly appeared completely normal, suggesting rapid repair of the damage to nerve membranes.

Leonard (1942) reported a decrease in amplitude of contraction to irreversible paralysis of active duodenal sections of isolated rat intestine when suspended in Locke's solution to which was added pyrethrin II

TABLE IV

Rat and Mouse Intraperitoneal and Chick Perivisceral LD_{50} Values and 95% Confidence Limits[a,b]

Grade of pyrethrum	Formulation dosed		Rat		Mouse		Chick	
	Diluent	Pyrethrins content (% w/v)	LD_{50} (mg/kg)	95% Confidence limits	LD_{50} (mg/kg)	95% Confidence limits	LD_{50} (mg/kg)	95% Confidence limits
PD	Neat	20.8	167	123–227	191	162–225	496	370–689
							502*	245–1030
Pale	Neat	20.9	189	142–251	185	143–239	382	65–1170
							409*	258–648
OR	Neat	27.2	341	266–437	317	258–389	1262	886–1795
	OPD	20.0	175	133–230	172	134–220	381	272–533
NP	Neat	77.8	798	630–1010	452	368–555	1565	974–3334
					333*	240–406		
	OPD	18.0	208	158–274	160	101–256	284	217–371
							240*	162–356
	Corn oil	25.7			226	164–314		

[a] Reprinted with permission of *Pyrethrum Post*.

[b] *Key:* All LD_{50} values are expressed in mg/kg of total pyrethrins. OPD, Odorless petroleum distillate; *, result of retest on same sample; PD, partially dewaxed oleoresin; Pale, pale extract; OR, pyrethrum oleoresin; NP, nitromethane concentrate of pyrethrins.

TABLE V

Rat and Mouse Oral LD_{50} Values and 95% Confidence Limits[a,b]

	Formulation dosed		Rat		Mouse	
Grade of pyrethrum	Diluent	Pyrethrins content (% w/v)	LD_{50} (mg/kg)	95% Confidence limits	LD_{50} (mg/kg)	95% Confidence limits
PD	Neat	20.3	794	604–1045	461	364–583
Pale	Neat	20.9	584	481–710	796	648–978
OR	Neat	27.2	634	495–812	344	283–432
	OPD	20.0	584	452–755	273	213–350
NP	Neat	77.8	900	733–1106	308	254–374
			715*	575–888	285*	203–401
	OPD	18.0	900	733–1106	786	564–918

[a] Reprinted with permission of *Pyrethrum Post.*

[b] *Key:* All LD_{50} values are expressed in mg/kg of total pyrethrins. OPD = Odorless petroleum distillate; * = result of retest on same sample; PD = partially dewaxed oleoresin; Pale = pale extract; OP = pyrethrum oleoresin; NP = nitromethane concentrate of pyrethrins.

at a concentration of 1:6000. This finding agreed with that of Perot and Gaudin (1933), who used rabbit intestine. Leonard stated that the finding suggests that the intense purgative action in the intact mouse or rat is not caused by local stimulation but is secondary to central nervous system stimulation. Leonard reported that the pyrethrin I mixture in sesame oil appeared less toxic than pyrethrin II. Two mice given 275 mg/kg survived, although they went through prolonged convulsions without prostration. Nine of 14 mice given 550 mg/kg died.

Lorber (1972) compared the effect of pyrethrins on intact and splenectomized dogs. He reported no effect from the dosage in normal dogs but did observe erythroid hyperplasia in the bone marrow and an increase in reticulocytes in peripheral blood in two recently splenectomized dogs. This finding would seem to indicate that splenectomy permits some of the pyrethrins, which would otherwise be metabolized in the liver, to reach the bone marrow where they interfered with hematoporesis.

Lutz-Ostertag and Lutz (1970) injected synergized pyrethrins into eggs and noted strong teratogenic effects on the chick embryos. Apparently, this works by a sterilizing effect on the gonads, they report. The females were not affected after 17 days' incubation, but the males were more sensitive. The finding posed, but did not answer, the question of whether nonsynergized pyrethrins would have the same effect. It should be mentioned that there is no evidence in mammalian systems, at least, that unmetabolized pyrethrins ever reach the gonads or any part of the reproductive system—they are simply metabolized too rapidly.

Springfield *et al.* (1971) have shown that pyrethrins cause significant liver enlargement in rats along with increased microsomal enzyme activity. This result is at variance with what many toxicologists have believed and again raises the question about the purity of the material these investigators worked with.

B. Chronic Toxicity

Some of the first reliable data on chronic toxicity of pyrethrum to animals was reported by Lehman in 1952. In a 104-week feeding study with white rats, he found the lowest feeding level of pyrethrins to produce gross effects to be 5000 ppm and the highest level without gross effects, 1000 ppm. His lowest level with tissue damage was 5000 ppm, and his highest level without tissue damage was 1000 ppm. From this work Lehman (1954) reported the chronic oral LD_{50} at 250 mg/kg for rats. In other words, the animal can ingest a greater amount of pyrethrins than the acute LD_{50} every day and still not exceed a 50% mortality over a period of 90 days.

A World Health Organization (WHO) working paper on pyrethrins (Anonymous, 1970) reported that Leong and Martin (1966), whose original report was unavailable, conducted a study of the chronic toxicity of pyrethrins. The working paper indicates two groups of dogs (three male and three female) were fed 0 to 5000 ppm of pyrethrins in the diet for 3 months. The animals getting the pyrethrins had poor appetite and lost weight. One died, but gross and microscopic examination showed no disorders that could be attributed to pyrethrins. All of the animals exhibited tremors, occasional salivation, and labored respiration. The same working paper cites a report by Weir (1966b) in which 5000 ppm pyrethrins fed to dogs showed definite toxic effect. Without the original reports, one cannot tell if these last two citations were two dog studies or one. The report goes on to indicate that at 5000 ppm no gross pathological or microscopic abnormalities were observed in the dog. Further, Weir (1966b) reported that in a reproduction study with dogs receiving 5000 ppm in the diet for 21 days after mating, there was a reduction in size of litters and in weight. No other effects were noted.

In the WHO working paper are cited the results of a reproduction study to assess teratogenicity with rabbits (Anonymous, 1970). Two groups of nine rabbits were administered 90 mg/kg/day of pyrethrins in corn oil from day 0 to day 16 of gestation. No pathological effects on animals or pups were noted, and no teratogenic effects from pyrethrins were observed.

Bond and DeFeo (1969) fed one quarter of the acute LD_{50} level to rats (or 380 mg/kg/day) for 90 days with only one death out of 20 animals. The controls showed one death in ten animals. Pathology in the animals fed those doses was minimal. Two of the animals showed in detail what might be some increase in eosinophilia of the cytoplasm of liver, indicating the beginning of cell necrosis. Two other animals showed slight kidney involvement. The rest of their study involved combinations of pyrethrins with other drugs, for instance ethyl alcohol, aspirin, caffeine, nicotine, and phenobarbital. As might be expected, toxic effects of pyrethrins were enhanced when pyrethrins were added to ethyl alcohol, caffeine, and aspirin, but not with phenobarbital or nicotine in acute studies. In chronic studies, including combinations with piperonyl butoxide, Tropital, and sulfoxide, some interesting conclusions were drawn. Pyrethrins with alcohol or caffeine produced many deaths in the treated animals. Piperonyl butoxide was less toxic when combined with pyrethrins, whereas sulfoxide was more lethal. The combination with Tropital produced fewest lethal effects. Likewise, the pyrethrins–Tropital mixture produced fewest histopathological changes. Most treatments resulted in reduced weight gains, which was not unexpected because the doses were very high. Except in the case of caffeine mixtures, the livers of the animals showed weight gains, with the piperonyl butoxide treatment producing the greatest liver enlargement.

Although the FDA has not been lenient in setting tolerances for chlorinated hydrocarbons and phosphates, it has recognized a high margin of safety in tolerances for pyrethrins. In a memorandum of February 6, 1956 from Robert S. Roe of the Division of Pharmacology to Commissioner George P. Larrick, the statement was made, "The requested tolerance of 2.7 ppm for pyrethrins is quite safe. Pyrethrins have so little cumulative toxicity that the Division of Pharmacology would be willing to see it exempted from the requirement of a tolerance when used with piperonyl butoxide. The Division of Foods recommends that pyrethrins be exempted."

C. Inhalation Toxicity

When the United States became involved in a war in the Pacific islands during World War II, the need to control mosquito-borne diseases was critical. Out of this need came the immediate acceptance of aerosol bombs as a means of insect control. The first bombs were high pressure, that is, of about 40 to 75 lb/in^2. As a result many fine, respirable particles were present. The urgent need for toxicological data on these insecticidal aerosols led to many investigations in a short period of time. Most of these studies have not appeared in the literature. In fact, when Carpenter *et al.* reported on their work with a new insecticide, allethrin, in 1950, they cited no previous work.

The first aerosol bombs were made by Westinghouse Corporation out of floats from their refrigeration condensers. The first tests on their possible effects on animals were conducted by Drs. Paul S. Neal and W. F. Van Oettingen of the United States Public Health Service. They sprayed the cages of test rats every day until the cages dripped with the oil from the bombs. Some rats died of pneumonia induced by inhalation of the oil, but no deaths could be attributed to the pyrethrins or to the pyrethrins–sesame oil solution used as a synergistic combination. As the exposure was repeated over many weeks, the military felt no qualms about using the aerosol bombs on personnel at only a few grams per 1000 cubic feet.

The only detailed work on inhalation studies found is that of Carpenter *et al.* (1950). Their inhalation studies with aerosols were particularly interesting. With 27 to 85 half-hour exposures at 0.5 gm pyrethrins in peanut oil per 1000 ft^3, no excess mortality or histopathological changes were observed in white rats compared with controls dosed with peanut oil alone. A minor diffuse thickening of the walls of the alveoli of the lungs was noted in some rats killed after 27 of the 30-min exposures, but because this was not unique to any one exposure group, it was attributed to the peanut oil. This was certainly a legitimate conclusion because rats killed after 85 exposures did not show this thickening nor did those killed after a 21-day period following last exposure. In the study 384 tissues were taken from 96 animals for histopathological examination. Out of this number, only 13 tissues were considered at all abnormal; five of these were in the peanut oil group, and only three were in the pyrethrins' group. Those were not considered evidence of significant effect. Hematologic studies showed no significant differences from normal. In inhalation studies of aerosols with dogs subjected to 40 30-min exposures of 0.5 gm pyrethrins per 1000 ft^3, the only effect noticed, and that was questionable, was minor congestion of the lungs in a few of the dogs. The effects were of dubious significance, because congestion was present in a few dogs in several different exposure groups, and these had been immunized against distemper a month before the study. Hematologic studies showed no significant abnormalities. Even in exposure of rats to 167 to 185 gm per 100 ft^3, no abnormal histopathological conditions were noted except minor lung congestion, and none of the five rats died. These studies are especially significant when one realizes that they were done at exposure levels of up to 10,000 times the recommended human exposure.

Jolly and Waterhouse (1962) covered inhalation toxicity of pyrethrins in detail in their report, but their data is taken from Carpenter *et al.* (1950).

Weir and Crews (1966) exposed rats to aerosols of 1% pyrethrins at 2 mg/liter for 1 hr and found no observable gross or microscopic changes except pinpoint hemorrhages in the lungs of a few rats.

Martin and Leong (1966) exposed rats to a fine mist of 1% pyrethrins at a nominal concentration of 2 mg/liter for 1 hr. No animals died, although all appeared hyperexcitable and exhibited extreme hostility toward each other. These signs disappeared in 24 hr. Gross examination showed all tissues normal except for pinpoint hemorrhages in the lungs.

D. Dermal Toxicity

Lehman (1954) reported the dermal LD_{50} of pyrethrins to rabbits as 200–400 mg/kg. This work was apparently done in the laboratories of the FDA's Division of Pharmacology.

Without indicating the amount used, Shimkin and Anderson (1936) rubbed 2.4% pyrethrins into the hair and skin of six animals (rats or guinea pigs), with no mortality but some hair loss. A 1% solution placed in a rabbit's eye caused no immediate observable irritation, but did cause a conjunctival bleb after 6 hr.

Carpenter *et al.* (1950) reported a LD_{50} value of 2060 mg/kg for dermal toxicity in a skin penetration test with male albino New Zealand strain white rabbits with pyrethrum oleoresin. Another commercial preparation of the same strength permitted survival of all five rabbits at a dose level of 3160 mg/kg.

Malone and Brown (1968) report toxicity at the maximum dose that could be applied in dermal toxicity studies with rats and rabbits. All formulations tested with rats showed a dermal LD_{50} greater than 1350 mg/kg with the nitromethane concentrate being 5400 mg/kg. With rabbits, the LD_{50} for odorless petroleum distillate solution of nitromethane extract was 4500 mg/kg, whereas the LD_{50} for the extract itself was 19,800 mg/kg. Other formulations gave a LD_{50} with rabbits at 5000 mg/kg.

These data, except for Lehman's report, indicate that pyrethrum and the pyrethrins are poorly absorbed through the intact skin. Jolly and Waterhouse (1962) refer to a value by Leonard for dermal toxicity, but Leonard does not mention dermal studies in the article cited (Leonard, 1942).

The WHO working paper (Anonymous, 1970) reported no skin sensitization in one study of two groups of nine male guinea pigs and a 1% pyrethrins formulation.

III. Toxicity to Humans

A. Poisoning

Despite the long period of time during which pyrethrum flowers, their powder, or extracts have been used for pest control, few documented cases of human poisoning from their use or misuse have been reported.

Hayes (1963) made note of one case of a 2-year-old child who ate about $\frac{1}{2}$ oz. of pyrethrum powder and subsequently died. No details were given in the original article (Anonymous, 1889). Hayes (1969), in his paper on "Pesticides and Human Toxicity," cites three references dealing with pyrethrins. All, however, are on dermatitis or allergic reactions.

Gnadinger, in his comprehensive review (1936, 1945) of pyrethrum covering 2364 references up through 1945, cites no authenticated cases of poisoning by pyrethrum. He refers to a number of old cases where illness or death was reported, but then states, "These instances are not well authenticated because the purity of the powder was not established." Gnadinger states that he took doses of 50 mg by mouth, and reports no effect other than numbness of the tongue and lips. Perhaps the greatest amount of information on the possible hazard of poisoning from the pyrethrins comes from the work of Chevalier (1930). In studies with children and adults he dosed the subjects with coated pills containing pyrethrins for control of intestinal worms (*Ascaris*, tapeworms, *Trichuris*, and *Oxyuris*). The coating on the pill protects the pill through the stomach, where otherwise metabolism would start destroying it, negating its use as a vermifuge, and dissolves in the intestine, setting the pyrethrins free where they are needed as a vermifuge. Chevalier used dosages of 5 mg daily for three consecutive days for children, with a double dosage for adults. These doses were insufficient in some cases, so the doses were doubled, that is, 10 mg for children and 20 mg for adults daily for three consecutive days. He found no untoward reactions in the many cases treated and was assured of the complete harmlessness of the treatment. One group on an archeological investigation in Syria used this treatment periodically for a year as a preventive.

They developed neither worms, which other treatments had not prevented, nor side effects. He further recommended an external rinse of pyrethrins in alcohol for parasites causing itching at the anus.

B. Allergic Responses

Reports on investigations of allergic responses caused by pyrethrins make up the bulk of the toxicological literature dealing with human exposure to pyrethrins.

One of the early records of allergic reaction is a report on pyrethrum dermatitis among factory workers by McCord *et al.* (1921). Their investigations reported that about 30% of the workers engaged in grinding, filling, weighing, and sealing pyrethrum powder were found to have erythema venenstum, skin roughening, and erythema with pruritus which cleared up on removal of exposure. One man, a plant engineer, had an anaphylactic type of reaction. Shortly after entering the dust ladened room, his facial skin turned red and burned and itched. His cheeks rapidly became swollen, also the eyelids to the extent that his eyes became closed. Pruritus was very intense. The entire condition disappeared in 2 days after removal from exposure to the dust.

Kesten and Laszlo (1931) reported two severe cases of dermatitis following exposure of their patients to commercial preparations of pyrethrum. One patient, an alcoholic, sprayed her bed 12 hr before sleeping in it. Within 2 days her eyelids were swollen shut and her face had an "oozing eczema." It took about a month for it to clear up following removal from exposure.

Sulzberger and Weinberg (1930) reported on an especially severe case of dermatitis from insect powder in a workroom. This patient, a woman, lost 40 lb. Her hands, arms, chest, thorax, back, thighs, and legs were covered with eczematous, vesicular, papular and, in places, weeping eruption. It took from November 19 to the following February for her condition to clear up after thorough cleaning of the workroom and discontinuance of the use of the pyrethrum insect powder.

Feinberg (1934) showed that of 225 patients sensitive to ragweed pollen, 104, or 46.2%, had cutaneous reactions with extracts of commercial pyrethrum.

In another investigation reported in the Public Health Bulletin (Anonymous, 1934), 20 cases of severe dermatitis occurred in 2 years in a factory making insect spray from Japanese insect flowers (pyrethrum). At the time of the report, 18 were still working with the flowers but, for the most part, they were not bothered any longer. Either they had developed an immunity or the original diagnosis was wrong. Two women workers were assigned elsewhere in the plant and were no longer exposed.

Sequeira (1936) studied workers in Kenya where much of the pyrethrum on the market today is grown. He found a common series of symptoms starting with itching of the hands and face, followed by erythema, which might be papular and vesicular, and, often, considerable edema. Painful deep cracks might appear in the flexures. Sweating and exposure to the sun aggravated the condition. In every case the symptoms rapidly cleared up on removal from exposure.

Lord and Johnson (1947) found that a pyrethrum-containing ointment used as a mosquito repellent produced skin sensitization in 9.7% of the men and 25.9% of the women using the preparation on a fairly regular daily basis. They found, however, that the skin sensitization was not related to the pyrethrins but to other materials present in the plant extracts. The commercial concentrates (40% pyrethrins) caused the most intense reactions of all the test preparations tried.

Zucker (1965) also found that patients sensitive to ragweed tended to be sensitive to pyrethrum. Again, it was found that the sensitization was caused by other materials present in the flower extract and not by the pyrethrins. The refined pyrethrins produced only mild or doubtful reactions. In Zucker's inhalation studies, seven of 14 patients with asthma failed to show symptoms following exposure to pyrethrum aerosol. The symptoms of those that reacted were mild and were limited to nose and throat discomfort. Two complained of eye irritation. Symptoms cleared up in a few minutes. No patients developed asthma.

Martin and Hester (1941) examined pyrethrum flowers for the causative agent for dermatitis. They found a volatile oil to be highly active in causing dermatitis. Also, a colorless extract obtained by petroleum ether extraction of flowers mixed with charcoal gave an intense reaction. Pyrethrins in 93% concentration gave no reaction.

According to the WHO working paper (Anonymous, 1970), 200 humans were patch-tested with 1% pyrethrins in water. Under the conditions of the test, it was concluded that pyrethrins were not a primary irritant or skin sensitizer.

C. Outlook for Pyrethrum Toxicity to Humans

Although much work has been done on the toxicology of pyrethrum and the pyrethrins, it is not complete enough to permit registration of the pyrethrins as a new insecticide if it were introduced to the market today. The long years of usage without ill effects other than dermatitis tell a story in themselves. Pyrethrum has demonstrated by usage that it is a safe insecticide.

Now we are beginning to get well-conducted and complete studies, such as that of Elliott *et al.* (1972), which prove conclusively that the pyrethrins are metabolized to water-soluble compounds that are in turn rapidly eliminated from the body without ill effects in the process.

Undoubtedly, the new emphasis for safer insecticides will stimulate more investigators to look into the toxicology of the pyrethrins. I do not believe such investigations will add much to what we now know; i.e., pyrethrins are a safe insecticide.

IV. Summary

The pyrethrins have an acute single dose oral LD_{50} against white rats variously reported at 200 to 1500 mg/kg, with 200 to 400 probably being the range for less pure pyrethrum preparations and 400 to 800 being the range for the more purified preparations. Dosages for guinea pigs and mice fall within these same ranges.

The pyrethrins show little evidence of causing chronic problems of toxicity. Experimental animals can be fed a large portion of an acute LD_{50} every day without harmful effects. Bond and DeFeo did not determine the chronic LD_{50}, whereas Lehman gave a chronic LD_{50} against rats of 250 mg/kg/day. His acute LD_{50} was 200 mg/kg for single doses.

The pyrethrins cause some inhalation problems because of impurities, resulting in asthma-like symptoms. Heavy inhalation dosages do not seem to produce harmful effects in animals.

Dermal toxicity of pyrethrins is negligible, because they seem to be poorly absorbed through the intact skin. The consensus of reported values for the dermal LD_{50} of the pyrethrins runs from 1350 mg/kg to about 5000 mg/kg, except for Lehmans report of 200–400 mg/kg. Why Lehman's values have been consistently lower than those of other investigators cannot be determined at this time.

There is little evidence of human toxicity problems from the pyrethrins. Long years of experience have produced no clear-cut cases of human poisoning from the pyrethrins.

Many humans have had dermal problems from exposure to pyrethrum. This dermatitis is an allergic manifestation and disappears rapidly on removal from exposure. There is some relationship between sensitivity to ragweed pollen and sensitivity to pyrethrum. The dermatitis is caused by impurities that are not present in the current commercial preparations of pyrethrins.

Pyrethrins do not pose a hazard of acute toxicity to humans.

Acknowledgment

The author wishes to thank Mrs. Claudia Lewis for her help in obtaining many of the references needed, Mrs. Vicki Roper for her translations of some of the original articles, and Mrs. Mary Daniel who helped throughout the preparation of this manuscript.

References

Ambrose, A. M., and Robbins, D. J. (1951). Comparative toxicity of pyrethrins and allethrin. *Fed. Proc., Fed. Amer. Soc. Exp. Biol.* **10**, 276–277.

Anonymous. (1889). *Chem. Drug.* **35**, 557.

Anonymous. (1934). Dermatitis due to pyrethrum contained in an insecticide. *Pub. Health Bull.* **215**, 51–54.

Anonymous. (1970). "Working Paper on Pyrethrins." World Health Organization, Rome.

Barthel, W. F., Haller, H. L., and LaForge, F. B. (1944). Pyrethrins for aerosols. The preparation of 98% pure pyrethrins for use in freon aerosol bombs. *Soap Sanit. Chem.* **20**, 121–122.

Beck, P. (1969). Memorandum to McLaughlin Gormley King, Minneapolis, Minnesota.

Bond, H. W., and DeFeo, J. J. (1969). "Toxicity of Pyrethrum in Combination with Certain Common Chemicals and Drugs." Report of University of Rhode Island to Kenya Pyrethrum Co.

Camougis, G., Davis, W. M., and Erickson, P. A. (1971). "An Evaluation of Potential Risks Resulting from the Use of Pyrethrins as Insecticides." Report of New England Research, Inc., Worcester, Massachusetts.

Carpenter, C. P., Weil, C. S., Pozzani, U. C., and Smythe, H. C., Jr. (1950). Comparative acute and subacute toxicities of allethrin and pyrethrin. *Arch. Ind. Hyg. Occup. Med.* **2**, 420–432.

Casida, J. E., Kimmel, E. C., Elliott, M., and Janes, N. F. (1971). Oxidative metabolism of pyrethrins in mammals. *Nature (London)* **230**, 326–327.

Chevalier, J. (1930). Lepyrethre (chrysantheme insecticide) activité pharmacodyamique et thérapeutique. *Bull. Sci. Pharmacol.* **37**, 154–165.

Elliott, M., Janes, N. F., Kimmel, E. C., and Casida, J. E. (1972). Metabolic fate of pyrethrin I, pyrethrin II, and allethrin administered orally to rats. *J. Agr. Food Chem.* **20**, 300–313.

Feinberg, S. M. (1934). Pyrethrum sensitization. *J. Amer. Med. Ass.* **102**, 1557–1558.

Gaudin, O. (1937). "Recherches sur l'action physiologique des pyrethrins." Vigot, Paris.

Gnadinger, C. B. (1936). "Pyrethrum Flowers," 2nd ed. McLaughlin Gormley King, Minneapolis, Minnesota.

Gnadinger, C. B. (1945). "Pyrethrum Flowers," Supplement to 2nd ed. (1936–1945). McLaughlin Gormley King, Minneapolis, Minnesota.

Godin, P. J., Sleeman, R. J., Snarey, M., and Thain, E. M. (1964). Jasmolin II: A new constituent of pyrethrum extract. *Chem. Ind. (London)* pp. 371–372.

Godin, P. J., Stevenson, J. H., and Sawicki, R. M. (1965). The insecticidal activity of jasmolin II and its isolation from pyrethrum. *J. Econ. Entomol.* **58**, 548–551.

Godin, P. J., Sleeman, R. J., Snarey, M., and Thain, E. M. (1966). The jasmolins, new insecticidally active constituents of *Chrysanthemum cinerariaefolium*. *J. Chem. Soc., C* pp. 332–334.

Hayes, W. J., Jr. (1963). "Clinical Handbook on Economic Poisons," Pub. Health Serv., Publ. No. 476. US Govt. Printing Office, Washington, D.C.

Hayes, W. J., Jr. (1965). Insecticides, rodenticides and other economic poisons. *In* "Drill's Pharmacology in Medicine" (J. P. DePalma, ed.), pp. 989–1004. McGraw-Hill, New York.

Hayes, W. J., Jr. (1969). Pesticides and human toxicity. *Ann. N.Y. Acad. Sci.* **160**, 40–54.

Jolly, D. W., and Waterhouse, C. E. (1962). Report of the Huntingdon Research Centre on the Mammalian Toxicity of Pyrethrum, Huntingdon, England.

Kesten, B., and Laszlo, E. (1931). Dermatitis due to sensitization to contact substances. *Arch. Dermatol. Syphilol.* **23**, 221–237.

Kimbrough, R. D., Gaines, T. B., and Hayes, W. J., Jr. (1968). Combined effect of DDT, pyrethrum and piperonyl butoxide on rat liver. *Arch. Environ. Health* **16**, 333–341.

LaForge, F. B., and Barthel, W. F. (1944). Constituents of pyrethrum flowers. XVI. Heterogeneous nature of pyrethrolone. *J. Org. Chem.* **9**, 242–249.

LaForge, F. B., and Barthel, W. F. (1945a). Constituents of pyrethrum flowers. XVII. The isolation of five pyrethrolone semicarbazones. *J. Org. Chem.* **10**, 106–113.

LaForge, F. B., and Barthel, W. F. (1945b). Constituents of pyrethrum flowers. XVIII. The structure and isomerism of pyrethrolone and cinerolone. *J. Org. Chem.* **10**, 114–120.

LaForge, F. B., and Barthel, W. F. (1945c). Constituents of pyrethrum flowers. XIX. The structure of cinerolone. *J. Org. Chem.* **10**, 222–227.

LaForge, F. B., and Barthel, W. F. (1947). Constituents of pyrethrum flowers. XX. The partial synthesis of pyrethrins and cinerins and their relative toxicities. *J. Org. Chem.* **12**, 199–202.

LaForge, F. B., and Soloway, S. B. (1947a). The structure of dihydrocinerolone. *J. Amer. Chem. Soc.* **69**, 186.

LaForge, F. B., and Soloway, S. B. (1947b). Constituents of pyrethrum flowers. XXI. Revision of structure of dihydrocinerolone. *J. Amer. Chem. Soc.* **69**, 2932–2935.

Lehman, A. J. (1952). Chemicals in food. Part II. Pesticides. Section III. Subacute and chronic toxicity. *Ass. Food Drug Off. U.S., Quart. Bull.* **16**, 47–53.

Lehman, A. J. (1954). A toxicological evaluation of household insecticides. *Ass. Food Drug Off. U.S., Quart. Bull.* **18**, 3–13.

Leonard, C. S. (1942). The effects of pyrethrins on certain mammals. *J. Econ. Entomol.* **35**, 261–264.

Leong, K. J., and Martin, A. R. (1966). Unpublished report from Hazelton Laboratories, Inc. to S. C. Johnson and Son, Inc.

Lorber, M. (1972). Hematoxicity of synergized pyrethrin insecticides and related chemicals in intact, totally and subtotally splenectomized dogs. *Acta Hepato-Gastroenterol.* **19**, 66–78.

Lord, K. A., and Johnson, C. G. (1947). The production of dermatitis by pyrethrum and attempts to produce a non-irritating extract. *Brit. J. Dermatol. Syph.* **59**, 367–375.

Lutz-Ostertag, Y., and Lutz, H. (1970). Action tératogène et stérilisante des pyréthrines synérgisées sur l'embryon de poulet. *C. R. Soc. Biol.* **164**, 777–779.

McCord, C. P., Kilker, C. H., and Minster, D. K. (1921). Pyrethrum dermatitis. *J. Amer. Med. Ass.* **77**, 448–449.

Malone, J. C., and Brown, N. C. (1968). Toxicity of various grades of pyrethrum to laboratory animals. *Pyrethrum Post* **9**(3), 3–8.

Martin, J. T., and Hester, K. H. C. (1941). Dermatitis caused by insecticidal pyrethrum flowers (*Chrysanthemum cinerariaefolium*). *Brit. J. Dermatol. Syph.* **53**, 127–142.

Martin, W. R., and Leong, K. J. (1966). Unpublished report from Hazelton Laboratories to S. C. Johnson and Son, Inc., March 31.

Metcalf, R. L. (1955). "Organic Insecticides," pp. 37–72. Wiley (Interscience), New York.

Moore, J. B. (1966). Chemistry and biochemistry of pyrethrins, natural pest control agents. *Advan. Chem. Ser.* **53**, 39–50.

Moore, J. B. (1972). Pyrethrum and pyrethrins. Unpublished manuscript from McLaughlin Gormley King, Minneapolis, Minnesota.

Negherbon, W. O. (1959). "Handbook of Toxicology," Vol. 3, pp. 637–638. Saunders, Philadelphia, Pennsylvania.

O'Brien, R. D. (1967). "Insecticides: Action and Metabolism," pp. 165–168. Academic Press, New York.

Perot, E., and Gaudin, O. (1933). Action des pyréthrins sur l'intestine isolé de lapin. *Bull. Sci. Pharmacol.* **40**, 7–13.

Sequeira, J. H. (1936). Pyrethrum dermatitis. *Brit. J. Dermatol. Syph.* **48**, 473–476.

Shimkin, M. B., and Anderson, H. H. (1936). Acute toxicities of rotenone and mixed pyrethrins in mammals. *Proc. Soc. Exp. Biol. Med.* **34**, 135–138.

Springfield, A. C., Carlson, G. P., and DeFeo, J. J. (1971). Increased microsomal enzyme activity due to pyrethrum administration. *Toxicol. Appl. Pharmacol.* **19**, 394.

Staudinger, H., and Ruzicka, L. (1924a). Insect poisons. I. Isolation and constitution of the active portion of dalmatian insect powder. *Helv. Chim. Acta* **7**, 177–201.

Staudinger, H., and Ruzicka, L. (1924b). Insect poisons. II. Constitution of chrysanthemummono- and dicarboxylic acids. *Helv. Chim. Acta* **7**, 201–211.

Staudinger, H., and Ruzicka, L. (1924c). Insect poisons. V. Synthesis of tetrahydropyrethrolone, the reduction product of pyrethrolone. *Helv. Chim. Acta* **7**, 245–259.

Staudinger, H., and Ruzicka, L. (1924d). Insect poisons. X. Synthesis of pyrethrins. *Helv. Chim. Acta* **7**, 448–458.

Sulzberger, M. B., and Weinberg, C. B. (1930). Dermatitis due to insect powder. *J. Amer. Med. Ass.* **95**, 111–112.

Verschoyle, R. D., and Barnes, J. M. (1972). Toxicity of natural and synthetic pyrethrins to rats. *Pestic. Biochem. Physiol.* **2**, 308–311.

Weir, R. J. (1966a). Unpublished report from Hazelton Laboratories to S. C. Johnson and Son, Inc., June 1.

Weir, R. J. (1966b). Unpublished report from Hazelton Laboratories, Inc. to S. C. Johnson and Son, Inc., August 2.

Weir, R. J., and Crews, L. M. (1966). Unpublished report from Hazelton Laboratories, Inc. to S. C. Johnson and Son, Inc., June 1 (supplement to report of March 1, 1966).

Zucker, A. (1965). Investigation of purified pyrethrum extracts. *Ann. Allergy* **23**, 335–339.

Chapter 7

Toxicity of Pyrethrum to Fish and Wildlife

RICHARD E. PILLMORE

I. Introduction

Assessing the field hazards of pyrethrum, or any other insecticide, to fish and wildlife is difficult. Certainly, determination of acute toxicity on a few selected nontarget species does not fully predict what happens to wildlife under the complex and varied conditions of actual field use. As Leedy (1959) suggests, the development of reliable methods of field appraisal has been a long-standing need. Field appraisals to determine the safety or impact of pesticides on a forest ecosystem must assess both the direct and the indirect effects of insecticide applications on both target and nontarget organisms and must be concerned both with the short- and long-term effects under varying conditions of field use. This paper describes the cooperative efforts of the United States Forest Service and the United States Bureau of Sport Fisheries and Wildlife, particularly in the 1970 field appraisal of pyrethrum, aerially applied at 0.1 and 0.2 lb/acre on the Pike National Forest, Colorado.

In 1964, the United States Forest Service established a research and development project to find effective insecticides that are less persistent and more selective than DDT and are without the objectionable side effects. The cooperative effort began under an existing agreement between the two agencies. Those candidate materials of the Forest Service Insecticide Evaluation Project that showed sufficient promise to merit field testing would be evaluated by biologists from the Denver Wildlife Research Center for their effects on forest birds and mammals. In the past, only short-term field studies were conducted on areas with high spruce budworm (*Choristoneura occidentalis*) populations. Insecticide applications were appraised both for efficacy against a target species and for possible effects on nontarget species, including fish, birds, and mammals. Initial appraisals of one material, Zectran, for effects on birds and mammals are described by Pillmore *et al.* (1972).

However, we believed a new approach was necessary utilizing more permanent study areas regardless of pest levels. We selected an experimental area where forest pest species were not currently abundant enough to justify spraying but where there was a variety of birds sufficiently abundant for an appraisal. The area was large enough for 12 treatment blocks that were spaced to minimize cross-contamination, and were in terrain accessible for both vehicle and foot travel. The selected area was not subject to the conflicting land uses of insect control, logging, excessive disturbance by livestock grazing, or human recreational use during the course of our study. Streams suitable for evaluating effects on fish or other aquatic life were also present.

For this volume, most of the data presented represent contributions of others, which I shall attempt to put together and interpret as coordinator for the 1970 cooperative field appraisals. For the Bureau of Sport Fisheries and Wildlife, Lawrence DeWeese and Merle Richmond of the Denver Wildlife Research Center conducted the bird studies. Crosswell Henderson, Fishery Management Services Division, assisted by Robert Hillen, Wildlife Services Division, made water chemistry determinations, conducted fish bioassays, and sampled streams for drifting and bottom organisms. Dr. Richard Schoettger, Fish Pesticide Laboratory, Columbia, Missouri, provided unpublished fish toxicity data. For the Forest Service, Patrick Shea assisted by Charles Tiernan sampled arboreal and terrestrial flying insects; their collections are still being processed with the assistance of Faye Shon and only the aquatic samples are completed. Charles Richmond and Ted Jones collected foliage samples for spray deposit analysis, and René Pieper processed foliage samples and determined pyrethrins content by gas chromatography. Ray Miskus provided technical assistance in formulating the pyrethrum spray and Donn Cahill calibrated spray equipment.

II. Experimental Area

The experimental area was located in the Pike National Forest, about 10 miles north of U.S. Highway 24 between Divide and Florissant, Teller County, Colorado, at an elevation of 8500 to 9000 ft. The soils of this region are shallow and well-drained and are derived from the Pikes Peak granite. A ponderosa pine–Douglas-fir (*Pinus ponderosa–Pseudotsuga menziesii*) forest stand type (Marr, 1961) is typical for this area. Douglas-fir usually predominates on more mesic north-facing slopes, where it is often mixed with blue spruce (*Picea pungens*). Aspen (*Populus tremuloides*) is present, usually in mixed stands of small trees on the slopes, but in the valley bottoms it may form groves with trees up to a foot or more in diameter. The main drainages typically have a willow-bottomland complex of stand types, with a bordering grassland complex on the north and west sides. The small streams draining this area are headwaters of West Creek, a tributary of the South Platte River.

A. Treatment Blocks and Study Plots

On the basis of our experience, we selected 160 acres as the minimum-sized treatment block for any field appraisal of insecticide effects on birds or mammals. Even this size area limited us to study of smaller species, whose home range would be confined within this area. Using a compass and chain, we laid out 12 treatment blocks of 160 acres each as shown in Fig. 1. Twelve blocks were established because our original design called for use of six blocks per year—two untreated and two each for two treatments. Because 160 acres is too large for one observer to cover in a bird census, we developed only the central 40 acres for intensive study, leaving a treated buffer strip around study plots.

The central 40 acres of all plots contained an orientation grid with marked orange reference stakes, 30 inches tall, every two chains. Additional

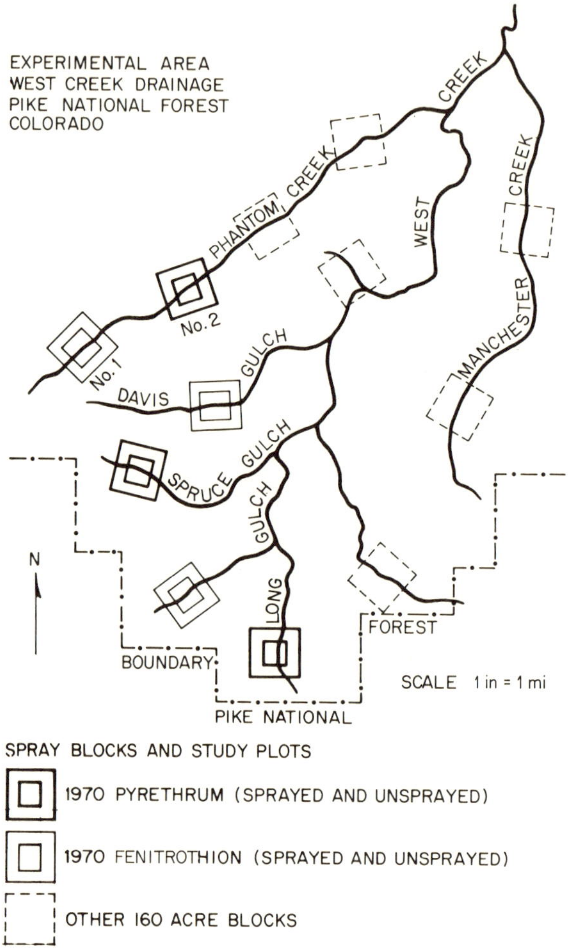

Fig. 1. Experimental area, spray blocks, and study plots for 1970 field appraisal of pyrethrum effects on fish and wildlife.

marking with surveyor's plastic flagging tape helped to keep observers properly oriented. Plots were laid out across drainages and most had small but permanent streams, some with active or abandoned beaver ponds.

B. Pyrethrum Applications

Pyrethrum was an attractive insecticide candidate because it is a natural material whose active components, pyrethrins, are highly toxic to a variety of forest insect pests. Pyrethrins are not persistent in the environment and are relatively nontoxic to warm-blooded animals (Miskus and Andrews, 1972); however, they were not suitable for field testing until these workers developed a stabilized formulation in mineral oil with antioxidant and ultraviolet screening materials added. Two treatment blocks were sprayed on the morning of June 18, 1970, from a Bell (G 3B 47) helicopter with ten spray nozzles (S.S. D6-45°). We calibrated this conventional spray system at a boom pressure of 55 psi, an air speed of 60 mph and a swath width of 100 ft to deliver 0.5 gal spray per acre. According to Akesson *et al.* (1970), these nozzles arranged perpendicular to the air stream will produce about 225 μm volume mean drop diameter (VMD) with 2.8% oil emulsified in 90–95% water.

Pyrethrum spraying began at 5:34 AM MDT at the treatment block in Long Gulch, which was covered twice to obtain the 0.2 lb/acre application. The second treatment block, in Spruce Gulch, was sprayed immediately following the first but was covered only once for the 0.1 lb/acre application and completed by 7:15 AM MDT.

Throughout the spraying the weather was clear and there was little or no breeze.

C. Other Insecticides Used

Two days after spraying pyrethrum, a fenitrothion formulation in oil was similarly applied on two additional treatment blocks. The application rate was 1.0 lb/acre in the Stone Gulch block and $\frac{3}{8}$ lb/acre in the Davis Creek block. The same helicopter and spray equipment were used after flushing the spray system with oil.

It is also necessary to mention the materials sprayed in 1971, because treatment blocks in Long Gulch and Spruce Gulch were again sprayed. The Spruce Gulch plot was sprayed with a pyrethroid, resmethrin (SBP 1382), at 0.05 lb/acre on June 19, 1971. The Long Gulch plot was sprayed with a suspension of trichlorfon in oil at 1.0 lb/acre on June 20, 1971. For the 1971 spraying, the spray nozzles were S.S. 80015, which produces a finer spray, approximately 180 μm VMD with low volatility oil according to Akesson *et al.* (1970).

Observers in the spray blocks reported that coarser spray droplets fell like rain with a slant in the direction of flight shortly after the helicopter passed overhead, and that the finer spray, visible in the sunlight, settled much more slowly and showed a tendency to drift down slopes toward the valley bottom.

III. Methods of Field Appraisal

Field appraisals are seldom completed as planned. What began as an appraisal of insecticides applied specifically to study their effects on birds and mammals and their insect food supply in 1968, was expanded in 1969 to include fish and their aquatic food supplies. The appraisal for effects on small mammals was dropped in 1969 because of their scarcity on the plots. The planned insecticide applications in 1969 were cancelled because of weather and formulation problems.

A. Estimating Bird Numbers and Survival

Trend counts, used with modification since 1965, reflect day-to-day activity of birds but have proved too variable to detect any but conspicuous short-term effects. In conducting a "trend count," an observer covers the 40-acre plot early each morning beginning about the same time, starting from the same point, traversing the same route at two-chain intervals, plotting on a grid map the locations of all birds seen or heard, and completing the count in approximately 2 hr. For the pyrethrum appraisal the same individual conducted all trend counts on the two treated plots the same day and made two corresponding trend counts of the untreated plot the following day. Another individual did the same for the fenitrothion appraisal. Four such counts were completed prior to spraying and again following spraying for a total of 32 counts, eight on each treated plot and 16 on the untreated. The following year, the same two observers were used and changed only on the Spruce Gulch plot. Different observers were used in 1968 and 1969 for censusing birds.

We had not expected the breeding bird census to provide more than a year-to-year trend; however, it was possible to estimate with considerable accuracy the number of pairs of several species present on each plot or near its boundary prior to spraying. Postspray counts could then be used to check on the continued occupancy of an identified pair territory. Breeding bird censusing was similar to trend counting. Plots were covered in much the same manner, but starting points, routes, and time to complete coverage of area were varied and not specified. The method is generally consistent with recommendations by the International Bird Census Committee (Robbins, 1970). The locations of nests found and observations of marked individuals helped in differentiating breeding pairs. Territories along the plot boundaries caused some problems, as they could be used in the appraisal but contributed only some fraction to the estimated density of breeding pairs per 40 acres. Renesting in some cases and suspected second nestings in others may have influenced the number of pairs estimated for some species in a given plot.

Survival was indicated by recapture or observations of previously banded birds or color-marked individuals following the spraying. Mist-nets, usually stretched across the valley bottoms in openings through the willows, were used to capture birds which were banded with Bureau of Sport Fisheries and Wildlife aluminum bands. Our method of marking was temporary, and consisted of red, green, and yellow dye applied to the plumage.

All nests located in each plot were checked before spraying to determine if they were active (containing eggs or young) at the time of treatment. The progress of these nests and nests found subsequently were checked during the 2 weeks following spraying. Information obtained provided specific and positive survival information, not only for adults attending nests, but also for young in the nests. Many of the nests found were in tree cavities; where possible, these were reached by ladders (up to 20 ft long) and examined with aid of a mirror and light device we developed for the purpose. We recorded as successful those nests found empty and apparently undisturbed when young should have fledged and those nests still active 2 weeks following the spraying. Abandoned nests with eggs, disturbed nests, or empty nests that should still have contained young were recorded as unsuccessful.

B. Checking Fish Survival

The Colorado Game, Fish, and Parks Department furnished about 1000 fingerling rainbow trout (3–5 in long) in 1969 and about 200 additional fish in 1970 from the State Hatchery at Salida. The fish were distributed in the streams and beaver ponds, with most released to swim free, but with some held in minnow traps and buckets on each study block. All fish in minnow traps were placed in streams at the lower end of each spray block. The mesh ($\frac{1}{4}$ in or larger) was large enough and the traps were placed so most aquatic organisms could drift through. Fish in the floating minnow buckets ($\frac{1}{4}$ in mesh) were placed several hundred yards downstream from the spray block, except in the Davis Creek plot, where they were within the spray treatment block. All buckets and traps were checked for mortality just prior to treatment.

In addition to checking confined fish after the treatments, investigators walked the streams looking for dead fish and fished the beaver ponds with rod and reel.

C. Sampling Terrestrial and Aquatic Insects

The primary method of sampling terrestrial insects was by Malaise trap, which is an effective means of capturing many flying insects. A single trap was placed in each of the five study plots where fish were confined. The traps were visited daily and the collecting jars replaced. Because the collections have not yet been processed, these data are not available. In 1969, we did some limited bird food habits examinations (33 birds collected, representing 23 different species) and made a gross comparison between what we found in bird crops and gizzards and what was in the collection jars. It was obvious that the trap was not sampling the same insects that the birds were eating, but the data should, when available, indicate general insect abundance in the treated and untreated plots.

There should also be some limited data from plastic insect drop cloths placed to collect falling insects beneath tree canopies.

The data from the aquatic insect sampling were available and consisted of counts, sorted by family, for the samples of drifting insects collected in 10 min by a plankton net placed in the stream before and at intervals after spraying. Stream bottom samples were taken with a square foot, or Surber, sampler. Streams were small enough that virtually the entire flow was sampled; however, the velocity of flow was inadequate to keep many organisms in suspension.

Water temperatures were taken and measurements of dissolved oxygen, pH, alkalinity, hardness, and electrical conductivity were made with Hach test kits.

D. Sampling for Pyrethrins Deposit

Aerial application does not insure a uniform deposit. The uneveness in the deposition pattern was evident in the foliage residues and drop deposit ratings obtained by personnel of the Insecticide Evaluation Project. They made foliage collections and exposed smoked glass slides at nine stations in each treated plot and collected unsprayed foliage samples as a check. Collections were completed as soon as possible following the spraying and also 1 and 5 days after spraying. There were three stations each for aspen, Douglas-fir, and willow foliage. At each station different plants and different sides of the same plants were sampled and combined for a composite sample. All composite samples were handled in the same fashion.

Dip samples of water were also taken from the upper, middle, and lower stream segments within the treatment plot, following spraying and again 5 days later. These samples and the vegetation were processed at Denver to stabilize pyrethrins before shipping to Berkeley, California, where gas chromatography was used for analysis.

IV. Results of Field Appraisal

A. Bird Data

Averages of the four pre- and four postspray bird trend counts are shown in Table I. Trend counts fail to show any appreciable changes in species composition or relative abundance of birds attributable to spraying. Some bird species, such as robins, appear to have increased during postspray counts, whereas gray-headed juncos increased on treated plots only. Species that decreased generally did so on both treated and untreated plots. The increases noted above are attributable to fledging young being included in the later counts. Some postspray decreases probably reflect reduced conspicuousness associated with pairs incubating eggs, and some increases were caused by the greater conspicuousness associated with increased foraging to feed nestling birds.

We reduced variability by expressing counts per unit of time instead of by area covered, and by having the same individual make all counts on assigned plots. Analysis of variance of the trend count totals, expressed in birds per minute, showed no significant ($\alpha = 0.1$) differences attributable to the pyrethrum treatment; however, a significant ($\alpha = 0.05$) difference was found, for the pretreatment but not the posttreatment period, between the early and late morning counts on the untreated plot. Variability was higher for first counts (early morning) than for second counts (late morning). Feeding activity later is perhaps more dependable than the earlier singing activity for trend counts.

Plotted trend count observations were combined with breeding bird censuses, and the composite map by species provides the estimate of breeding pairs for many species present on the plots. Some species may be overlooked or be difficult to characterize as territorial or breeding pairs unless the nest can be located. Pairs with territories along the periphery of the 40 acres are estimated to the nearest half-territory per 40-acre plot for the data in Table II but some may be included as full-pair territories in appraisal (Fig. 2). Effects of spraying in 1970

TABLE I

Average Number of Birds per Trend Count by Species, on Pyrethrum-Treated and Untreated Plots[a]

Species	Early morning counts				Late morning counts			
	Sprayed, 0.2 lb/acre		Unsprayed		Sprayed, 0.1 lb/acre		Unsprayed	
	Before	After	Before	After	Before	After	Before	After
Mallard duck and green-winged teal (*Anas platyrhynchos* and (*A. carolinensis*	0	0.25	2.25	0	0	0	0.5	0
Broad-tailed hummingbird (*Selasphorus platycercus*)	2.5	2.0	2.5	2.0	2.0	0.75	2.25	2.0
Red-shafted flicker (*Colaptes cafer*)	0.5	1.0	2.25	0.75	0.5	1.0	0.5	0.5
Yellow-bellied sapsucker (*Sphyrapicus varius*)	0.5	1.25	1.5	2.0	0	1.0	0.5	1.5
Williamson's sapsucker (*S. throides*)	0	0	0.5	1.75	0	0	0.5	0.5
Hairy woodpecker (*Dendrocopos villosus*)	0	0	0	0	0	0.25	0	0
Empidonax spp.	0.75	1.0	0	0	0.5	0.5	0	0
Mountain chickadee (*Parus gambeli*)	2.5	2.75	3.0	2.75	4.25	2.25	5.0	4.0
Red-breasted nuthatch (*Sitta canadensis*)	0.5	0.25	0.75	0.5	0.25	0.5	0	0
Pigmy nuthatch (*S. pygmaea*)	0	0	0	0	0	0	0.5	1.75
Brown creeper (*Certhia familiaris*)	0	0	2.5	0	0	0	1.75	1.75
Violet-green swallow (*Tachycineta thalassina*)	0	0	2.5	0	2.0	1.0	2.25	0.5
Steller's jay (*Cyanocitta stelleri*)	0.25	0.5	0.5	0	0	0	0	0
Gray jay (*Perisorius canadensis*)	0.25	0.75	1.0	1.75	0	1.25	1.25	0.75
House wren (*Troglodytes aedon*)	0.25	0.5	3.5	1.25	0.75	1.0	1.25	1.0

Robin (*Turdus migratorius*)	2.75	4.25	6.25	5.25	2.0	4.0	3.25	4.0
Townsend's solitaire (*Myadestes townsendi*)	1.0	1.0	1.75	0	0.5	0	0	0
Mountain bluebird (*Sialia currucoides*)	0	0.5	0	0	0.5	0.75	0	0.25
Golden-crowned kinglet (*Regulus satrapa*)	0	0	0.25	0.5	0	1.25	0	0.5
Ruby-crowned kinglet (*R. calendula*)	1.75	3.0	1.25	1.25	3.0	2.75	1.0	0.5
Warbling vireo (*Vireo gilvus*)	0.25	0	3.75	2.75	0.5	0.5	0.5	2.75
Audubon's warbler (*Dendroica auduboni*)	1.0	0.25	1.75	0.5	1.75	0	2.25	0
MacGillivray's warbler (*Oporornis tolmiei*)	0.25	0	0.5	0.25	0	0.25	0	0.75
Wilson's warbler (*Wilsonia pusilla*)	2.5	0.5	1.5	2.25	1.25	1.0	0.5	1.75
Red-winged blackbird (*Agelaius cyanocephalus*)	0	0	0.25	0	0	0	0.5	0
Pine siskin (*Spinus pinus*)	0.5	1.25	1.75	1.0	0.5	0.25	1.5	0.5
Green-tailed towhee (*Chlorura chlorura*)	1.0	0.5	0.25	0	0.5	0	0.5	0
Gray-headed junco (*Junco caniceps*)	4.25	8.75	3.5	1.5	2.75	3.75	2.5	1.0
Chipping sparrow (*Spizella passerina*)	1.25	2.25	0	0	0	0.25	0	0
White-crowned sparrow (*Zonotrichia leucophrys*)	0.25	0	0	0	0	0	0	0
Lincoln's sparrow (*Melospiza lincolnii*)	1.5	2.25	4.25	4.25	0.25	1.25	0.25	3.25
Song sparrow (*M. melodia*)	0	0	0.75	0.5	0	0	0	0.5
Totals	26.25	34.75	50.50	32.75	23.75	24.50	29.00	30.00

[a] Four counts before and 4 counts after treatment.

TABLE II

Estimated Pairs of Selected Breeding Bird Species and a Summation of Total Species on Pyrethrum-Treated and Untreated 40-Acre Plots in the Pike National Forest, Colorado

Species	Phantom Creek (control, not sprayed)			Spruce Gulch (0.1 lb/acre)				Long Gulch (0.2 lb/acre)			
				Prespray years		Spray year	Post-spray year	Prespray years		Spray year	Post-spray year
	1969	1970	1971	1968	1969	1970	1971	1968	1969	1970	1971
Gray-headed junco	5.0	2.0	2.0	6.0	4.0	9.0	3.0	6.0	5.0	5.5	6.0
Mountain chickadee	5.5	6.5	6.0	3.5	4.5	3.5	2.0	3.5	4.0	3.5	4.0
Ruby-crowned kinglet	4.5	3.0	4.0	6.5	5.0	5.0	3.0	2.5	3.5	3.5	5.0
Audubon's warbler	2.5	4.0	3.0	4.0	2.5	2.5	0.0	4.0	3.0	4.5	1.0
Warbling vireo	3.5	2.0	3.0	2.5	1.5	2.0	1.0	3.5	2.5	P[a]	1.0
Lincoln's and song sparrows	4.5	3.5	4.0	3.0	3.0	2.0	3.0	5.0	4.5	3.5	5.0
Total (all species)	44	44	48	48	38	53	34	48	47	48	42

[a] P = This species present but no nests found or pair territory defined by plotted observations.

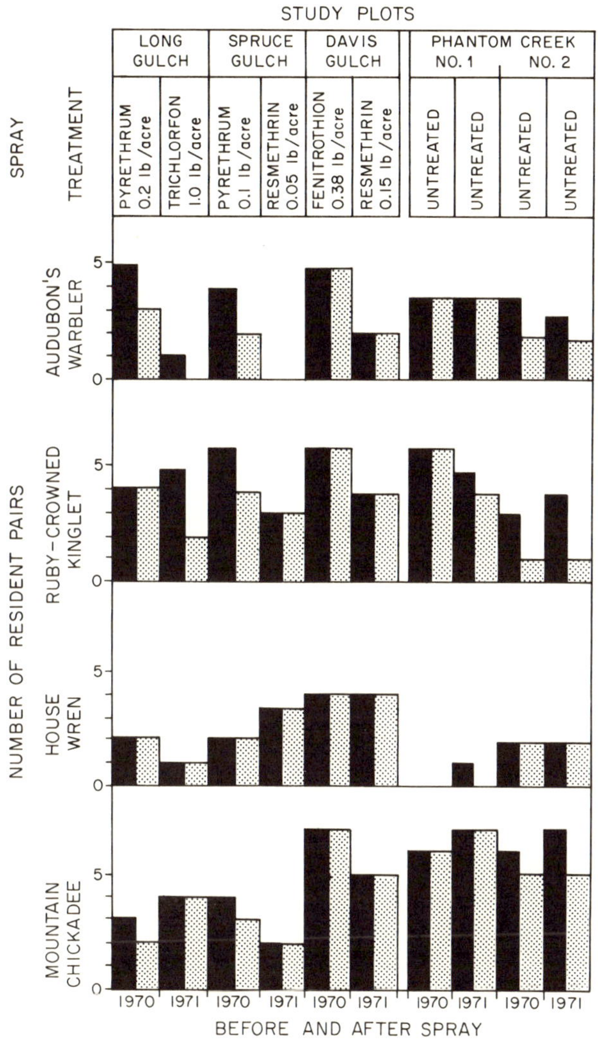

Fig. 2. Histogram for four insectivorous bird species showing number of pairs per treated or untreated plot indicated by censuses before (black) and still present after (gray) aerial spraying treatments involving four insecticides.

may be reflected by an apparent reduction of pairs of Audubon's warblers present in 1971, prior to the 1971 treatments. Most observers were aware of a seasonal trend of decreased singing activity of the male Audubon's warbler. This could have affected censuses by different observers. Observer effects are expected, but the decrease in numbers of

Audubon's warblers from 1970 to 1971 was more than expected on three treated plots when these were compared with two untreated plots censused by two observers. (In both 1970 and 1971, one observer censused Long Gulch and Phantom Creek No. 2, and the other censused Davis Creek and Phantom Creek No. 1; Spruce Gulch was censused in 1970 by the first observer and in 1971 by the second observer). Similar data for three other species of insectivorous birds in Fig. 2 do not show the same pattern as for Audubon's warblers.

On the treated Long Gulch plot, ten of the 15 nests found, representing nine species, were active (containing eggs or young) at the time of spraying. Of the ten nests, eight were still active 2 weeks later or were known to have fledged young birds. One robin nest was abandoned, and a Townsend's solitaire nest was destroyed by predators. On the treated Spruce Gulch plot, all 16 nests found, representing eight species, were active at the time of spraying. Eleven of these were active or known to have fledged young 2 weeks later, a junco nest was destroyed by predators, and the fate of four could not be determined. On the untreated Phantom Creek plot, 21 of the 27 nests found, representing 13 species, were active at the time of spraying, and 17 of these were still active or had fledged young 2 weeks later. The fate of four nests could not be determined.

Eighteen birds were individually marked with colored dye on their plumage before spraying on the Long Gulch plot. Five of these were repeatedly seen or recaptured during the 10-day posttreatment period. On the Spruce Gulch and Phantom Creek plots, the capturing and marking followed the spraying. On the Spruce Gulch plot, four of 12 birds marked were seen or recaptured one or more times during the postspray period. On the Phantom Creek plot, seven of the 13 marked birds were recaptured or seen after the date of their first capture.

In addition to the five marked birds seen after spraying in the Long Gulch plot, three birds banded in 1969 were recaptured on the same plot in 1970. One of the three was again recaptured on this same plot in 1971, along with seven of the 22 birds banded there in 1970.

B. Fish Data

None of the rainbow trout exposed to the water and to drifting food organisms were killed by either the pyrethrum or fenitrothion spraying in 1970. There were noticeable effects on fish in only one plot, the $\frac{3}{8}$ lb/acre fenitrothion-treated plot in Davis Creek. Here fish were gulping at the surface and had partially lost equilibrium in both the minnow bucket and trap about an hour after spraying, and their apparent distress continued most of the day. Dissolved oxygen was measured and found to be 8.0 ppm, which is normal for all streams on the area. All fish had apparently recovered by the next morning (24 hours postspray). No effects were observed among any of the other confined fish, though some affected insects were seen drifting into the minnow traps and being eaten by fish.

Water temperatures, pH, alkalinity, and hardness (in ppm $CaCO_3$), and specific conductivity (in μmho), measured over 3 years, were similar in all drainages in the area; for Phantom Creek, Spruce Gulch, and Long Gulch the average ($n = 10$) values (June 11 through 23) were about 10°C (5°–16°C), 7.2 (7.0–7.3), 61.1 (50–75) ppm $CaCO_3$, 63.4 (40–85) ppm $CaCO_3$, and 143 (110–170) μmho, respectively. Water temperatures tended to be slightly higher in and below beaver ponds than above them.

Wild fish were present in beaver ponds just a few hundred feet downstream from the treatment block in both Long and Spruce Gulches, and some fish were likely present within the treatment blocks but this was not confirmed. Because of the larger number of fish stocked there in 1969, the beaver ponds just below the Long Gulch spray block yielded good catches of trout (9–14 in long) after spraying and well into fall of 1970 but not in 1971.

C. Insect Data

Other than the observed knockdown of insects during and shortly after spraying, the data on terrestrial insects are not yet available.

The most abundant aquatic organisms in the streams were immature insects belonging to the four orders: (*1*) Diptera (midges, black flies, and mosquitoes), (*2*) Plecoptera (stoneflies), (*3*) Ephemeroptera (mayflies), and (*4*) Trichoptera (caddisflies). The numbers of these taken in the 10-min drift samples and the square-foot bottom or Surber samples are shown in Table III for the two pyrethrum-treated plots.

Henderson reported that about an hour after spraying in Long Gulch a few terrestrial and some mature aquatic insects were floating on the surface and many immature aquatic insects were noted to be affected by the spray. Numbers of caddisfly larvae were crawling from their cases, and drift samples contained mostly affected or dead organisms during the remainder of the day. A Surber sample taken about 7 hr postspray contained only dead organisms; one taken 31 hr postspray contained all dead organisms except for one black fly and one midge larva.

Because the current was not great enough to carry most of the affected organisms, these settled in the pool areas, where the large white caddisfly

TABLE III

Numbers of Immature Aquatic Insects Collected in Drift and Bottom Samples before and at Intervals after Pyrethrum Spraying on June 18, 1970.[a]

		Samples										
		Drift net (10 min)								Bottom (ft²)		
			Hours postspray								Hours postspray	
Plot	Insect orders	Pre-spray	1	2	4	7	10	31	56	Pre-spray	7	31
Long Gulch 0.2 lb/acre	Diptera	0	7	11	15	11	3	0	2	5	10	4
	Ephemeroptera	3	0	1	10	41	34	2	1	5	9	2
	Plecoptera	1	6	50	0	16	5	0	4	4	8	6
	Trichoptera	4	2	6	2	2	4	2	0	19	10	0
	Totals	8	15	68	27	70	46	4	7	33	37	12
			Hours postspray								Hours postspray	
		Pre-spray	1	2	4	6	8	27	55	Pre-spray	8	27
Spruce Gulch 0.1 lb/acre	Diptera	1	2	0	0	1	0	1	0	6	6	10
	Ephemeroptera	2	16	20	5	3	2	2	1	24	23	24
	Plecoptera	0	17	113	6	8	2	3	1	4	22	19
	Trichoptera	3	4	8	3	3	3	1	1	1	5	8
	Totals	6	39	141	14	15	7	7	3	35	56	61

[a] Unpublished data, Insecticide Evaluation Project, U.S. Forest Service, Berkeley, California.

larvae were very conspicuous. Only the drift from the short riffle area above the collecting station was well represented in the 10-min collections.

The effect of the pyrethrum on organisms in the Spruce Gulch stream was noticeably less than in Long Gulch. The 8- and 27-hr postspray Surber samples contained many live mayfly, stonefly, and caddisfly larvae.

D. Pyrethrins Deposit Data

Residue results are summarized in Table IV. There was a considerable range in pyrethrins deposit, 0 to 6.1 ppm, which appeared heaviest on

TABLE IV

Pyrethrins Deposit as Residues in Foliage and Water Samples and Mean Rating (on Scale of 1–10, Low–High) of Drop Deposit on Smoked Glass Slides[a]

Plot, treatment, and mean drop deposit rating	Composite sample (no. stations in parentheses)	Residues (ppm)	
		Spray day mean (range)	Day after spray mean (range)
Long Gulch	Douglas-fir (3)	1.84 (0.18–4.7)	0
0.2 lb/acre	Aspen (3)	5.96 (5.7–6.1)	0.97 (0.22–1.6)
(rating 5.4)	Willow (3)	1.68 (0.54–3.7)	0.53 (0.52–0.61)
	Water (3)	0.035 (0.003–0.070)	0
Spruce Gulch	Douglas-fir (3)	0.43 (0–1.2)	0
0.1 lb/acre	Aspen (3)	2.38 (0.7–3.4)	0.31 (0–0.47)
(rating 3.7)	Willow (2)	<0.27	0
	Water (3)	<0.001 (<0.001–0.002)	0
Untreated	Douglas-fir (1)	0	
Check	Aspen (1)	0	
Samples	Willow (1)	0	
	Water (1)	0	
Recovery—88% at 0.1 ppm			

[a] Unpublished data, Insecticide Evaluation Project, United States Forest Service, Berkeley, California.

aspen foliage. Residues were much reduced after 1 day, and no appreciable residues were present after 5 days, although pyrethrins were still detectable in four of the nine samples from the 0.2 lb/acre application.

Water analysis results are also shown in Table IV. These results show no trace of pyrethrins after 5 days.

Certainly, weather, type of aircraft, flight patterns, or spray swaths influence deposit along with the size, range, frequency distribution, and behavior of spray droplets impinging on varied environmental substrates.

V. Use of Laboratory Toxicity Data

Toxicity information is helpful in interpreting the results of a field test; and it also helps in planning a field appraisal because it can give indications of the kinds of effects that may be expected under some conditions of field use. Rudd and Genelly (1956) reviewed the hazards of pyrethrum to wildlife, and they stated that no hazard to warm-blooded animals exists. The mammalian toxicity of pyrethrins is covered in considerable detail by others in this volume. In general, from what is known about pyrethrins toxicity it appears extremely unlikely that the pyrethrum spraying could have had any direct toxicity to forest birds and mammals.

A. Toxicity of Pyrethrum to Birds

Some preliminary avian toxicity tests were conducted at the Denver Wildlife Research Center. Acute oral dosing with the 20% pyrethrins extract yielded the following results: (*1*) Three female mallard ducks (*Anas platyrhynchos*) survived doses of 10,000 mg/kg (Tucker and Crabtree, 1970). (*2*) Eight male Japanese quail (*Coturnix coturnix*) were dosed, two each at 1250, 2500, 5000, and 10,000 mg/kg; one died at the 5000 mg/kg dose and one at the 10,000 mg/kg dose, but all others survived and appeared normal when the test was terminated after 14 days and the birds were examined for any gross internal pathology (unpublished data). These results did not permit calculation of a median lethal dose, or LD_{50}, but did not indicate a need for further testing.

B. Toxicity of Pyrethrum to Fish and Fish Food

For fish, both direct fish toxicity and toxicity to other aquatic organisms on which fish feed are important considerations in a field appraisal. Rudd and Genelly (1956) implied that fish were more sensitive to pyrethrins in an emulsifiable formulation than to those contained in the powdered dry flowers. Negherbon (1959) gave the harmful and lethal concentration of pyrethrins (as pyrethrum flowers) for carp (*Cyprinus carpio*) as 2 and 5–10 ppm.

Bridges and Cope (1965) compared the toxicity of 24.6% pyrethrins extract mixed with equal amounts of piperonyl butoxide, a pyrethrum synergist, with that of rotenone and DDT on three species of fish—rainbow trout (*Salmo gairdneri*), channel catfish (*Ictalurus punctatus*), and bluegill (*Lepomis macrochirus*). They described their procedure as using reconstituted water (deionized water with appropriate amounts of different salts added), hatchery stock fish (ten fish per test concentration), controlled temperatures, and the method of Litchfield and Wilcoxon (1949) for analyzing the fish mortality to obtain a median lethal concentration (LC_{50}) value for specific time periods. Essentially the same procedures are still used for the static fish bioassay at the Fish Pesticide Laboratory, now located at Columbia, Missouri. Unpublished data from this laboratory are presented with those of Bridges and Cope in Table V.

TABLE V

Toxicity of Pyrethrum Extract to Aquatic Organisms[a]

Organism[b]	Bioassay	Temperature (°C)	LC_{50} or TL_{50}[c] (μg/liter) and 95% confidence intervals		
			24 hr	48 hr	96 hr
Rainbow trout (av. 0.3 gm)	static	12	56 (40–64)	54 (48–60)	54 (48–60)
Channel catfish (av. 0.5 gm)	static	25	96 (80–108)	82 (71–93)	80 (70–92)
Bluegill (0.6 gm)	static	25	80 (72–89)	74 (64–86)	74 (64–86)
Stonefly	static	14	10 (7.1–14)	6.4 (4.8–8.6)	1.0 (0.7–1.4)
Steelhead trout (av. 0.2 gm)	static	12	—	—	25 (20–30)
Lake trout (av. 0.5 gm)	static	12	—	—	37 (32–44)
Coho salmon (av. 0.7 gm)	static	12	—	—	39 (33–45)
Scud	static	21	28 (20–39)[c]	—	11 (8–15)
Waterfleas					
Daphnia	static	21	—	25 (18–32)[c]	—
Simocephalus	static	21	—	42 (34–55)[c]	—
Coho salmon (av. 0.8 gm)	intermittent flow		—	—	23 (18–29)
Northern pike (av. 0.8 gm)	intermittent flow		—	—	17 (13–22)
Yellow perch (av. 2.6 gm)	intermittent flow		—	—	45 (36–54)

[a] Published and unpublished data of Fish Pesticide Laboratory, U.S. Bureau of Sport Fisheries and Wildlife, Columbia, Missouri.
[b] First four values from Bridges and Cope (1965), others unpublished data.
[c] TL_{50} are immobilization values.

Recalculating these data on the basis of the amount of pyrethrins contained in our tests, we have the following 96-hr LC_{50} values: (*1*) 24.6% with synergist: rainbow trout, 2.7 μg/liter (0.0027 ppm); channel catfish, 3.9 μg/liter (0.0039 ppm); and bluegill, 3.7 μg/liter (0.0037 ppm). (*2*) Twenty percent without synergist (static bioassay): steelhead (*Salmo gairdneri*), 5.0 μg/liter (0.005 ppm); lake trout (*Salvelinus namaycush*), 7.4 μg/liter (0.0074 ppm); and coho salmon (*Oncorhynchus kisutch*), 7.8 μg/liter (0.0078 ppm). (*3*) Twenty percent without synergist (intermittent flow, raw water bioassay): coho salmon, 4.6 μg/liter (0.0046 ppm); northern pike (*Esox lucius*), 3.4 μg/liter (0.0034 ppm); and yellow perch (*Perca flavesecus*), 9.0 μg/liter (0.009 ppm). These data indicate a possible increased toxicity for pyrethrum with synergist. With the intermittent flow bioassay, the degradation of pyrethrins as in the static bioassay is less a factor, so a somewhat higher toxicity is expected. Of the fish species tested, the northern pike would appear to be the most sensitive to pyrethrins poisoning.

The immature stonefly (*Pteronarcys californica*) represents an important item in the diet of fish, and synergized pyrethrins are five or more times more toxic to aquatic stonefly nymphs than to the rainbow trout, according to Bridges and Cope (1965). Additional unpublished data supplied by the Fish Pesticide Laboratory indicate that some aquatic crustaceans, the scud (*Gammarus lacustris*) and two waterfleas (*Daphnia pulex* and *Simocephalus serrutatus*), are slightly less sensitive to pyrethrins than are the aquatic insects (Table V).

VI. Interpretation of Results

The objectives of this study are to identify possible effects and problem areas for consideration in the development of future projects, and to increase the sensitivity and reliability of our field appraisal for detecting direct and indirect effects over both the short- and the long-term study.

Pyrethrum, with its low acute and chronic oral toxicity to warm-blooded animals, was not expected to cause direct effects on birds, and none were seen. From experience we expected a temporary increase in insect vulnerability to bird predation to be followed in time with increased foraging effort, depending upon the effectiveness of the insecticide and the abilities of the birds to find alternative prey species. Without data on insect prey populations and specific feeding habits of birds, we can only speculate.

We found nothing in our data to indicate that resident birds on the study plots were adversely affected by the pyrethrum spraying. There appeared to be no differences that we would attribute to the spraying in the success of nesting pairs in raising young, in the territories they occupied, or in the observed activities of marked and unmarked individuals. The short-term impact of the spraying was certainly less than that of weather. There was a snow storm on June 11, 1970, that forced juncos, robins, and probably some other species nesting at that time to renest. The high number of eight pairs of gray-headed juncos in the Spruce Gulch plot (Table II) probably represents problems of interpreting the plotted census data arising from a shifting and renesting population rather than a true increase in numbers of pairs. Robins are so mobile within plots that our estimate of breeding pairs was based more on nest locations than upon plotted observations. The fact that nesting success appeared as good in the treated as in the untreated plots is an indication that foraging activity of birds was successful, but we are not certain that it was in all cases restricted to within treatment blocks.

With respect to possible long-term effects, the data in Table II do not show much year-to-year variation in total pairs for the treated Long Gulch and the untreated Phantom Creek No. 2 plots, and we suspect that some of the differences in the Spruce Gulch plot must be explained by the difference in observers. The differences most likely to be related to treatment involve Audubon's warblers. We suspect that the apparent reduction in the number of pairs on treated plots, while numbers on the untreated plots remained stable, was caused by a reduction in some unidentified prey species, probably one inhabiting the forest canopy.

Pyrethrins, as seen in Table V, are quite toxic to fish and even more toxic to some of the aquatic organisms on which fish feed. Some effects on fish might be expected but none were seen, although the residue measured (Table IV) in the upper part of the stream in Long Gulch following the spraying was 70 ppb (μg/liter) and averaged 35 ppb of pyrethrins—well above the 96-hr LC_{50} values for fish, which ranged from 2.7 to 9 ppb. The more sensitive insects, however, were affected severely in this plot, as seen in Table III. In both treated streams there was a great increase in the numbers of drifting insects in the first 8 to 10 hr after spraying. What the tables do not show is that most drifting organisms in the Long Gulch stream were dead or immobile in the drift or bottom and net collections made one or more hours after spraying, whereas in the Spruce Gulch stream live organisms were prevalent in all postspray collections. Had the streams been flowing faster, the drift would apparently have been much heavier, because the affected and dead organisms were visible on the bottoms of all pools or ponds examined within the spray block.

We suspect that the reason rainbow trout were not affected was that most of the pyrethrins were on the water surface in oil and did not enter solution in the water to affect the fish. In contrast to the fish in the pyrethrum-treated plots, the fish in one of the fenitrothion-treated plots were visibly affected over a period of hours, but there were no fish killed. The effects were noted only in the Davis Creek plot; none were seen in the Stone Gulch plot, which had a heavier application. The apparent explanation is that the stream is screened by tree and shrub canopies in Stone Gulch, and in the Davis Creek plot beaver ponds exposed considerable water surface to collect the spray. Such ponds also provided a reservoir for the contamination and tended to release it over a greater period of time than in a free-flowing stream where the contaminated segment of the stream would soon move through and out of the plot. In the Long Gulch plot, there was also a beaver pond within the treatment block at the upper end of the stream segment; this pond could have provided a reservoir for the pyrethrins contamination, which was also increased by larger spray droplets, but this would have depended upon life of the pyrethrins. It is possible that pyrethrum in an emulsifiable formulation would have affected fish.

VII. Summary

Pyrethrum is toxic to fish and fish food (aquatic insects and crustaceans) but not to birds, according to limited laboratory studies. These findings and the known low hazard of pyrethrum for warm-blooded animals indicate that pyrethrum spraying is extremely unlikely to produce any direct toxic effects to forest birds and mammals. Pyrethrum was aerially applied to a forest ecosystem at 0.1 and 0.2 lb/acre to evaluate the short- and long-term effects on nontarget organisms. Sensitive insects were severely affected but fish were not. Birds suffered neither direct nor long-term effects attributable to pyrethrum. Residues of pyrethrins on tree foliage and in water were greatly reduced or not detectable at 1 to 5 days after treatment. Insecticidal dosages of pyrethrum appear to have a minimal direct impact on fish and wildlife and no long-term effects in the forest ecosystem.

References

Akesson, N. B., Yates, W. E., and Wilce, S. E. (1970). Key to safer and effective aerial application-controlling spray atomization. *Agrichem. Age* pp. 10–17.

Bridges, W. R., and Cope, O. B. (1965). The relative toxicities of similar formulations of pyrethrum and rotenone to fish and immature stoneflies. *Pyrethrum Post* 8(1), 3–5.

Leedy, D. (1959). Pesticide-wildlife problems and research needs. *Trans. N. Amer. Wildl. Conf.* **24**, 150, 156, and 164.

Litchfield, J. T., Jr., and Wilcoxon, F. (1949). A simplified method of evaluating dose-effect experiments. *J. Pharmacol. Exp. Ther.* **96**, 99–113.

Marr, J. W. (1961). Ecosystems of the east slope of the Front Range in Colorado. "Institute of Arctic and Alpine Research Contributions," No. 4. Univ. of Colo. Press, Boulder.

Miskus, R. P., and Andrews, T. L. (1972). Stabilization of thin films of pyrethrins and allethrins. *J. Agr. Food Chem.* **20**, 313–315.

Negherbon, W. O. (1959). "Handbook of Toxicology," Vol. III, pp. 636–651. Saunders, Philadelphia, Pennsylvania.

Pillmore, R. E., Flickinger, E. L., and Richmond, M. L. (1972). Forest spraying of Zectran and its safety to wildlife. *J. Forest.* **69**, 721–727.

Robbins, C. S. (1970). Recommendations for an international standard for a mapping method in bird census work. *Audubon Field Notes* **24**, 723–726.

Rudd, R. L., and Genelly, R. E. (1956). Pesticides: Their use and toxicity in relation to wildlife. *Game Bull.* **7**, 48–49.

Tucker, R. K., and Crabtree, D. G. (1970). Handbook of toxicity of pesticides to wildlife. *U.S., Fish Wildl. Ser., Resource Publ.* **84**, 1–131.

Chapter 8

Tests for Possible Teratogenic, Carcinogenic, Mutagenic, and Allergenic Effects of Pyrethrum

CLARA H. WILLIAMS

I. Introduction

Based upon substantial data presented as evidence at safety hearings held in 1950, the Secretary of Health, Education, and Welfare published in the Federal Register March 11, 1955 a list of tolerances and exemptions from tolerances for pesticide chemicals in or on a number of raw agricultural commodities. This was known as the Miller Amendment and represented the establishment of Sec. 408 of the Food, Drug, and Cosmetic (FD and C) Act. Included in the list was the exemption of pyrethrins and pyrethrum flowers when applied to growing crops. The notice stated that: "no health hazards are to be anticipated as pyrethrins, with or without synergists, are used today." On growing crops pyrethrins deteriorate so rapidly that no attention has been given to measuring residues from such use and this was one of the reasons for exempting the pyrethrins from a tolerance for this purpose.

In December 1955 a petition was received by the Food and Drug Administration (FDA) requesting a tolerance for 20 ppm piperonyl butoxide on wheat, corn, rye, barley, popcorn, rough rice, and buckwheat from a postharvest application in combination with pyrethrins. An exemption from a tolerance for the pyrethrins component was also sought. Final action on the petition resulted in the establishment of a tolerance of 3 ppm for pyrethrins on the above crops (Code of Federal Regulations §120.128, Federal Register February 15, 1956). The toxicity data used in evaluating the safety of the pyrethrum and its synergist were a chronic toxicity study using a combination of piperonyl butoxide and pyrethrum, at levels of 1000 ppm of the former and 164 ppm of the latter. Changes noted at these levels were similar to those resulting from the administration of 1000 ppm piperonyl butoxide alone, indicating no potentiation of toxicity of the pyrethrum by the synergist.

In the Federal Register of August 9, 1956 a tolerance was granted for residues of 1 ppm pyrethrins from postharvest application on 33 crops including fruits, vegetables, and grain. The toxicity data upon which the tolerance was established were the same as those used in establishing the 3 ppm tolerance on the grains. In this instance the pyrethrins were used in a spray or dust formulation on freshly picked fruits and vegetables while in storage or in processing plants for the control of *Drosophila* and other insects. Pyrethrum was also used directly on dried fruit, tree nuts, grains, and oil seeds as a protective treatment against insect infestation during storage.

In 1968, FDA received a petition requesting a negligible residue tolerance for pyrethrins at 0.5 ppm in milk fat (reflecting negligible residues of 0.02 ppm in whole milk) and for 0.1 ppm in meat, fat, and meat byproducts of cattle, goats, hogs, horses, and sheep. The tolerance was subsequently granted.

These now comprise the pyrethrins pesticide residues tolerances established under Section 408 of the FD and C Act to date. There are several food additive tolerances for pyrethrins which were established under Section 409 of the FD and C Act to cover pyrethrins residues in food from treated cellophane or multiwall paper bags.

In recent months the Environmental Protection Agency (EPA) scientists involved in evaluating the toxicity data in petitions for pesticide tolerances have formulated guidelines for toxicity testing procedures. For tolerances for residues at the nonnegligible level (levels of pesticide above 0.1 or 0.2 ppm) the guidelines for carcinogenicity tests specify lifespan studies in two rodent species. Therefore, the tolerances for 3 ppm post harvest residues of pyrethrins on a number of raw agricultural commodities would be considered nonnegligible. These toxicologically significant amounts would require adequate chronic toxicity tests. A teratogenic study and a three-generation reproduction study are also required for nonnegligible residues.

A description of the available carcinogenic, teratogenic, mutagenic, and allergenic data on pyrethrins follows, together with a description of the protocols, if any, currently being suggested for such tests.

II. Teratogenic Tests

The Report of the Secretary's Commission on Pesticides and Their Relationship to Environmental Health (Mrak, 1969) refers to a number of compounds that have been tested for teratogenicity, but pyrethrins are not mentioned. In a recent survey of the pesticide literature by Durham and Williams (1972) no evidence of teratogenicity of pyrethrins was found. One test for measuring teratogenic potential is known as the Somers test and was originally carried out in rabbits. It is also possible to estimate teratogenic potential by use of the multigeneration reproduction test. This latter test is less likely to give a positive response, because teratogenicity appears more frequently when a challenging dose is given during the sensitive period of organogenesis, as is the case in the Somers test.

A. Reproduction Studies

A reproduction study with pyrethrins has been performed in rats. Hazelton Laboratories (Weir, 1966b) conducted a two-litter one-generation reproduction study with 12 female rats fed 5000 ppm pyrethrins in the diet and 20 female rats that served as controls. The results showed that the ingestion of pyrethrins at a level of 5000 ppm resulted in significant growth suppression in the nursing first (F_{1a}) litter pups and, to an increased degree, in the nursing second (F_{1b}) litter pups. The reproductive performance of the parental generation rats (P_1) in the three test groups, reflected by the various indices determined for these groups, was not adversely affected. The indices, the mean litter size and weights of the pups at 24 hr, were comparable among the test and control groups and among the first and second litters in each group.

Necropsies performed on representative F_{1a} and F_{1b} weanlings from the control and test groups revealed no compound-related gross pathological changes in the organs.

Although this study indicates that pyrethrins have no effect on rat reproductive capacity, it does not provide conclusive evidence that pyrethrins possess no teratogenic potential. More modern toxicological protocols for reproductive tests require a three generation study, thus giving more opportunity for the teratogenic properties of a chemical to appear. More than one dose level should also be used in order to establish a dose–response relationship.

Hazelton Laboratories (Weir, 1966b) also performed a study in rabbits to evaluate and characterize the teratogenic potential of pyrethrins. Nine proven breeding does of the New Zealand strain were given 90 mg/kg of pyrethrins by capsule from day 8 through day 16 of the gestation period. A control group was given 0.45 ml/kg vegetable oil on day 29 and day 30 of gestation. Four does from each group were sacrificed by intravenous air embolism and the young were delivered by caesarian section. The remaining does in each group were allowed to complete full term with normal parturition; however, one control doe was sacrificed on day 36 of the gestation period. Each doe was sacrificed by intravenous air embolism and necropsied. At necropsy the uteri of all does were closely scrutinized for sites of fetal attachment and evidence of resorption. All litters were examined for stillbirths and anatomical malformations and all fetuses were examined grossly and fixed in formalin. Subsequent skeletal clearing and alizarin staining were performed on two fetuses chosen at random from each litter. The control and the pyrethrins-treated animals gave similar findings in the parameters measured.

A more satisfactory test for teratogenic potential would include several levels of administration of the test compound to enable a dose–response relationship to be measured.

B. Guidelines for Testing

Guidelines for teratogenicity testing have been proposed by the Toxicology Branch of the Pesticides Tolerances Division of the EPA. They are:

(a) *Species.* Rat, mouse, hamster, rabbit, or monkey. If the rat is used for the reproduction study, it is recommended that a different species be used for the teratogenicity study.

(b) *Number of animals.* Twenty pregnant female rodents per level, ten pregnant female nonrodents per level.

(c) *Dose levels.* Three levels plus a control, with levels selected so that a "no-effect" level will be ascertained even though an effect is found at the highest level.

(d) *Route of administration.* Orally.

(e) *Period of dosing.* Daily through the sensitive period of organogenesis. For example, days 6 through 15 for rats, days 6 through 10 for hamsters.

(f) *Method of birth.* Except for monkeys, deliver fetuses by caesarian section 1 day prior to term.

(g) *Observations.* Number and position of resorption sites, number of corpora lutea and implantation sites, fetal weights, external malformations, and skeletal anomalies after clearing and staining with Alizarin red.

III. Carcinogenic Tests

The Mrak report (1969) examined the available toxicity data on about 100 pesticidal chemicals. The report that the panel used for the pyrethrins carcinogenicity evaluation was referenced as an FDA unpublished report (1951). On the basis of this report, the panel placed pyrethrins in priority group C3 which states "the following compounds yielded an increased tumor incidence in comparison with negative controls but the level of significance was less than 0.02. . . . These compounds have third priority for additional testing." The study is the same one described in Lehman's Summaries of Pesticide Toxicity (1965) under "Pyrethrins," and in which the statement is made that "pyrethrins produced no effect on tumor production."

A. Long-Term Toxicity Studies in Rats

This long-term toxicity test was actually performed in 1951 but the pathology was not evaluated until 1959 by Long in collaboration with Fitzhugh. Three groups of Osborne-Mendel rats, 24 per group and approximately evenly divided as to sex, received 0.5%, 0.1%, or 0.02%, respectively, of pyrethrum in the diet for 2 years. There was also a control group of 24 rats, likewise equally divided as to sex, and also serving as controls for a series of rats fed rotenone. Of these 96 rats, 78 were received in the pathology laboratory and 55 were sectioned and studied microscopically. Of the 96 animals started on the experiment, 37 survived 102 or more weeks. Survival was normal in all the test groups, but mortality was high among the controls, only three surviving and the other controls largely succumbing to infections of the respiratory tract.

Pathological changes attributable to pyrethrum were relatively minimal. There was a small total amount of hepatic damage consisting of slight bile duct proliferation plus occasional miscellaneous lesions at 0.5% and 0.1% but not at 0.02%. A tendency to smaller follicles in the thyroid gland was moderate at 0.5%; unfortunately, too few thyroids were sectioned at lower dosage levels to determine at what dose this mild effect faded out. Pyrethrum had no effect on tumor production.

In another study, Ambrose and Robbins (1951) fed rats pyrethrum in their diet for 2 years at 0, 200, 1000, and 5000 ppm. At 1000 ppm there was no tissue damage but at 5000 ppm tissue damage and gross toxic signs were noted. The effect noted was upon liver.

Lehman (1954), in referring to the repeated administration of pyrethrum, states that the detoxification of pyrethrins in rats is so rapid that they can ingest more than an acutely fatal dose every day throughout their life with little or no injury. In this case the pyrethrum was uniformly mixed with the food and the intake intermittent.

In the 1970 FAO/WHO evaluation of pyrethrins, 200 ppm of pyrethrum in the diet was set as the level causing no toxicological effect. This is equivalent to 10 mg/kg body weight per day. The FAO/WHO (1967) report recommends for pyrethrins: "Long term studies should be made in at least one more species with special emphasis on the effect on liver." Such studies are now underway at a commercial laboratory.

B. Methods for Testing

The Toxicology Branch of the Pesticides Tolerances Division has also drawn up guidelines for carcinogenicity testing:

a. Species. Usually two rodents.

b. Selection of animals. Weanling random-bred animals. At the option of the experimenter, these animals may be derived from a normal stock colony and maintained on test diet for their lifespan or they may be the F_{1a} progeny from a reproduction study in which the parents have been treated with the pesticide.

c. Dosage. At least three dosage levels plus a control; the highest dosage should elicit any specific pharmacological action the test compound might have. One dose level should show no observable differences from control animals.

d. Duration. Two years for rats and hamsters, 18 months for mice.

e. Number of animals. The groups should be large enough to assure an adequate number of survivors at the end of the study. Depending upon the strain of rodent, it is suggested to use from 25 to 50 animals of each sex per dose level.

f. Observations. (*i*) Growth, food consumption, general appearance, signs of local and systemic toxicity, terminal organ weights, and gross and histopathological examination of both rodent species. (*ii*) In one species (usually the rat), clinical laboratory tests should be performed at 3, 6, 12, 18, and 24 months. These should include routine hematological measurements, biochemical parameters, qualitative urinalysis, and cholinesterase activity at intervals on plasma and red cells and terminally in the brain.

IV. Mutagenic Tests

In Recommendation 10 of *The Secretary's Commission on Pesticides and Their Relationship to Environmental Health* (Mrak, 1969) it is stated that research should be expanded relative to the metabolism and degradation of pesticides and their effects on the integrated system by which organisms derive energy, build protoplasm, and reproduce. Additional studies on teratogenesis, mutagenesis, and carcinogenesis must be supported. Neither pyrethrins nor their commonly used synergist piperonyl butoxide has been reported as having mutagenic potential but it is possible that it has never been tested.

Geneticists are focusing their interest upon pesticides because of their ubiquitous presence in our environment. A number of chemicals cause mutations in nonmammalian systems such as bacteria and *Drosophila*. To evaluate the possible mutagenic potential of a pesticide used on food crops, these nonmammalian tests are not relevant. A mammalian test, such as the dominant lethal test, is more appropriate. Evaluation of tests for mutagenicity is difficult, as the art of mutagenicity testing is in its infancy. In the EPA toxicology guidelines, it is suggested that testing laboratories become familiar with the various tests but the EPA is not requiring them routinely. Questions about such matters as dose–response relationships, positive findings in one test and negative findings in others, and the proper test species cannot be satisfactorily answered at this time.

V. Allergenic Effects

Injury to man from pyrethrum has most frequently resulted from the allergenic properties of the material rather than from its direct toxicity. Although the allergy has been associated with occupational or therapeutic contact, it is impossible to exclude any importance of it in connection with food residues.

Pyrethrum sensitivity may manifest itself in several forms in man. Contact dermatitis is by far the most common. Pyrethrum flowers are closely related to ragweed. Refined pyrethrins are not at all allergenic.

A. Clinical Reports

A study of the literature reveals that allergic skin reactions have been the most important effects from pyrethrum in man. Much of the clinical data consists of reports of dermatitis. Feinberg (1934) has found that

46.2% of 225 individuals who were sensitive to ragweed (belonging to the plant family Compositae as does pyrethrum) proved to be sensitive to unrefined or ground pyrethrum flowers. Martin and Hester (1941) investigating a subject sensitive to pyrethrum found that the sensitizing agent was not pyrethrins but a volatile oil not present in the purified or refined extracts.

In skin tests of a refined pyrethrum extract used in modern aerosol preparations (Pyrocide 175 RC containing 20% pyrethrins) on atopic individuals who had shown a positive reaction either to ragweed or unrefined pyrethrum, Zucker (1965) found negative reactions in 88 out of 110 cases and definitely positive reactions in only four out of 110. He found approximately 43% cross-reactions between ragweed and unrefined pyrethrum similar to Feinberg's results. In these studies Zucker also studied 14 allergic individuals who gave histories of adverse reactions to household insecticide sprays. These individuals were tested by being placed for 6 min in an average sized room after 6 sec of spraying with pyrethrum aerosol containing 0.3% pyrethrins. Seven of the 14 had no reactions; five had itching or dryness of nose or throat. Two complained of eye irritation, which was mild and lasted only a few minutes. None of the patients developed asthma or serious reactions and no exposures had to be interrupted. Only two of the seven who had such symptoms had mild reactions when subjected to further exposures consisting of only pyrethrins and freon. Only one of these reacted when the pyrethrins–freon exposure was repeated. Zucker concluded that there were no definite allergic reactions to refined pyrethrins in his study and that a positive skin test to unrefined pyrethrum does not seem to indicate allergy to insecticides that contain present-day refined pyrethrins.

Leyton (1968), using the passive transfer technique involving the transfer of human serum to the skin of monkeys, confirmed Zucker's observations. Sera from individuals highly sensitive to unrefined pyrethrum in Zucker's studies were used and showed no reactions when challenged with the refined pyrethrins (Pyrocide 175 RC).

B. Human Patch Test

Hazelton Laboratories (Weir, 1966a) conducted a patch test with pyrethrins in human volunteers. In this study 200 individuals, comprised of 160 white females, 20 white males, 16 colored females, and 3 colored males, were tested for skin sensitization by application of a mixture containing 1.47% pyrethrins and 2.94% technical piperonyl butoxide. The volunteers were divided into three groups as follows: (*a*) 70 subjects

21–80 years old (mostly middle aged white females); (*b*) 106 subjects ages 19–71 (heterogenous groups of hospital workers and nurses); (*c*) 24 subjects ages 15 to 60 years (mostly young pregnant females, four caseworkers).

Patches were cotton gauze squares secured at two extremities. An initial test was conducted in which the treated patch remained in place for 4 days and was then removed. Twelve days later another patch was applied for a 3-day interval. No effects upon the treated skin areas were noted at any time in any of the volunteers.

C. Industrial Exposure

A confidential report from one commercial manufacturer of pyrethrins-containing products indicates that in the 18-year history of manufacturing and selling such products no medically documented case of allergy was noted. During this period, the following pattern of dermal experience was reported:

Alleged allergy	1 per 19.7 million units
Contact dermatitis	1 per 8.2 million units
Unclassified	1 per 6.7 million units

These tests are important because they provide considerable medical evidence that refined pyrethrins are not allergenic, although it is clear that ground pyrethrum powders have allergenic properties. Moreover, they add a note of caution against the use of skin tests with unrefined pyrethrum, as the results of such tests do not indicate a potential allergy to insecticides that contain modern, refined pyrethrins.

VI. Summary

From the toxicity data presently available, there is no evidence that pyrethrins are carcinogenic, teratogenic, or mutagenic in test animals or, by extrapolation, in man. The proven allergenic properties of pyrethrins have been conclusively shown to be caused by impurities in the pyrethrum flowers. The carcinogenic potential of pyrethrins has not been satisfactorily investigated by modern-day standards. Because the established tolerances for pyrethrins are above the negligible level, an adequate lifespan feeding study of pyrethrins should be made. This is particularly important because of the frequent presence of pyrethrins in the ecosystem, not only on some stored products but also from use of pyrethrins dispensers. Tolerances or exemptions from a tolerance have been petitioned for the use of pyrethrins dispensers in food processing and food storage areas. For these latter uses inhalation toxicity studies have been suggested by EPA toxicologists.

References

Ambrose, A. M., and Robbins, D. J. (1951). Comparative toxicity of pyrethrins and allethrin. *Fed. Proc., Fed. Amer. Soc. Exp. Biol.* **10**, 276.

Durham, W. F., and Williams, C. H. (1972). Mutagenic, teratogenic, and carcinogenic properties of pesticides. *Annu. Rev. Entomol.* **17**, 123–148.

FAO/WHO (1967). Evaluation of some pesticide residues in food. FAO/PL: CP 15; WHO/FOOD ADD/67.32.

FAO/WHO (1970). Evaluations of some pesticide residues in food. AGP: 1970/M/12/1.

Feinberg, S. M. (1934). Pyrethrum sensitization. *J. Amer. Med. Assoc.* **102**, 1557–1558.

Lehman, A. J. (1954). Toxicological evaluation of household insecticides. *Ass. Food Drug Off. U.S., Quart. Bull.* **18**, 3–13.

Lehman, A. J. (1965). Summaries of pesticide toxicity. *Ass. Food Drug Off. U.S.* p. 126.

Leyton, L. (1968). Cited as personal communication in "Toxicology and Pharmacology of Pyrethrum," unpublished report by Dr. Joseph Moore.

Long, E., and Fitzhugh, O. G. (1959). Pathological changes produced in rats by the feeding of pyrethrum for two years. Unpublished FDA study.

Martin, J. T., and Hester, K. H. C. (1941). Dermatitis caused by insecticidal pyrethrum flowers (*Chrysanthemum cinerariaefolium*). *Brit. J. Dermatol. Syph.* **53**, 127–142.

Mrak, E. M., Chairman. (1969). "Report of the Secretary's Commission on Pesticides and their Relationship to Environmental Health." U.S. Department of Health, Education and Welfare, Washington, D.C.

Weir, R. J. (1966a). Human patch test. Unpublished report from Hazelton Laboratories to S. C. Johnson & Son, Inc.

Weir, R. J. (1966b). Reproduction study—rats neopynamin and pyrethrin. (report submitted by Hazelton Laboratories to S. C. Johnson & Son, Inc.).

Zucker, A. (1965). Investigation of purified pyrethrum extracts. *Ann. Allergy* **23**, 335–339.

Chapter 9

Interactions in the Toxicity of Pyrethrum, Synergists, and Other Chemicals to Mammals

HOWARD BOND, KAREN MAUGER, and JOHN J. DeFEO

I. Introduction

Pyrethrum, an extract of the flower heads of *Chrysanthemum cinerariaefolium*, is a very effective insecticide with a low mammalian toxicity, as discussed elsewhere in this volume. In order to minimize the amount of pyrethrum used and to enhance the insecticidal effects of many preparations, synergistic chemicals are usually mixed with pyrethrum. These insecticidal substances are being used more and more in our modern society so that humans are being exposed to these chemicals in ever increasing quantities in homes, shopping areas, places of work, dining facilities, and other public places where insect removal is desirable. Information concerning the possible effects of these insecticidal substances on mammalian systems is of importance in determining how these chemicals might be used more efficiently and in as safe a manner as possible.

Many people are also taking drugs of various types in their everyday life the actions of which conceivably could be influenced by simultaneous administration of the insecticides. It is therefore important to determine whether interactions might occur with some of the more commonly used drugs, such as alcohol, acetylsalicylic acid (aspirin), caffeine, nicotine, and phenobarbital. This study was designed to obtain information concerning toxicity and possible interactions resulting from the actions of the insecticides and drugs acutely and over a 90-day period. Information obtained would therefore be a point of departure for more detailed studies concerning reactions and interactions that would be of help in elucidating the mechanisms by which these substances produce their effects in the mammalian system.

II. Materials and Methods

A. Animals and Chemicals

The animals employed in this study were albino rats obtained from the Charles River Breeding Laboratories, Inc. (Wilmington, Mass.).

The chemicals used were: Premium Pyrocide 175, a 20% solution of pyrethrins in petroleum distillate, referred to in this report as pyrethrum (McLaughlin Gormley King Co., Minneapolis, Minn.); tropital, a synergist containing 90% piperonal bis[2-(2-butoxyethoxy)ethyl]acetal and 10% related compounds (McLaughlin Gormley King Co.); piperonyl butoxide, a synergist consisting of 80% α-[2-(2-butoxyethoxy)ethoxy]-4,5-methylenedioxy-2-propyltoluene and 20% related compounds (FMC Corp., Middleport, New York); sulfoxide, a synergist containing 88% 1,2-methylenedioxy-4-[2-(octylsulfinyl)propyl]benzene and 12% related compounds (S. B. Penick Co., New York, N.Y.). All other drugs and chemicals were purchased from scientific supply houses.

B. Acute Toxicity

Exploratory studies were carried out on three groups of two rats, male and female. The results permitted estimation of an approximate LD_{50} value as a basis for further study. The signs of effects produced by the various drugs and chemicals were observed. From the results of these studies doses were calculated for actual LD_{50} studies for each substance. The LD_{50} studies were carried out on five groups of 10 rats (five male, five female). Each animal in each group of ten was administered one dose of the substance orally by means of a metal stomach tube. The animals were then carefully observed throughout the first hour, then at 3, 6, and 24 hr. The survivors were again checked at least once daily for 2 weeks. The LD_{50} values are based on 24-hr mortality and the values were calculated by the method of Litchfield and Wilcoxon (1949). These results were evaluated and then used as the basis for the acute interaction studies, which were carried out by administering a constant dose of pyrethrum (LD_{50} dose) and varying doses of the other substances. When necessary, because of excess mortality, the dose of pyrethrum was reduced to either $\frac{1}{2}$ or $\frac{1}{4}$ its calculated LD_{50} value and the other substances were again tested in various doses. The animals were observed through the same time periods as with the single substances, the new LD_{50} values were calculated, and the curves were compared for parallelism and significance, again by the method of Litchfield and Wilcoxon (1949).

C. Subacute Toxicity

The subacute toxicity study was designed so that most of the animals would survive during 90 days of daily administration of the drugs and chemicals, so that some pathological changes would occur which would indicate a minimal toxic effect as well as the type of changes produced, and if possible so that an upper limit (maximum effective dose) for safe usage would be established.

1. *Testing Schedule*

A dose of one-fourth of each LD_{50} value as determined above was selected for the subacute toxicity studies. Rats that weighed approximately 100 g (1 month old), when purchased, were allowed to adapt to their new quarters in community cages for 1 week. Twenty-four hours prior to starting the actual tests, the animals were moved to a special room where they were housed in pairs (same sex) in metabolism cages. Each substance or combination of substances was tested in groups of 20 rats (ten male, ten female).

Throughout the study each rat was weighed every morning and the test substances were administered orally. The animals were then returned to their metabolism cages where they were permitted water *ad libitum* and 20 gm/day of Rockland rat chow (Tekland Inc., Monmouth, Ill.) for each animal.

2. Tests Performed

The urine of each group was tested at 12, 28, 58, and 88 days. Care was taken to prevent contamination and to assure minimal evaporation in the special collecting units. Prior to the tests, urine was collected for 24 hr; measurements for volume, pH, and specific gravity were taken; and special tests for protein were carried out. Blood obtained from the tail veins was tested on days 13, 30, 59 and 91. The hematocrit value, hemoglobin content, white and red blood cell counts, and differential counts were determined.

The animals that survived were sacrificed on day 91 when organ weights were obtained and various tissues were removed and prepared for gross observations and histological study. These latter studies were performed by Merle C. Loveless, D.V.M., Associate Professor of Animal Science Technology, State University, Agricultural and Technical College, Delhi, New York 13753.

III. Results

A. Acute Toxicity

The results (Tables I and II) indicated a general pattern of increased toxicity (decreased LD_{50} dosage) when the drugs were combined with

TABLE I

LD_{50} Values for Orally Administered Drugs

Drug	LD_{50} (mg/kg)	95% confidence limits	Comments
Pyrethrum	1440	(1171–1771)	20% solution of pyrethrins in petroleum distillate
Alcohol	13,800	(10,698–17,802)	absolute
Aspirin	1200	(1000–1440)	5% aspirin in 10% sodium citrate
Caffeine	330	(289–376)	10% solution citrated
Nicotine	90	(74–110)	nicotine sulfate solution (40%)
Phenobarbital	330	(244–446)	sodium phenobarbital (40%)
Piperonyl butoxide	7500	(6098–9225)	commercial grade
Sulfoxide	2000	(1000–4000)	commercial grade
Tropital	4200	(3621–4872)	commercial grade

TABLE II

Effect of Pyrethrum on Acute Toxicity (LD_{50}) of Orally Administered Drugs

Drug	Pyrethrum (mg/kg)	LD_{50} (mg/kg)	95% confidence limits	Percent of single LD_{50} values[a]
Alcohol	720	11,040[b]	(9798–12,420)	50 and 80[b]
Aspirin	720	900	(504–1620)	50 and 75[c]
Caffeine	720	211	(132–337)	50 and 64[c]
Nicotine	1440	11[d]	(6–24)	100 and 12[b]
Phenobarbital	1440	182[d]	(125–261)	100 and 55[b]
Piperonyl butoxide	720	3000	(1200–7500)	50 and 40[c]
Sulfoxide	720	600[d]	(360–1020)	50 and 30[b]
Tropital	360	2520[d]	(2184–2940)	25 and 60[b]

[a] First value is that for pyrethrum, second value is that for test substance.

[b] Curves are parallel to single curve. LD_{50} dose of alcohol not significantly different from single LD_{50} ($P < 0.05$).

[c] Curves deviate from parallelism ($P < 0.05$).

[d] Significantly different from single value ($P < 0.05$).

pyrethrum. The LD_{50} for alcohol did not differ significantly from the LD_{50} obtained for alcohol and pyrethrum. Aspirin, caffeine, and piperonyl butoxide, when combined with pyrethrum, produced LD_{50} curves that deviated from parallelism. The LD_{50} of the other substances, nicotine, phenobarbital, sulfoxide, and tropital, differ significantly from those obtained when pyrethrum was added. It should be pointed out that the combined LD_{50} values result from the interaction of a significant dose of pyrethrum with the particular drug or chemical. Thus, nicotine and phenobarbital are combined with a dose of pyrethrum equal to its LD_{50} and this may be interpreted as a protective or inhibitory interaction of the two drugs on pyrethrum. The greatest increase in the acute toxicity was produced by the three synergists when combined with pyrethrum.

B. Subacute Toxicity

The doses used for daily oral administration for 90 days are listed in Table III. These are obviously extremely high doses that were necessary to produce minimal effects for the purposes of this study. Alcohol, caffeine, and piperonyl butoxide produced the most deaths during the 90-day study (Table IV). When pyrethrum was added, alcohol, caffeine, and sulfoxide produced the highest rate of mortality. The most interest-

TABLE III

Daily Oral Dose Used in Subacute Studies (90 Day)

Drug group	Daily dose (mg/kg)	Drug group	Daily dose (mg/kg)
Pyrethrum	360	pyrethrum (P)	360
Alcohol	3450	P + alcohol	180 + 2800
Aspirin	300	P + aspirin	180 + 225
Caffeine	82.5	P + caffeine	180 + 52.8
Nicotine	22.5	P + nicotone	360 + 3
Phenobarbital	82.5	P + phenobarbital	360 + 45.4
Piperonyl butoxide	1875	P + piperonyl butoxide	180 + 750
Sulfoxide	500	P + sulfoxide	180 + 150
Tropital	1050	P + tropital	90 + 630

TABLE IV

Mortality Resulting from Subacute Studies (90 Day)

Treatment	Animals	Number dead	Mortality[a] (%)
Controls	10	1	10
Pyrethrum (P)	20	1	5
Alcohol	20	4	20
Aspirin	20	2	10
Caffeine	20	5	25
Nicotine	20	1	5
Phenobarbital	20	2	10
Piperonyl butoxide	20	8	40[b]
Sulfoxide	20	1	5
Tropital	20	0	0
P + alcohol	20	6	30[c]
P + aspirin	20	0	0
P + caffeine	20	6	30[c]
P + nicotine	20	1	5
P + phenobarbital	20	0	0
P + piperonyl butoxide	19	1	5[c]
P + sulfoxide	20	6	30[c]
P + tropital	20	1	5

[a] Compared by Fisher's exact method (McNemar, 1954).
[b,c] Significantly different from control, or from drug tested alone.

ing results are those which indicate that piperonyl butoxide is more toxic when used alone, and sulfoxide becomes more toxic when combined with pyrethrum. Tropital and pyrethrum, when administered singly or in combination, produced the lowest mortality of the insecticidal substances.

C. Weight Gains

Tropital and alcohol were the only two substances that did not significantly affect cumulative weight gains in males when administered alone. All other substances reduced this gain (Table V). In the females, all the substances tended to reduce gains although once again the effects of alcohol and tropital are not statistically different from the controls. Pyrethrum, itself, produced a significant decrease in weight gains in both sexes. When administered with pyrethrum, all of the substances, except alcohol, reduced gains significantly in the male animals (Table VI). Sulfoxide produced a greater effect than the other synergists when combined with pyrethrum. In both sexes aspirin, nicotine, and sulfoxide produced the greatest effect when combined with pyrethrum. The weight gains in percent also indicate that alcohol and tropital affect this parameter least, although there is a general trend to lower total gains. When combined with pyrethrum, alcohol and phenobarbital have the least effect.

TABLE V

Cumulative Weight Gains in Subacute Study of Single Drugs (90 Day)

Treatment	Male		Female	
	Grams ± S.E.	Percent	Grams ± S.E.	Percent
Controls	276 ± 16	236	174 ± 16	147
Pyrethrum	141 ± 18[a]	95	105 ± 6[a]	71
Alcohol	303 ± 22	209	142 ± 9	114
Aspirin	120 ± 11[a]	53	106 ± 6[a]	63
Caffeine	151 ± 22[a]	83	92 ± 7[a]	46
Nicotine	171 ± 10[a]	86	88 ± 5[a]	58
Phenobarbital	175 ± 20[a]	86	106 ± 6[a]	54
Piperonyl butoxide	181 ± 30[a]	119	129 ± 10[a]	75
Sulfoxide	166 ± 13[a]	77	140 ± 6[a]	92
Tropital	267 ± 4	212	147 ± 5	128

[a] Significantly different from controls of same sex at 95% level of confidence.

TABLE VI

Cumulative Weight Gains in Subacute Study of Drug Interactions (90 Days)

Treatment	Male		Female	
	Grams ± S.E.	Percent	Grams ± S.E.	Percent
Controls	276 ± 16	236	174 ± 16	147
Pyrethrum (P)	141 ± 18[a]	95	105 ± 6[a]	71
P + alcohol	255 ± 15	209	173 ± 8	142
P + aspirin	147 ± 8[a]	104	133 ± 12[a]	99
P + caffeine	187 ± 7[a]	126	159 ± 11	112
P + nicotine	160 ± 13[a]	137	139 ± 5[a]	128
P + phenobarbital	215 ± 8[a]	215	173 ± 12	175
P + piperonyl butoxide	170 ± 14[a]	140	161 ± 9	139
P + sulfoxide	135 ± 16[a]	113	121 ± 11[a]	95
P + tropital	199 ± 19[a]	167	160 ± 7	132

[a] Significantly different from controls of same sex at 95% level of confidence.

D. Organ Weights

The weights of the various organs are recorded in Tables VII through XII. Some of the interesting changes produced by single drugs are those by pyrethrum and nicotine on the heart and spleen of males and females (Table VII) and gonads and adrenals of the females (Table VIII)

TABLE VII

Comparative Organ Weights Resulting from Subacute Studies (90 Days)[a] (Single Drugs)

Treatment group	Heart		Spleen	
	Male	Female	Male	Female
Controls	1.20 ± 0.06	0.96 ± 0.06	0.77 ± 0.05	0.68 ± 0.03
Pyrethrum	0.89 ± 0.05[b]	0.83 ± 0.03[b]	0.48 ± 0.03[b]	0.54 ± 0.03[b]
Alcohol	1.25 ± 0.05	0.84 ± 0.04[b]	0.80 ± 0.06	0.55 ± 0.04
Aspirin	1.10 ± 0.05	0.89 ± 0.02	0.79 ± 0.07	0.60 ± 0.04
Caffeine	1.13 ± 0.05	0.94 ± 0.05	0.64 ± 0.05[b]	0.51 ± 0.03
Nicotine	1.04 ± 0.03[b]	0.77 ± 0.04[b]	0.59 ± 0.03[b]	0.55 ± 0.03[b]
Phenobarbital	1.04 ± 0.07	0.95 ± 0.03	0.84 ± 0.07	0.62 ± 0.04
Piperonyl butoxide	0.99 ± 0.04[b]	0.92 ± 0.03	0.67 ± 0.02	0.59 ± 0.04
Sulfoxide	1.12 ± 0.03	0.97 ± 0.02	0.68 ± 0.03	0.70 ± 0.05
Tropital	1.22 ± 0.03	0.84 ± 0.02[b]	0.70 ± 0.03	0.64 ± 0.03

[a] Grams, mean ± S.E.

[b] Significantly different from controls of same sex at 95% level of confidence.

TABLE VIII

Comparative Organ Weights Resulting from Subacute Studies (90 Days)[a] (Single Drugs)

Treatment group	Gonads		Adrenals	
	Male	Female	Male	Female
Controls	3.37 ± 0.21	0.20 ± 0.01	0.09 ± 0.01	0.12 ± 0.01
Pyrethrum	3.46 ± 0.009	0.13 ± 0.009[b]	0.08 ± 0.003	0.09 ± 0.005[b]
Alcohol	3.68 ± 0.10	0.17 ± 0.02	0.09 ± 0.007	0.10 ± 0.002[b]
Aspirin	3.80 ± 0.11	0.15 ± 0.01[b]	0.11 ± 0.012	0.10 ± 0.007
Caffeine	3.18 ± 0.11	0.17 ± 0.01	0.09 ± 0.005	0.11 ± 0.007
Nicotine	3.64 ± 0.19	0.14 ± 0.005[b]	0.08 ± 0.003	0.09 ± 0.004[b]
Phenobarbital	3.60 ± 0.13	0.17 ± 0.01	0.09 ± 0.005	0.11 ± 0.006
Piperonyl butoxide	3.52 ± 0.08	0.17 ± 0.01	0.09 ± 0.004	0.12 ± 0.006
Sulfoxide	3.56 ± 0.05	0.19 ± 0.02	0.09 ± 0.005	0.10 ± 0.005
Tropital	3.72 ± 0.12	0.17 ± 0.01	0.09 ± 0.005	0.09 ± 0.005

[a] Grams, mean ± S.E.

[b] Significantly different from controls of same sex at 95% level of confidence.

which show a decrease in size. However when compared on a weight to 100 gm body weight basis this effect is minimized because of the general decrease in overall body weight. The most significant effects are seen in the liver, which was enlarged by pyrethrum, alcohol, phenobarbital, piperonyl butoxide, sulfoxide, and tropital (Table IX).

TABLE IX

Comparative Organ Weights Resulting from Subacute Studies (90 Days)[a] (Single Drugs)

Treatment group	Liver		Kidneys	
	Male	Female	Male	Female
Controls	11.56 ± 0.99	8.61 ± 0.94	2.71 ± 0.11	2.13 ± 0.12
Pyrethrum	14.37 ± 1.08[b]	12.61 ± 0.68[b]	2.72 ± 0.21	2.00 ± 0.10
Alcohol	14.83 ± 0.71[b]	8.95 ± 0.42	3.06 ± 0.13	1.88 ± 0.07
Aspirin	12.22 ± 0.14	9.44 ± 0.32	3.22 ± 0.14[b]	2.11 ± 0.08
Caffeine	9.43 ± 0.48	8.01 ± 0.33	2.40 ± 0.09	1.99 ± 0.04
Nicotine	12.60 ± 0.54	9.22 ± 0.69	2.96 ± 0.07	1.84 ± 0.07[b]
Phenobarbital	16.67 ± 1.25[b]	13.94 ± 0.54[b]	2.82 ± 0.15	2.25 ± 0.07
Piperonyl butoxide	26.28 ± 2.05[b]	22.0 ± 1.30[b]	2.85 ± 0.15	2.45 ± 0.10
Sulfoxide	16.83 ± 0 64[b]	14.81 ± 0.48[b]	3.18 ± 0.12[b]	2.47 ± 0.08[b]
Tropital	14.80 ± 0.27[b]	10.90 ± 0.50[b]	3.12 ± 0.05[b]	2.24 ± 0.10

[a] Grams, mean ± S.E.

[b] Significantly different from controls of same sex at 95% level of confidence.

TABLE X

Comparative Organ Weights Resulting from Subacute Studies (90 Days)[a] (Drug Combinations)

Treatment group	Heart		Spleen	
	Male	Female	Male	Female
Controls	1.20 ± 0.06	0.96 ± 0.06	0.77 ± 0.05	0.68 ± 0.03
Pyrethrum (P)	0.89 ± 0.05[b]	0.83 ± 0.03[b]	0.48 ± 0.03[b]	0.54 ± 0.03[b]
P + alcohol	1.09 ± 0.05	0.97 ± 0.03	0.65 ± 0.06	0.60 ± 0.04
P + aspirin	0.89 ± 0.03[b]	0.87 ± 0.03	0.52 ± 0.04[b]	0.56 ± 0.04
P + caffeine	1.07 ± 0.03	1.02 ± 0.03	0.51 ± 0.02[b]	0.68 ± 0.05
P + nicotine	0.90 ± 0.03[b]	0.85 ± 0.02[b]	0.52 ± 0.04[b]	0.55 ± 0.03[b]
P + phenobarbital	0.91 ± 0.03[b]	0.85 ± 0.04	0.66 ± 0.04	0.56 ± 0.05
P + piperonyl butoxide	0.94 ± 0.05[b]	0.89 ± 0.03	0.67 ± 0.05	0.61 ± 0.05
P + sulfoxide	0.82 ± 0.05[b]	0.85 ± 0.06	0.42 ± 0.05[b]	0.49 ± 0.05[b]
P + tropital	0.97 ± 0.05[b]	0.94 ± 0.04	0.69 ± 0.05	0.77 ± 0.05

[a] Grams, mean ± S.E.

[b] Significantly different from controls of same sex at 95% level of confidence.

The drug combinations also produced some interesting effects. As before, the apparent decrease in size of the heart and spleen (Table X) is minimized when the loss of body weight gains is taken into account. Effects on the gonads and adrenals are not significant, except the reduction in gonad weights with females receiving pyrethrum only (Table XI). The greatest overall increases in organ size were those of the liver (Table XII). Piperonyl butoxide alone produced the greatest increase in liver size of all the substances. However, when combined with pyrethrum, the combination with phenobarbital showed a greater increase than did that with piperonyl butoxide.

E. Pathology

The pathological changes are reported according to positive reactions and with abstracts of the actual descriptions provided by the pathologist. Only those tissues actually showing any pathological effect are mentioned.

In the test series using each substance alone, alcohol and phenobarbital produced a significant effect on the liver and a low incidence of effects on the kidney (Tables XIII and XIV). Pyrethrum also produced a low-level effect on liver and kidney. No apparent pathological changes

TABLE XI

Comparative Organ Weights Resulting from Subacute Studies (90 Days)[a] (Drug Combinations)

Treatment group	Gonads		Adrenals	
	Male	Female	Male	Female
Controls	3.37 ± 0.21	0.20 ± 0.01	0.09 ± 0.01	0.12 ± 0.01
Pyrethrum (P)	3.46 ± 0.12	0.13 ± 0.009[b]	0.08 ± 0.003	0.09 ± 0.005[b]
P + alcohol	3.47 ± 0.11	0.18 ± 0.01	0.06 ± 0.007	0.12 ± 0.01
P + aspirin	3.52 ± 0.12	0.21 ± 0.01	0.08 ± 0.006	0.12 ± 0.006
P + caffeine	3.32 ± 0.08	0.21 ± 0.01	0.09 ± 0.004	0.13 ± 0.009
P + nicotine	3.22 ± 0.07	0.36 ± 0.18	0.08 ± 0.005	0.12 ± 0.007
P + phenobarbital	3.51 ± 0.08	0.18 ± 0.01	0.09 ± 0.003	0.11 ± 0.006
P + piperonyl butoxide	3.53 ± 0.09	0.23 ± 0.02	0.08 ± 0.006	0.11 ± 0.007
P + sulfoxide	3.15 ± 0.16	0.17 ± 0.01	0.08 ± 0.004	0.10 ± 0.006
P + tropital	3.38 ± 0.12	0.18 ± 0.01	0.08 ± 0.004	0.12 ± 0.007

[a] Grams, mean ± S.E.

[b] Significantly different from controls of same sex at 95% level of confidence.

TABLE XII

Comparative Organ Weights Resulting from Subacute Studies (90 Days)[a] (Drug Combinations)

Treatment group	Liver		Kidneys	
	Male	Female	Male	Female
Controls	11.56 ± 0.99	8.61 ± 0.94	2.71 ± 0.11	2.13 ± 0.12
Pyrethrum (P)	14.37 ± 1.08[b]	12.61 ± 0.68[b]	2.72 ± 0.21	2.00 ± 0.10
P + alcohol	17.02 ± 1.23[b]	13.44 ± 0.64[b]	2.95 ± 0.12	2.25 ± 0.10
P + aspirin	11.57 ± 0.37	10.86 ± 0.42[b]	2.64 ± 0.10	2.17 ± 0.07
P + caffeine	13.66 ± 0.57	14.32 ± 0.93[b]	2.76 ± 0.09	2.28 ± 0.13
P + nicotine	13.27 ± 0.63	13.18 ± 0.58[b]	2.79 ± 0.19	2.05 ± 0.06
P + phenobarbital	16.61 ± 0.67[b]	16.07 ± 1.54[b]	2.75 ± 0.14	2.09 ± 0.10
P + piperonyl butoxide	15.26 ± 0.77[b]	14.89 ± 0.60[b]	2.71 ± 0.14	1.99 ± 0.14
P + sulfoxide	11.99 ± 0.84	11.10 ± 2.14	2.43 ± 0.13	1.79 ± 0.27
P + tropital	11.99 ± 0.86	12.22 ± 0.51[b]	2.81 ± 0.11	2.21 ± 0.06

[a] Grams, mean ± S.E.

[b] Significantly different from controls of same sex at 95% level of confidence.

TABLE XIII

Incidence and Description of Pathological Changes

Treatment	Liver	N^a
Pyrethrum (P)	"eosinophilia with beginning cell necrosis"	2/18
Alcohol	"cloudy swelling—beginning liver necrosis"	8/8
Phenobarbital	"cloudy swelling of parenchymal cells—cells enlarged—cytoplasm granular"	11/11
P + alcohol	degeneration of parenchymal cells—cells swollen—early stage of necrosis	1/13
P + aspirin	no apparent changes	
P + caffeine	congested—centrolobular necrosis	5/16
P + nicotine	acute toxic hepatitis—coagulative necrosis	4/16
P + phenobarbital	beginning necrosis	1/10
P + piperonyl butoxide	no apparent changes	
P + sulfoxide	no apparent changes	
P + tropital	no apparent changes	

[a] Number positive/total number in group.

TABLE XIV

Incidence and Description of Pathological Changes

Treatment	Kidney	N^a
Pyrethrum (P)	"some protein in proximal tubules"—"complete tubular degeneration"	1/17
Alcohol	"some interstitial nephritis"—"advanced tubular necrosis"	1/7
Phenobarbital	no apparent changes	
P + alcohol	toxic tubular necrosis "textbook picture"	1/19
P + aspirin	amyloidosis of proximal tubule—low to moderate	8/15
P + caffeine	toxic nephrosis	1/15
P + nicotine	tubular necrosis	1/17
P + phenobarbital	small focal area of lymphocyte infiltration	1/14
P + piperonyl butoxide	no apparent changes	
P + sulfoxide	tubular nephrosis	3/16
P + tropital	no apparent changes	

[a] Number positive/total number in group.

were reported for any of the three synergists. In the combination series the most significant changes were those brought about by caffeine and nicotine in the liver, by aspirin and sulfoxide in the kidney, and by nicotine in the pancreas (Table XV). Some important points should be emphasized here. The pathological effects of both alcohol and pheno-

TABLE XV

Incidence and Description of Pathological Changes

Treatment	Pancreas	N[a]
Pyrethrum (P)	no apparent changes	
Alcohol	no apparent changes	
Phenobarbital	no apparent changes	
P + alcohol	large areas of necrosis	1/18
P + aspirin	no apparent changes	
P + caffeine	necrosis	1/15
P + nicotine	necrosis	3/12
P + phenobarbital	no apparent changes	
P + piperonyl butoxide	no apparent changes	
P + sulfoxide	no apparent changes	
P + tropital	no apparent changes	

[a] Number positive/total number in group.

barbital were practically eliminated when they were combined with pyrethrum. Although some changes occurred in the urine and blood, these did not appear to be significant in terms of range of values or in overall pattern.

F. Other Effects

Information from studies by other laboratories is also relevant to the consideration of interactions involving pyrethrum synergists.

The possibility of carcinogenic activity by any of the substances used in insecticidal preparations is of obvious importance. A great deal of work covering this topic has been reviewed by Mrak (1969) and Brown (1971).

The synergists that are most widely employed are methylenedioxyphenyl derivatives. Piperonyl butoxide is prepared from dihydrosafrole, sulfoxide from isosafrole, and tropital from piperonal. Several methylenedioxyphenyl derivatives have been reported to be carcinogenic when fed at high levels over a long period. They have also been shown to be cocarcinogens when administered with known carcinogens. For example, safrole is a hepatocarcinogen in rats (Long *et al.*, 1963) and dihydrosafrole has been reported to initiate esophageal cancer in rats (Long and Jenner, 1963). Falk and co-workers (1965) reported on the possible cocarcinogenic activity of piperonyl butoxide and sulfoxide and

TABLE XVI

Calculated Daily Intake (mg/kg) of Piperonyl Butoxide in Long-Term Feeding Trials[a]

Animal	Stated dose	Duration	Conversion factor 1 ppm in food (mg/kg/day)	Calculated dose (mg/kg/day)	Tumors	Reference
Mouse	300 ppm	18 months	0.150	45	no statistical effect	Innes *et al.*, 1969
Mouse[b]	890 ppm	18 months	0.150	133	no statistical effect	Innes *et al.*, 1969
Rat	5000 ppm	17 weeks	0.100	500	liver and kidney damage	Lehman, 1951, 1952
Rat	10,000 ppm	24 months	0.050	500	liver and kidney damage, benign tumors	Sarles and Vandergrift, 1952
Dog	32 mg/kg/day	52 weeks			reduction in weight gains, no tumors	Sarles and Vandergrift, 1952

[a] Brown (1970).

[b] 1112 ppm Butacide containing 80% piperonyl butoxide.

their interference with the detoxification and elimination of the known carcinogen, 3,4-benzpyrene, in rats. The two synergists were administered by the oral, intraperitoneal, and intravenous routes. These workers also established that the retention and greater activity of 3,4-benzpyrene was caused by hepatic damage that was related to the alteration of specific detoxification enzymes. However, it is also important to note that piperonyl butoxide, when fed to rats for 18 months at high doses up to 20,000 ppm, did not produce liver tumors.

There is other evidence to show that piperonyl butoxide has additive and synergistic effects when combined with pyrethrum and DDT in rats (Kimbrough *et al.*, 1968) and with freon in mice (Epstein *et al.*, 1967). However, once again, the synergist alone did not produce the high mortality in mice or hepatic damage which was produced by freon alone.

In the review by Brown (1971) there is a table, which is reproduced here as Table XVI, that summarizes the results of several workers who examined animals following the feeding of various dose levels of piperonyl butoxide. These results, which show some damage produced by this compound, led him to report: "Long-term feeding studies have failed to indicate that piperonyl butoxide is a carcinogen, although liver damage may occur." He does recognize the body of evidence showing the inhibition of oxidative detoxification of several drugs by piperonyl butoxide, which might also inhibit the oxidation of substances such as benzpyrene and thus point to its possible role as a cocarcinogen.

IV. Discussion

The acute toxicity studies were carried out primarily to detect the types of effects and levels of activity that might be produced by the test substances. There obviously was an interaction between pyrethrum and some of the drugs. It was interesting to note the minimal interaction between alcohol and pyrethrum. Although the LD_{50} doses of nicotine and phenobarbital combined with pyrethrum are significantly smaller than their single doses, the fact that pyrethrum is administered in its full LD_{50} amount indicates some inhibitory interaction on pyrethrum.

The subacute toxicity studies were carried out with relatively high quantities of each drug and chemical. The insecticidal substances would never be encountered in such large amounts under normal usage. However, even with the high doses, the mortality frequency was excessive only for alcohol, caffeine, and piperonyl butoxide when used alone. In combination with pyrethrum the highest mortality was produced by alcohol and caffeine. The ability of pyrethrum to increase the lethality of sulfoxide bears further study. Tropital produced the lowest frequency of mortality of the synergists tested. Pyrethrum also demonstrates a minimal lethal effect.

The fact that body weight gains are generally reduced should not be too surprising. The large quantities of drugs and chemicals that were administered could very easily affect the digestive and metabolic mechanisms of the animals, interfering with the proper efficient absorption of foods.

The most significant change in organ size was obviously that of the livers. Pyrethrum and the three synergists all were active in this regard. Others have reported on this also (Kimbrough *et al.*, 1968). In contrast to such reports describing definite pathological changes, our studies of 90 days duration did not show such changes produced by pyrethrum or the synergists, although there was a definite enlargement of the livers.

The ability of pyrethrum to reduce the pathological changes induced by alcohol and phenobarbital, alone, is of considerable interest. This ability has led to a study of the mechanism of the pyrethrum effect on hepatic microsomal drug metabolism in rats (Springfield *et al.*, 1971). The results indicate that the liver enlargement began within 4 days following daily doses of the substance; the liver remained enlarged throughout the continued administration but returned to control values within 7 days when the treatment was stopped. This indicates a reversible effect is possible. This study also indicates that the enlargement of the liver occurs along with an increased drug metabolism and is dose related. The lowest dose level tested, 85 mg/kg, produces only minimal changes, giving an approximation of minimum effect. However, this should not be interpreted as an actual no-effect level. The evidence accumulated over the past few years tends to show that the insecticidal substances considered in this report can produce certain types of toxic reactions in various mammalian systems. However, in every case the dose levels administered are admittedly extremely high. They are doses in fact that would be almost impossible to attain by any normal means of dispersal. It is also obvious that many reactions occur only after long-term ingestion of the substance.

In attempting to relate experimental studies to man, Brown (1971) discusses the possible uptake by man, in his food, of the residues of aerosols. He reports that it is possible to ingest about 2.03 mg of piperonyl butoxide in a day which approximates 0.029 mg/kg/day in a 70-kg man. Even under conditions of minimal air changes and with twice the recommended amount applied by aerosol (1.33 mg pyrethrins and 9.0 mg piperonyl butoxide in 3000 ft^3 every 15 min), the amount of the synergist inhaled would be 1.0 mg in 8 hr. Brown also reports that an oral dose of 50 mg to man has produced no effect on metabolism.

V. Summary

There are certain real toxic effects produced by pyrethrum and the synergists as reported by several workers that must be accepted. However, all emphasize the high doses used, and the length of time under test. Some of the data is inconclusive and further studies are necessary, especially in the area of drug interactions.

Based on the results of our studies and those of others, it can be stated that, when used in quantities normally recommended, the insecticide pyrethrum is extremely nontoxic. Because of their ability to interfere with microsomal drug metabolism the effect of the synergists cannot be overlooked, but these also can be considered to have an extremely low toxic potential when used in recommended amounts.

References

Brown, N. C. (1970). Report A 28/52. Wellcome Foundation Ltd.

Brown, N. C. (1971). A review of the toxicology of piperonyl butoxide. *Pyrethrum Post* **11**(2), 66–68.

Epstein, S. S., Joshi, S., Andrea, J., Clapp, P., Falk, H., and Mantel, N. (1967). The synergistic toxicity and carcinogenicity of freons and piperonyl butoxide. *Nature (London)* **214**, 526–528.

Falk, H. L., Thompson, S. J., and Kotin, P. (1965). Carcinogenic potential of pesticides. *Arch. Environ. Health* **10**, 847–858.

Innes, J. R. M., Ulland, B. M., Valerio, M. G., Petrucelli, L., Fishbein, L., Hart, E. R., Pallotta, A. J., Bates, R. R., Falk, H. L., Gart, J. J., Klein, M., Mitchell, I., and Peters, J. (1969). Bioassay of pesticides and industrial chemicals for tumorigenicity in mice: A preliminary note. *J. Nat. Cancer Inst.* **42**, 1101–1114.

Kimbrough, R. D., Gaines, T. B., and Hayes, W. J., Jr. (1968). Combined effects of DDT, pyrethrum and piperonyl butoxide on rat liver. *Arch. Environ. Health* **16**, 333–341.

Lehman, A. (1951). Chemicals in foods: A report to the Association of Food & Drug Officials on current developments. Part II. Pesticides. *Ass. Food Drug Off. U.S., Quart. Bull.* **15**, 112.

Lehman, A. (1952). Section II. Dermal toxicity. Section III. Subacute and chronic toxicity. *Ass. Food Drug Off. U.S., Quart. Bull.* **16**, 3 and 47.

Litchfield, J. T., Jr., and Wilcoxon, F. (1949). A simplified method of evaluating dose effect experiments. *J. Pharmacol. Exp. Ther.* **96**, 99–113.

Long, E. L., and Jenner, P. M. (1963). Esophageal tumors produced in rats by feeding of dihydrosafrole. *Fed. Proc., Fed. Amer. Soc. Exp. Biol.* **22**, 275–281.

Long, E. L., Nelson, A. A., Fitzhugh, O. G., and Hansen, W. H. (1963). Liver tumors produced in rats by feeding safrole. *Arch. Pathol.* **75**, 595–604.

McNemar, Q. (1954). "Psychological Studies," pp. 240–242. Wiley, New York.

Mrak, E. M., Chairman. (1969). "Report of the Secretary's Commission on Pesticides and Their Relationship to Environmental Health," pp. 507–564. U.S. Dept. of Health, Education and Welfare, Washington, D.C.

Sarles, M. P., and Vandergrift, W. B. (1952). Chronic oral toxicity and related studies on animals with the insecticide and pyrethrum synergist, piperonyl butoxide. *Amer. J. Trop. Med.* **1**, 862–883.

Springfield, A. C., Carlson, G. P., and DeFeo, J. J. (1971). Increased microsomal enzyme activity due to pyrethrum administration. *Toxicol. Appl. Pharmacol.* **19**, 394.

Chapter 10

Mode of Action of Synergists in Enhancing the Insecticidal Activity of Pyrethrum and Pyrethroids

IZURU YAMAMOTO

I. Introduction

Pesticide synergist is a term usually used for compounds that are nontoxic or negligibly toxic at the dosage employed but which serve to enhance the toxicity of a pesticide chemical when combined with it. *Synergism,* in a wider sense, is related to greater-than-additive toxicity, but the additive toxicity is somewhat difficult to assess when both compounds involved are toxic. Insecticide synergists are important both practically and from the academic standpoint; so, they have been extensively investigated. The effectiveness of an insecticide synergist is commonly expressed by the ratio of the LD_{50} of insecticide alone to the LD_{50} of insecticide with the synergist. Synergism of pyrethroids is considered in reviews by Casida (1970), Dahm (1957), Hayashi (1972), Hewlett (1960), Kearns (1956), Metcalf (1955, 1967), and Wilkinson (1968, 1971). In the present review, emphasis is placed on the sites and mechanisms of detoxication of pyrethroids in insects in relation to the action of synergists.

Synergists were originally sought as a means to increase the insecticidal activity of pyrethrum and to reduce the cost of insect control with this natural product. In 1938 *N*-isobutylundecyleneamide was recognized as a possible practical synergist for pyrethrum. In 1940 the synergistic action of sesame oil was discovered and the active components were found to be sesamin and sesamolin, each having the methylenedioxyphenyl (MDP) moiety in the molecule. Recognition of the importance of the MDP moiety for synergism led to the testing for synergistic potency of hundreds of MDP compounds that were synthesized or isolated from natural sources. Synergism is not only found for pyrethrum, but also for other pyrethroids, nicotine, rotenone, organophosphorus compounds, chlorinated hydrocarbons, and carbamates, whereas many types of compounds other than MDP derivatives are found to be synergistic to certain other types of insecticides. At the present time, however, synergists find their only practical use in combination with pyrethrum and pyrethroids. The structures and names of selected pyrethroid synergists are given in Fig. 1.

Many investigators have attempted to define the best combination and ratio of pyrethroid and synergist. However, the synergistic effect is greatly influenced by the test insect, formulation, pyrethroid–synergist ratio, method of administration, and choice of response, either kill or knockdown; these experimental differences complicate interpretation of the results. Piperonyl butoxide is generally quite effective in combination with natural pyrethrins, but this is not always the optimal synergist with synthetic pyrethroids. In the United States, natural pyrethrins are combined with one of four MDP synergists, piperonyl butoxide, sulfoxide, propyl isome, and Tropital, among which piperonyl butoxide is used most frequently. Commercial compounds lacking the MDP moiety are MGK 264, Synepirin 500, and octachlorodipropyl ether (Casida, 1970). Pyrethroids are commonly used in Japan and Southeast Asia for mosquito coils. Allethrin is particularly suitable for this use because of its heat stability. However, in an oil-based formulation or in an aerosol, the knockdown and kill effects of allethrin, even when synergized with piperonyl butoxide, are inferior to those of pyrethrins (Nishizawa, 1971). The search for effective synergists for allethrin and several new synthetic pyrethroids (such as tetramethrin, resmethrin, prothrin, and proparthrin) has led to the synergists currently used in Japan: MGK 264 predominates, followed by octachlorodipropyl ether, piperonyl butoxide, and Synepirin 500 (reviewed by Hayashi, 1972).

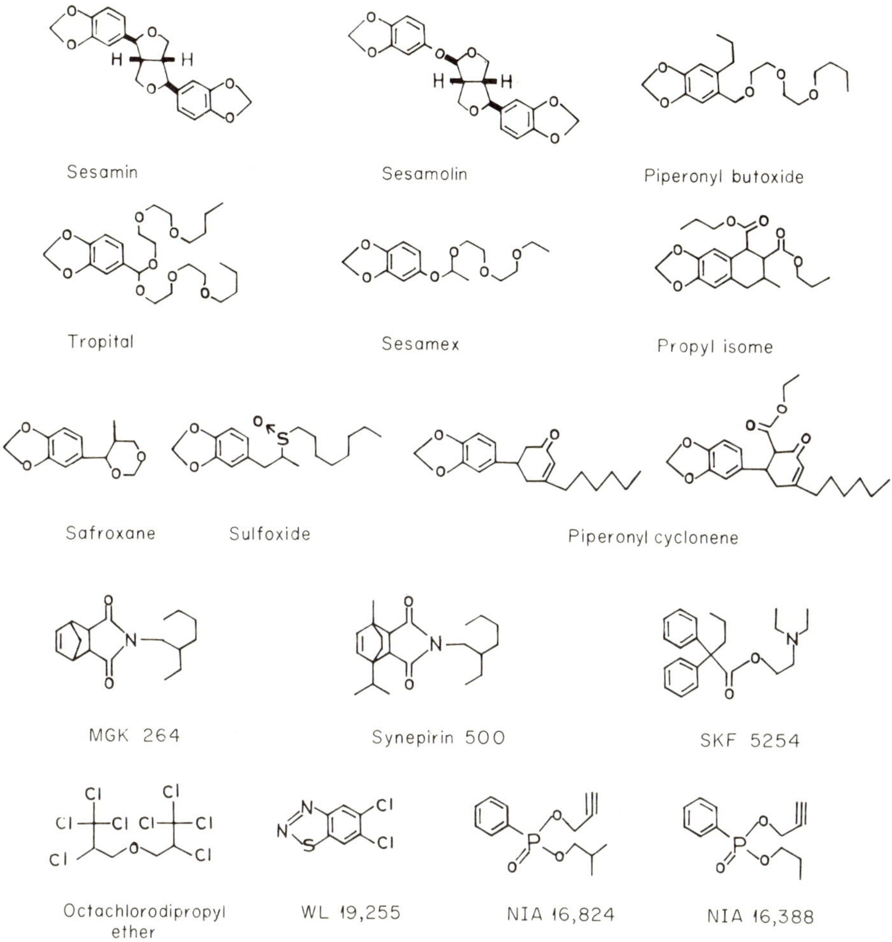

Fig. 1. Structures and names of selected pyrethroid synergists.

II. Mode of Action and Metabolism of Synergists

Many early theories on the mode of action of synergists (reviewed by Blackith, 1953; Metcalf, 1955) have been discarded for lack of experimental support. Synergists frequently decrease the penetration rate of the insecticide into the insect body (Chang and Kearns, 1964; Hayashi *et al.*, 1968; Winteringham *et al.*, 1955). Therefore, an inappropriate combination of a pyrethroid with a synergist may lead to decreased insecticidal activity, particularly in the knockdown effect. It is not necessary for the synergist to be applied along with the insecticide in order to obtain a synergistic action. Synergism also results on treatment of an insect with the synergist before the application of pyrethrins (Lindquist *et al.*, 1947), or when separate sites of application are involved for the synergist and pyrethrins (Wilson, 1949). These findings suggest the involvement of a biochemical interaction in the insect body.

It is now well established that the synergists act, in the most part, by interfering with the *in vivo* detoxication of the insecticide. The current knowledge of the biochemical basis for synergist action is adequately considered in recent reviews (Casida, 1970; Wilkinson, 1968, 1971) and will be only briefly mentioned here. MDP compounds exert their synergistic action by inhibiting the oxidative metabolism of insecticides catalyzed by the mixed-function oxidase (mfo) system of microsomes. The detailed mechanism of the interaction of the synergist with the microsomal components is not well understood, but many synergists combine with cytochrome P-450 directly or at least with sites that alter the structure around the heme group, resulting in an alteration of function. It is possible that they act by serving as alternative substrates, being converted to catechol compounds, thereby sparing the insecticide chemical from detoxication (Casida *et al.*, 1966; Esaac and Casida, 1969; Wilkinson and Hicks, 1969), but they may also react with another site of the mfo system as the cations or free radicals formed by subtraction of the hydride ion or the hydrogen radical from the methylene group of the MDP moiety, preventing oxidative detoxication (Hansch, 1968; Hennessy, 1965, 1970; Wilkinson, 1971) (Fig. 2). Therefore, the methylene group of the MDP moiety is closely associated with the mode of action of the synergist. Knowledge of the biochemical basis for the action of other types of synergists is very limited but most of the compounds shown in Fig. 1 are likely to be substrates for the mfo system.

The MDP synergists are metabolized not only by attack at the methylene position of the benzodioxole but also by oxidation of various sites in the aliphatic side chains (Fig. 3). It is suggested that the importance of the long side chains containing oxygen or sulfur to the synergistic action with pyrethrins may result from improved fit on the active site of the pyrethrins metabolizing enzymes or from the availability of additional sites on the synergist that are metabolized, thereby reducing attack on the pyrethroid molecule (Esaac and Casida, 1969).

There is evidence that not all synergists are substrates for and inhibit the mfo system. Certain organic thiocyanates are synergists for carbamates and pyrethrins. However, these compounds are metabolized not by the mfo system but by glutathione S-transferases liberating hydrogen cyanide (Ohkawa and Casida, 1971). An *O*-(2-propynyl) phos-

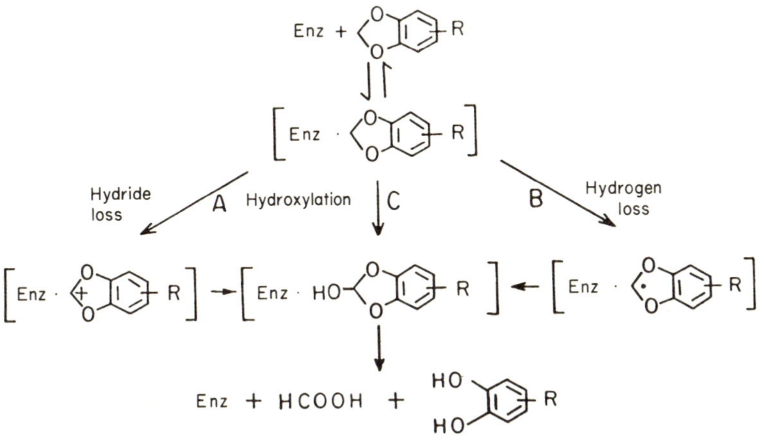

Fig. 2. Possible reaction mechanisms for methylenedioxyphenyl compounds at the active site of the mixed-function oxidase system (reproduced from Casida, 1970).

phonate acts, in part, by inhibiting ester cleavage of tetramethrin, as discussed in Section IV.

There appears to be another facet of synergism in addition to the normal inhibition of pyrethroid metabolism by synergists. The time required for pyrethrins to block nerve conduction in the exposed nerve cord of cockroaches is shortened by the direct application of sesamex or piperonyl butoxide to the nerve, and this effect apparently correlates with the concentration of the synergist, on the one hand, and with the increased kill effect, on the other (Burt and Goodchild, 1967). Possibly,

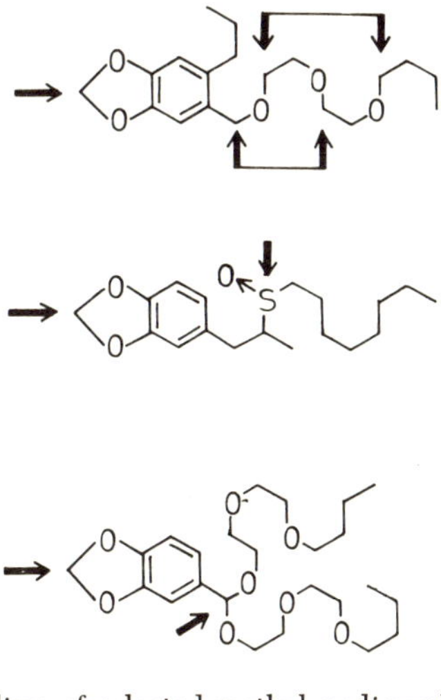

Fig. 3. Sites of metabolism of selected methylenedioxyphenyl synergists (piperonyl butoxide, sulfoxide, and Tropital).

the synergist alters the rate of pyrethrins penetration into the nerve or reduces the rate of oxidation of the penetrated pyrethrins.

III. Metabolism of Pyrethroids in Insects

For many years it was considered that the major detoxication mechanism for pyrethrins involved hydrolysis of the ester linkage, based on the following criteria: the ester nature of pyrethrins; the insecticidal inactivity of the component acid and alcohol; the production of an unspecified acid from incubation of pyrethrins *in vitro* with cockroach "lipase" preparation (Chamberlain 1950); dubious characterization of chrysanthemic acid and the alcohols from *in vivo* metabolism of [^{14}C]pyrethrins by the cockroach (Zeid *et al.*, 1953).

Later studies showed little or no production of chrysanthemic acid on *in vivo* metabolism in houseflies of allethrin (^{14}C labeled, Hopkins and Robbins, 1957; ^{3}H labeled, Hayashi *et al.*, 1968), pyrethrins (^{14}C labeled, Winteringham *et al.*, 1955), pyrethrin I and cinerin I (both ^{14}C labeled, Chang and Kearns, 1964) and tetramethrin (^{3}H labeled, Hayashi *et al.*, 1968). [^{14}C]Allethrin decomposed only to a small degree on incubation with a fly "lipase" preparation, whereas high concentrations of benzaldehyde, a lipase inhibitor, were required to reduce the *in vivo* metabolism of allethrin (Bridges, 1957). These findings led to the proposal that hydrolysis did not constitute the major detoxication mechanism. However, in these studies considerable amounts of metabolites were produced with the same R_f value on paper chromatography as the corresponding alcohols formed on ester cleavage. If these were indeed the alcohols, then a hydrolytic mechanism with subsequent degradation or conjugation of the free chrysanthemic acid could not be excluded, in spite of the claimed nonhydrolytic mechanism. This product was not the alcohol, at least in one case (Hopkins and Robbins, 1957), because the radioactive metabolite(s) with the same R_f value as allethrolone was derived from ^{14}C-acid-labeled allethrin. In any case, the evidence was not convincing because of the incomplete characterization of the products based on the low resolution capabilities of the chromatographic system used.

A proposal that pyrethrin I and cinerin I undergo detoxication by modification of the alcohol moiety while the acid moiety and the ester linkage remained intact (Chang and Kearns, 1964) was based on the finding that three out of five metabolites plus a small quantity of chrysanthemic acid gave a positive Denigès color reaction for chrysanthemic acid after saponification; therefore, these metabolites were assumed to be chrysanthemic acid esters. Two other products, including major me-

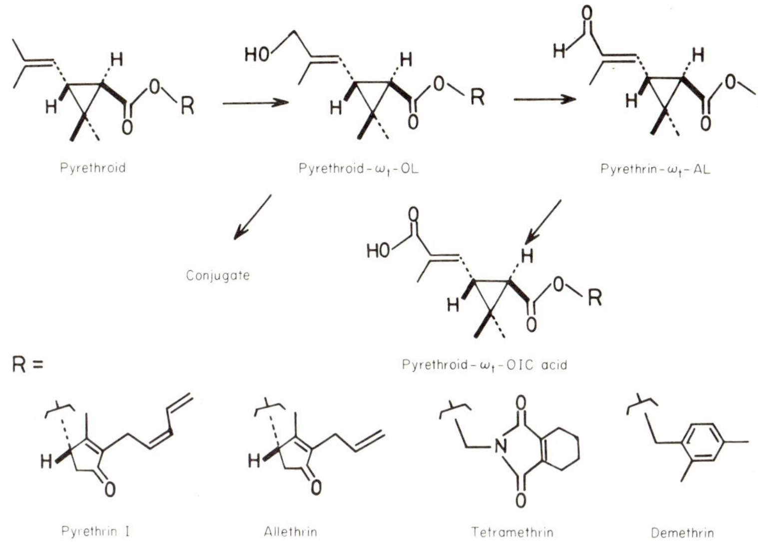

Fig. 4. Oxidative metabolism in insects of pyrethrin I, allethrin, tetramethrin, and dimethrin.

tabolites, did not seem to be chrysanthemic acid esters, based on the color reaction.

The first definitive demonstration of a nonhydrolytic mechanism for metabolism of ^{14}C-labeled pyrethrin I, allethrin, dimethrin, and tetramethrin involved the use of an *in vitro* enzyme system consisting of housefly abdomen homogenates fortified with reduced nicotinamide–adenine dinucleotide phosphate (NADPH) and living houseflies of the R_{Baygon} or R_{Hokota} resistant strain (Yamamoto and Casida, 1966; Yamamoto *et al.*, 1969). In each case, the metabolic pathway was established to be as shown in Fig. 4, involving oxidation of the *trans*-methyl of the isobutenyl grouping of the chrysanthemic acid moiety, by the microsomal mfo system, to the hydroxymethyl compound, which is then converted by the action of other enzymes to a conjugate or is oxidized to the corresponding aldehyde and acid compounds. The pyrethroid-ω_t-oic acids were characterized by conversion to chrysanthemum dicarboxylic acid on saponification and to the corresponding rethrin II- or pyrethroid II-type compounds by methylation. Oxo- and hydroxychrysanthemic acids were obtained on saponification of the corresponding ester metabolites from allethrin. These metabolites gave the 2,4-dinitrophenylhydrazone of allethrolone methyl ether on treatment with the hydrazine reagent in acidic methanol.

The following types of evidence established that, for allethrin, oxidation of this type is a major metabolic pathway. Each of the more than ten metabolites produced by the enzyme system is found also in treated houseflies, although the hydroxy intermediate is conjugated in the fly but is oxidized to the carboxyl compound in the enzyme system. Hydrolysis does not occur and all metabolites have an intact ester linkage because the same products are detected *in vitro* and *in vivo,* whether the acid-labeled or the alcohol-labeled allethrin is metabolized. Little if any free chrysanthemic acid or allethrolone is produced by the enzyme system. Modification on the alcohol moiety is a minor or nonexistent pathway because the metabolite mixture gives almost all of the radiocarbon from the metabolized alcohol-labeled allethrin as derivatives with the intact allethrolone moiety.

With other pyrethroids, there are varying degrees of evidence on the nature of the metabolic changes in houseflies. For pyrethrin I and dimethrin, only the acid-labeled preparations were used, and the corresponding ω_t-oic acids were identified. Other metabolites from pyrethrin I are probably not the hydrolysis products because the metabolites from pyrethrin I and allethrin, which differ only in the side chain of the alcohol moiety by two carbons, are almost comparable in thin-layer chromatography R_f values. The metabolites from dimethrin give a somewhat similar chromatographic pattern. The metabolites of cinerin I in living houseflies produce a chromatographic pattern similar to that of pyrethrin I, suggesting that there are analogous sites of metabolic attack on cinerin I and pyrethrin I (Chang and Kearns, 1964). Tetramethrin is also metabolized without significant ester cleavage because most of the products are the same both *in vitro* and *in vivo,* from either the acid- or alcohol-labeled preparation, and tetramethrin-ω_t-oic acid was identified as one among the many metabolites. These studies used housefly strains with high oxidase activity; a different result obtained by the use of a susceptible housefly strain will be discussed below.

In addition to the pathways of pyrethroid metabolism in houseflies shown in Fig. 4, there are other metabolites formed from allethrin, pyrethrin I, tetramethrin, and dimethrin that are not the oxo- or hydroxypyrethroids resulting from oxidative attack at the isobutenyl group. Although minor in quantity *in vitro,* and slightly more *in vivo,* they appear to result from other types of modifications in the molecules.

IV. Synergists and the Sites of Detoxication of Pyrethroids

A. Synergism by Inhibiting Oxidation of the Acid Moiety

The insecticidal activity of allethrin for houseflies is enhanced by piperonyl cyclonene and other synergists, including non-MDP com-

TABLE I

Toxicity to Houseflies of Allethrins and Certain of Their Metabolites[a]

Compound	LD_{50}(μg/fly) by:	
	Topical application	Injection
Allethrin	1.3	0.16
Allethrin II	—	1.8
Allethrin-ω_t-ol	770	15(12–24% kill)
Allethrin-ω_t-al	460	13(16–20% kill)
Allethrin-ω_t-oic acid	320(60% kill)	>5

[a] From Kobayashi *et al.* (1971).

pounds (Gersdorff *et al.*, 1951). This appears to result from *in vivo* inhibition of the allethrin-detoxifying enzymes (Bridges, 1957; Winteringham *et al.*, 1955). The major metabolic pathway of allethrin involves oxidation at the isobutenyl grouping of the chrysanthemic acid moiety. The enzyme activity of houseflies, as assayed by the ability to metabolize allethrin *in vitro,* decreases following treatment with piperonyl butoxide to a minimum after 4 hr but returns to nearly normal after 3 or 4 days. The same metabolites are produced with or without a synergist or at different times after treatment (Yamamoto *et al.*, 1969). The initial oxidation step, producing hydroxyallethrin or allethrin-ω_t-ol, is sensitive to inhibition by the synergists. This alcohol is converted to more polar conjugates or the allethrin-ω_t-oic acid, which are unlikely to penetrate the nerve or be insecticidally active. When topically applied or injected, allethrin-ω_t-ol, allethrin-ω_t-al, and allethrin-ω_t-oic acid show greatly reduced toxicity to houseflies (Kobayashi *et al.*, 1971) (Table I).

B. Synergism by Inhibiting Oxidation of the Alcohol Moiety

Oxidative attack on the isobutenyl grouping is important in pyrethroid detoxication. However, detoxication at this site may not always be the limiting factor in determining the *in vivo* stability of pyrethrins and pyrethroids. There is some evidence for the involvement of other pathways from studies of the effect of synergists on the insecticidal activity of various pyrethroids, particularly with modified alcohol groupings.

The insecticidal activity of pyrethroid-like compounds lacking the isobutenyl grouping, such as tetramethylcyclopropanecarboxylates, is synergized by high levels of piperonyl butoxide even though this acid structure is not expected to be significantly oxidized, so it is likely that, in this series, detoxication occurs primarily at the alcohol moiety (Berteau and Casida, 1969). At a 1:10 ratio of pyrethroid and piperonyl butoxide, pyrethrins, allethrin, and allethronyl tetramethylcyclopropanecarboxylate give the synergistic ratios of 6.3, 3.7, and 5.3, respectively (A. Hayashi and I. Yamamoto, unpublished results, 1972).

There are wide variations in the degree of synergism of pyrethrin I, pyrethrin II, cinerin I, cinerin II, and allethrin. With each of the pyrethrins, combination with piperonyl butoxide results in more rapid knockdown of flies; the knockdown effect lasts longer and the kill effect is increased. The degree of synergism is lowest during the first few minutes after application, probably because it takes time for the synergist to reach and/or react with the site of action; then it increases rapidly for 2 hr (Sawicki, 1962a,b,c). Table II shows the relative toxicity of each of the pyrethrins and allethrin with or without synergist (Chang and Kearns, 1962; Incho and Greenberg, 1952; Sawicki, 1962a). The degree of synergism is quite different among pyrethrin I, cinerin I, and allethrin, which differ only in the side chain of the alcohol moiety, and is highest for cinerin I and the least for allethrin. The stereochemistry of the acid and alcohol moieties is not of importance in determining the degree of synergism because it varies only to a small extent with the various allethrin isomers (Incho and Greenberg, 1952). Compounds formed by replacing the allyl group of allethrin with substituents other than those tabulated are also synergized to a varying degree (Barthel,

TABLE II

Relative Toxicity to Houseflies of Pyrethrins, Cinerins, and Allethrin with or without Synergists

	Relative toxicity[a]						Synergistic ratio[a]		
	Alone			With synergist					
	A	B	C	A	B	C	A	B	C
Pyrethrin I	1.00	1.00	1.00	1.00	1.00	1.00	9.1	16.1	18.0
Pyrethrin II	0.39	0.58	1.61	0.39	0.32	0.69	9.1	8.8	6.0
Cinerin I	0.45	0.25	0.62	0.73	0.94	0.98	14.5	60.2	27.0
Cinerin II	0.26	0.23	0.58	0.34	0.32	0.44	11.9	22.1	14.0
Pyrethrins	0.57		1.17	0.98		0.76	15.3		10.0
Allethrin	1.24			0.30			3.0		

[a] (A) Spray with piperonyl butoxide (1:10). From Incho and Greenberg (1952). (B) Topical application with sesamex (1:10). From Chang and Kearns (1962). (C) Topical application with piperonyl butoxide (1:8). From Sawicki (1962a).

1961; Hewlett, 1960). For example, the allethrin analogs with furfuryl, thienyl, and benzyl side chains instead of the allyl side chain of the alcohol moiety show less toxicity to houseflies than does allethrin, but in combination with piperonyl butoxide and its MDP analogs the furfuryl and thienyl compounds are more effective than allethrin (Incho and Ault, 1954). Data of this type is useful in establishing the potential toxicity of the pyrethroids, when detoxication is suppressed to a minimal degree, and indicates the primary site at which metabolic attack can occur.

On the assumption that the oxidation of the acid moiety is the critical detoxication step for all pyrethroids regardless of the alcohol moiety, these synergism results are interpretable on the basis that a change of the side chain on the alcohol moiety could, without being directly involved in the enzymic attack, affect the interaction of the isobutenyl portion of the molecule with the enzyme (Wilkinson, 1968). The alcohol moiety of allethrin is not extensively metabolized in houseflies but this has not been experimentally established for pyrethrin I, cinerin I, or related compounds.

A more plausible assumption is that, in addition to the acid moiety, the unsaturated side chain of the alcohol moiety of pyrethrin I and cinerin I is also involved to a major degree in the detoxication. However, that of allethrin is not. The pyrethric acid esters, pyrethrin II and cinerin II, are synergized, but to a degree half to less than half of the corresponding chrysanthemic acid esters. The synergized toxicity of pyrethrin II and cinerin II is almost equal, but is only one half to one third of the corresponding chrysanthemic acid esters. With the rethrin II compounds, possibly both cleavage of the carbomethoxyl grouping of the acid moiety by a nonoxidative mechanism and oxidation on the alcohol moiety occur, with MDP synergism only involving the latter case.

The nature of the metabolic modification on the alcohol moiety can be inferred from the result of pyrethrin I metabolism in mammals because the mfo systems of insects and mammals appear to work by a similar mechanism. Pyrethrin I is converted to metabolites with an intact ester linkage and with modification on both the isobutenyl grouping and the pentadienyl side chain (Elliott *et al.*, 1972). Probably in insects the initial stage of detoxication involves oxidation independently either on the acid or the alcohol moiety of pyrethrin I and cinerin I, but only on the acid moiety of allethrin, and either carbomethoxy cleavage of the acid moiety or oxidation of the alcohol moiety occurs with pyrethrin II and cinerin II. These reactions are illustrated with the pyrethrins in Fig. 5.

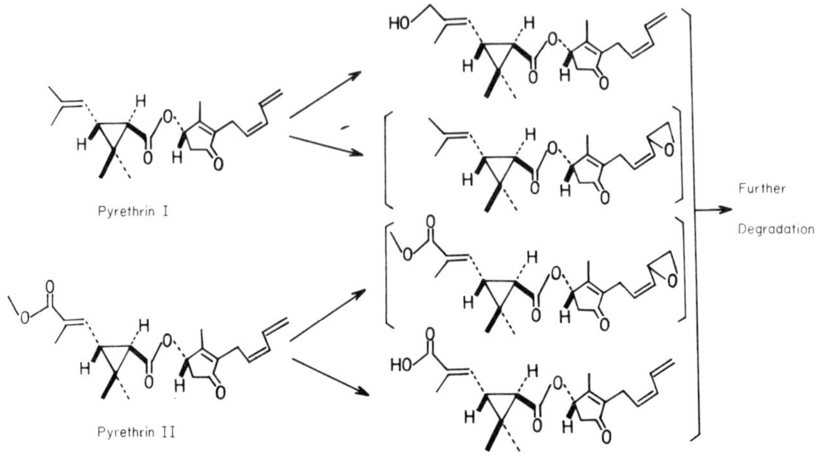

Fig. 5. Possible metabolic pathways in insects of pyrethrin I and pyrethrin II.

C. Synergism by Inhibiting Hydrolysis of the Ester Group

The metabolism of the highly insecticidal resmethrin possibly involves a different site or balance of attack on the molecule because little synergism occurs with the usual synergist levels, although high levels of piperonyl butoxide increase the insecticidal activity (Okuno *et al.*, 1969). Perhaps oxidation on the acid moiety occurs, but hydrolysis of the parent resmethrin and its oxidized products is more prominent; thus, chrysanthemic acid and its oxidized forms are detected on incubation of resmethrin with housefly and mammalian liver microsome–NADPH systems (C. O. Abernathy and J. E. Casida, personal communication, 1972) (Fig. 6).

This type of metabolism has recently been confirmed with tetramethrin (Miyamoto and Suzuki, 1973). Living houseflies metabolize ^{14}C-labeled tetramethrin mostly by cleavage to chrysanthemic acid and the component alcohol, *N*-(hydroxymethyl)-3,4,5,6-tetrahydrophthalimide, whereas the metabolites oxidized at the isobutenyl grouping of the acid moiety, with an intact ester linkage, as well as the oxidized forms of chrysanthemic acid are found in lesser amounts. There may be a competition between the ester cleavage and the oxidation because, at 15 min after injection, chrysanthemic acid and the cleaved alcohol far exceed the total oxidized ester products, and it seems that the latter increase without much increase of chrysanthemic acid after 30 to 60 min to reach the same level. An apparent discrepancy between this result and the previous one (Yamamoto *et al.*, 1969) is probably caused by the difference in the fly strain used. With the R_{Hokota} strain, which

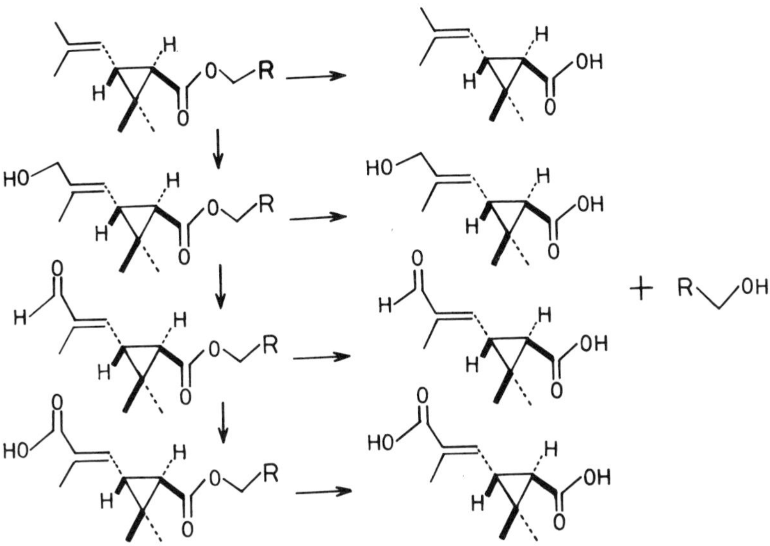

Fig. 6. General scheme for metabolic reactions of pyrethroids in insects involving oxidation on the acid moiety and cleavage at the ester linkage.

exhibits high oxidative metabolism of many foreign compounds, oxidation of tetramethrin is predominant, whereas in the SK strain, which is susceptible to pyrethroids and organophosphorus insecticides, both oxidation and ester cleavage occur, with the latter predominating.

It is interesting that, with tetramethrin, piperonyl butoxide, a typical mfo inhibitor, has little affect on the formation of chrysanthemic acid and the alcohol or the extent of the metabolism, although it decreases the formation of the oxidized metabolites slightly, whereas another synergist, NIA 16388 (Fig. 1), dramatically inhibits the metabolism. The analogous synergist, NIA 16824, also strongly synergizes tetramethrin (Niagara Chemical Division, 1968). The ester-cleaving activity is present mostly in the microsomal fraction (J. Miyamoto, personal communication, 1972). Therefore, the propynyl phosphonates probably act on the microsomes to inhibit both their hydrolase and their oxidase activities when the microsomes act on appropriate substrates, such as tetramethrin.

V. Summary

The stability of pyrethroids in houseflies is limited by oxidative and hydrolytic reactions occurring at various sites on the molecule, depending on the particular compound involved. Each of the chrysanthemates that has been studied, including pyrethrin I, allethrin, dimethrin, and tetramethrin, undergoes oxidation in living houseflies at the *trans*-methyl group of the isobutenyl side chain, producing alcohols that undergo subsequent oxidation to the corresponding aldehydes and acids. This site of oxidation is also involved on incubating these compounds with the housefly mfo system. Studies on the degree of synergism with varying alcohol substituents on the pyrethroids suggest that oxidase attack is also likely to occur on the alcohol moiety, with the ester linkage intact, although this remains to be verified with appropriate metabolic studies. The ester group is also cleaved, at least with tetramethrin. The type of detoxication, oxidation or ester hydrolysis, depends not only on the compound but probably also on the strain of houseflies because some have a low and others a high mfo activity.

Such pyrethroid synergists as MDP compounds appear to act by inhibiting the mfo system, minimizing the rate of formation of less toxic oxidation products. Other synergists may also inhibit cleavage of the ester group, at least with tetramethrin. Which is the optimal synergist for each pyrethroid depends on the relative rate of entry of the insecticide and synergist into the insect, the importance of various sites of metabolic attack in detoxifying the pyrethroid, and the potency of the synergist for inhibiting the enzyme(s) involved in attacking the biodegradable sites of the pyrethroid molecule.

References

Barthel, W. F. (1961). Synthetic pyrethroids. *Advan. Pest Contr. Res.* **4**, 33–74.

Berteau, P. E., and Casida, J. E. (1969). Synthesis and insecticidal activity of some pyrethroid-like compounds including ones lacking cyclopropane or ester groupings. *J. Agr. Food Chem.* **17**, 931–938.

Blackith, R. E. (1953). Pyrethrum synergists—a review. *Pyrethrum Post* **3**(2), 20–29.

Bridges, P. M. (1957). Absorption and metabolism of [^{14}C] allethrin by the adult housefly, *Musca domestica* L. *Biochem. J.* **66**, 316–320.

Burt, P. E., and Goodchild, R. E. (1967). Action of pyrethrum and synergised pyrethrum on the nervous system of *Periplaneta americana* L. *Rothamsted Exp. Sta., Rep.* pp. 167–168.

Casida, J. E. (1970). Mixed-function oxidase involvement in the biochemistry of insecticide synergists. *J. Agr. Food Chem.* **18**, 753–772.

Casida, J. E., Engel, J. L., Esaac, E. G., Kamienski, F. X., and Kuwatsuka, S. (1966). Methylene-C^{14}-dioxyphenyl compounds: Metabolism in relation to their synergistic action. *Science* **153**, 1130–1133.

Chamberlain, R. W. (1950). An investigation on the action of piperonyl butoxide with pyrethrum. *Amer. J. Hyg.* **52**, 153–183.

Chang, S. C., and Kearns, C. W. (1962). Effect of sesamex on toxicities of individual pyrethrins. *J. Econ. Entomol.* **55**, 919–922.

Chang, S. C., and Kearns, C. W. (1964). Metabolism *in vivo* of C^{14}-labelled pyrethrin I and cinerin I by house flies with special reference to the synergistic mechanism. *J. Econ. Entomol.* **57**, 397–404.

Dahm, P. A. (1957). The mode of action of insecticides exclusive of organic phosphorus compounds. *Annu. Rev. Entomol.* **2**, 247–260.

Elliott, M., Janes, N. F., Kimmel, E. C., and Casida, J. E. (1972). Metabolic fate of pyrethrin I, pyrethrin II, and allethrin administered orally to rats. *J. Agr. Food Chem.* **20**, 300–313.

Esaac, E. G., and Casida, J. E. (1969). Metabolism in relation to mode of action of methylenedioxyphenyl synergists in houseflies. *J. Agr. Food Chem.* **17**, 539–550.

Gersdorff, W. A., Nelson, R. H., and Mitlin, N. (1951). The relative effect of several pyrethrum synergists in fly sprays containing allethrin. *J. Econ. Entomol.* **44**, 921–927.

Hansch, C. (1968). The use of homolytic, steric, and hydrophobic constants in a structure-activity study of 1,3-benzodioxole synergists. *J. Med. Chem.* **11**, 920–924.

Hayashi, A. (1972). Pyrethroid synergists—effect and mode of action. *Kagaku (Tokyo)* **42**, 376–385.

Hayashi, A., Saito, T., and Iyatomi, K. (1968). Studies on the increment of the efficacy of insecticides (VIII). Metabolism of ^{3}H-pyrethroids in the adult house fly, *Musca domestica vicina* Macq. *Botyu-Kagaku* **33**, 90–95.

Hennessy, D. J. (1965). Hydride-transferring ability of methylenedioxybenzenes as a basis of synergistic activity. *J. Agr. Food Chem.* **13**, 218–220.

Hennessy, D. J. (1970). The potential of carbamate synergists as pest control agents. *In* "Biochemical Toxicology of Insecticides" (R. D. O'Brien and I. Yamamoto, eds.), pp. 105–114. Academic Press, New York.

Hewlett, P. S. (1960). Joint action in insecticides. *Advan. Pest Contr. Res.* **3**, 27–74.

Hopkins, T. L., and Robbins, W. E. (1957). The absorption, metabolism, and excretion of C^{14}-labeled allethrin by house flies. *J. Econ. Entomol.* **50**, 684–687.

Incho, H. H., and Ault, A. K. (1954). The toxicity to house flies of allethrin analogs in combination with piperonyl butoxide analogs. *J. Econ. Entomol.* **47**, 664–672.

Incho, H. H., and Greenberg, H. (1952). Synergistic effect of piperonyl butoxide with the active principles of pyrethrum and with allethrolone esters of chrysanthemum acids. *J. Econ. Entomol.* **45**, 794–799.

Kearns, C. W. (1956). The mode of action of insecticides. *Annu. Rev. Entomol.* **1**, 123–147.

Kobayashi, A., Yamashita, K., Ohshima, K., and Yamamoto, I. (1971). Chemical studies on pyrethroids. Part I. Syntheses of allethrin metabolites and their toxicities. *Agr. Biol. Chem.* **35**, 1961–1965.

Lindquist, A. W., Madden, A., and Wilson, H. (1947). Pre-treating house flies with synergists before applying pyrethrum sprays. *J. Econ. Entomol.* **40**, 426–427.

Metcalf, R. L. (1955). "Organic Insecticides," pp. 94–99. Wiley (Interscience), New York.

Metcalf, R. L. (1967). Mode of action of insecticide synergists. *Annu. Rev. Entomol.* **12**, 229–256.

Miyamoto, J., and Suzuki, T. (1973). Metabolism of tetramethrin in houseflies *in vivo*. *Pestic. Biochem. Physiol.* **3**, 30–41.

Niagara Chemical Division. (1968). Tech. Data Sheet on NIA 16824. FMC Corp., Middleport, New York.

Nishizawa, Y. (1971). Development of new synthetic pyrethroids. *Bull. W.H.O.* **44**, 325–336.

Ohkawa, H., and Casida, J. E. (1971). Glutathione S-transferases liberate hydrogen cyanide from organic thiocyanates. *Biochem. Pharmacol.* **20**, 1708–1711.

Okuno, Y., Fujimoto, K., Kadota, T., Miyamoto, J., and Hamuro, K. (1969). Insecticidal activity of a new synthetic pyrethroidal compound, 5-benzyl-3-furylmethyl-*dl-cis,trans*-chrysanthemate (NRDC-104, Chrysron). *Botyu-Kagaku* **34**, 157–165.

Sawicki, R. M. (1962a). Insecticidal activity of pyrethrum extract and its four insecticidal constituents against house flies. II. Synergistic activity of piperonyl butoxide with the four constituents. *J. Sci. Food Agr.* **13**, 260–264.

Sawicki, R. M. (1962b). Insecticidal activity of pyrethrum extract and its four insecticidal constituents against house flies. III. Knockdown and recovery of flies treated with pyrethrum extract with and without piperonyl butoxide. *J. Sci. Food Agr.* **13**, 283–292.

Sawicki, R. M. (1962c). Insecticidal activity of pyrethrum extract and its four insecticidal constituents against house flies. V. Knockdown activity of the four constituents with piperonyl butoxide. *J. Sci. Food Agr.* **13**, 591–598.

Wilkinson, C. F. (1968). Detoxication of pesticides and the mechanism of synergism. *In* "Enzymatic Oxidations of Toxicants" (E. Hodgson, ed.) pp. 113–149. North Carolina State Univ. Press, Raleigh.

Wilkinson, C. F. (1971). Insecticide synergists and their mode of action. *Proc. Int. Congr. Pestic. Chem., 2nd 1971* Vol. 2, pp. 117–159.

Wilkinson, C. F., and Hicks, L. J. (1969). Microsomal metabolism of the 1,3-benzodioxole ring and its possible significance in synergistic action. *J. Agr. Food Chem.* **17**, 829–836.

Wilson, C. S. (1949). Piperonyl butoxide, piperonyl cyclonene, and pyrethrum applied to selected parts of individual flies. *J. Econ. Entomol.* **42**, 423–428.

Winteringham, F. P. W., Harrison, A., and Bridges, P. M. (1955). Absorption and metabolism of [^{14}C]pyrethroids by the adult housefly, *Musca domestica* L., *in vivo*. *Biochem. J.* **61**, 359–367.

Yamamoto, I., and Casida, J. E. (1966). *O*-Demethyl pyrethrin II analogs from oxidation of pyrethrin I, allethrin, dimethrin, and phthalthrin by a house fly enzyme system. *J. Econ. Entomol.* **59**, 1542–1543.

Yamamoto, I., Kimmel, E. C., and Casida, J. E. (1969). Oxidative metabolism of pyrethroids in houseflies. *J. Agr. Food Chem.* **17**, 1227–1236.

Zeid, M. M. I., Dahm, P. A., Hein, R. E., and McFarland, R. H. (1953). Tissue distribution, excretion of $C^{14}O_2$ and degradation of radioactive pyrethrins administered to the American cockroach. *J. Econ. Entomol.* **46**, 324–336.

Chapter 11

Mode of Action of Pyrethrum on Arthropod Nerves

GEORGE CAMOUGIS

Introduction

The potent actions of pyrethrum as an insecticide have been well known for more than 100 years. During more recent years, remarkable progress has been made on the chemistry and toxicology of the active principles of pyrethrum; much of that progress is reviewed in this volume.

There are, however, still some notable gaps in our knowledge of pyrethrum. These gaps include our lack of knowledge of how the pyrethrins actually act on insects. For example, even though we know quite a bit about their various actions, the nature of the fatal lesions is not understood.

As with most potent insecticides, the toxic effects of pyrethrins are usually associated with direct actions on the nervous systems of insects. Actions on the nervous system have been implied by numerous field observations and confirmed by laboratory experiments. Indeed, many monographs and reviews assert that the mode of action of pyrethrins is to interfere with the normal functions of the nervous system (Shepard, 1951; Roeder, 1953; Metcalf, 1955; O'Brien, 1967; Yamamoto, 1970a,b). The object of this paper is an attempt to put this assertion into current perspective.

II. Actions of Pyrethrins on Arthropod Nerves

When a flying insect encounters an effective spray of pyrethrins, it is rapidly knocked down and paralyzed. This knockdown action is one of the more remarkable properties of pyrethrins. An insect on solid substrate is usually paralyzed after a preliminary period of excitation. In fact, Shepard (1951) points out that pyrethrum has practical value in this latter regard, as it may excite insects to abandon a hiding place. Much of the evidence supports the premise that pyrethrins have at least two actions on the arthropod nervous system. There is an initial excitatory action that is followed eventually by a blocking phase. This section will consider these two important actions.

A. Excitatory Actions

A number of studies present evidence for the excitatory action of pyrethrins on arthropod nerve. In early electrophysiological studies, Lowenstein (1942) demonstrated a strong excitatory phase on the action potentials of the abdominal nerve cord of the cockroach *Blatta orientalis.* Direct applications of varying concentrations of pyrethrins produced excitation followed by a decay of activity and finally by a block. The time from the application of the insecticide to the blocking phase showed a clear relationship to the concentration of the pyrethrum extract used. The higher concentrations required less time to block. Therefore, the initial excitatory phase gives way to the subsequent blocking phase, and the duration of both events varies inversely with concentration.

Similar results have been observed with crustacean nerve preparations. Ellis *et al.* (1942) studied the actions of pyrethrum on a crayfish nerve–muscle system in a claw preparation. Pyrethrum injection caused spontaneous contractions of the claw, and the magnitude and duration of contraction increased as the dosage was increased. Also, pyrethrum produced repetitive discharge in nerve fibers following single-shock stimulation. These results were also observed in crayfish preparations by Welsh and Gordon (1947). Pyrethrum was observed to cause spontaneous discharge within 1 min after application in low concentration. Higher concentrations produced block. In contrast, DDT caused the same type of repetitive discharge, but a longer period of time was required for the excitatory actions to occur.

More recently, Camougis and Davis (1971) studied the actions of pyrethrins on a variety of isolated nerve preparations with electrophysiological methods. These studies included the actions on the isolated abdominal ganglia of the crayfish. Isolated ganglia repeatedly showed marked excitation following exposure to pyrethrins. Typically, the initial excitatory response was followed several minutes later by a blocking phase. These observations were supplemented with *in vivo* experiments in which injections of pyrethrum or suspensions of pyrethrum in the aquarium were used in tests with freely moving crayfish. Results included immediate excitation, convulsions, and eventually paralysis and death.

Also in recent studies, Burt and Goodchild (1971a) were able to show an increase in the spontaneous electrical activity in the exposed sixth abdominal ganglion of the cockroach following application of pyrethrin I. In summary, then, we can conclude that pyrethrins cause a rapid and pronounced excitatory action on the arthropod nervous system, as shown by direct evidence from a number of studies.

B. Blocking Actions

Perhaps better known are the blocking actions of pyrethrins on arthropod nerve. These blocking actions are inferred by the paralysis of insects and known more directly from the experimental evidence.

Several early electrophysiological studies established the blocking activity of pyrethrum on arthropod nerves when administered directly to exposed nerve tissue. Lowenstein (1942) was able to show block in the cockroach abdominal nerve cord after the initial excitatory phase. As the concentration was increased, the time in minutes required for block was decreased. Generally comparable results were seen with crustacean nerve preparations. Ellis *et al.* (1942) and Welsh and Gordon (1947) reported block of crustacean peripheral nerve with low concentrations of pyrethrum. Schallek and Wiersma (1948) studied the crayfish central nervous system (CNS). They found that pyrethrum was effective in a concentration of 10^{-7} gm/cm^3, and it achieved a synaptic block in 2 min.

More recent studies with both cockroaches (Burt and Goodchild, 1971a,b) and crayfish (Camougis and Davis, 1971) have confirmed potent blocking activity. Burt and Goodchild (1971a,b) reported block of the electrical nerve discharges from the CNS of the cockroach after application of pyrethrin I in low concentration. Camougis and Davis (1971) observed block of both peripheral nerve and the CNS ganglia in the crayfish. All investigators observed the blocking effects to follow initial stimulation.

A repeated observation is that recovery following block is not readily achieved after washing with appropriate saline solution. This is seen with both pyrethrum and the synthetic pyrethroids (see Section IV). The physical-chemical properties of pyrethroids are probably responsible for this observation. Further speculations on reversibility will be made in Section V of this chapter.

III. Actions of Pyrethrins on Other Species

Despite the relatively numerous studies of pyrethrins on arthropod nerve, very little information exists on their effects on the nervous systems of other animals. Much of the information that is available is a consequence of toxicological studies. For example, reviews (Negherbon, 1959; Hayes, 1965) include hyperexcitability, tremor, convulsions, clonic spasm, and tetanic muscular paralysis as symptoms of pyrethrum poisoning in mammals. From this information we may assume an action on the nervous system.

In their comparative study on the neuropharmacology of pyrethrins, Camougis and Davis (1971) included tests with frog and rat nerve. Lidocaine, which is known to depress excitability and block conduction in all nerves generally (Takman and Camougis, 1970), was used as a standard. Thirty-four frog nerves and 16 rat nerve preparations were tested with varying concentrations of lidocaine and pyrethrins in a variety of testing sequences. Pyrethrins, even in concentrations up to 0.2%, were either inactive or produced slight or partial blocks. In contrast, lidocaine produced the expected block in conduction in concentrations that are known to block. As mentioned earlier, pyrethrins were very effective on crayfish nerve. These results established fundamental differences in the actions of pyrethrins on arthropod and vertebrate nerves.

These studies were extended to include tests with desheathed frog nerves. The rationale for tests with desheathed nerves is that many neurotropic substances often do not penetrate the nerve sheath, which acts as a diffusion barrier. Lidocaine is much more active when applied to desheathed nerve; approximately one-tenth the concentration required for intact nerves produces equal block in desheathed nerve. Because it is important to establish the success of the desheathing process, which

TABLE I

Summary of Results of Tests with Nerve Preparations from Various Species[a]

Species	Isolated nerve preparation	Action of pyrethrins	Action of lidocaine
Crayfish	ganglia (CNS)	excitation followed by block	block
Crayfish	peripheral nerve	block	not tested in sufficient concentration; presumed to block
Frog	peripheral nerve	no block or poor block	block
Frog	desheathed peripheral nerve	no block or poor block	more active block than with nondesheathed nerve
Rat	peripheral nerve	no block or poor block	block

[a] Modified from Camougis and Davis (1971).

is often a difficult procedure, the use of lidocaine as a standard is again critical. Therefore, lidocaine and pyrethrins were tested alternately on desheathed portions of frog nerves. Intact (nondesheathed) nerves were also tested as controls. Lidocaine showed predictably greater blocking activity on the desheathed nerves. In contrast, pyrethrins were about equally active on both types of nerve. The various tests on the three species are summarized in Table I.

IV. Neuropharmacology of Other Pyrethroid Insecticides

Some valuable information on the neuropharmacological actions of other pyrethroid insecticides has become available in recent years. Much of this information has resulted from the able efforts of Narahashi and his co-workers with the synthetic pyrethroid, allethrin. Allethrin has insecticidal powers and acute toxicity generally comparable to the natural pyrethrins (Negherbon, 1959).

Using intracellular microelectrode recording techniques, Narahashi and his co-workers have been able to study the actions of allethrin on various arthropod nerve giant fibers (Narahashi, 1962a,b; Berteau *et al.*, 1968). The results with microelectrode studies on cockroach giant nerve fibers showed the following actions: (*1*) slight depolarization, (*2*) an increase in the negative afterpotential, (*3*) repetitive afterdischarge, and (*4*) eventual block in conduction (Narahashi, 1962a,b).

These findings were generally confirmed when allethrin was tested on crayfish giant fibers (Berteau *et al.*, 1968). In addition, decreases were observed in the rate of rise of the nerve action potential prior to block. Recovery from block after washing in saline was poor.

Allethrin has also been studied on the giant axons of the squid (Narahashi and Anderson, 1967). These studies were designed to investigate the site and mechanism of action. The squid giant axon preparation, because of its size, lends itself to the study of drug perfusion on both sides of the excitable nerve membrane. Allethrin was found to act on both sides of the nerve membrane. Moreover, the findings confirmed the results of the tests on cockroach nerves. Allethrin was able to delay and inhibit both the early transient current and the late steady-state conductance change associated with the action potential. Again, poor reversibility was observed.

In addition to allethrin, the studies of Berteau *et al.* (1968) on crayfish giant fibers included tests with several synthetic compounds that lacked cyclopropane and ester groupings. Although the potencies of these compounds varied, they had qualitatively similar pyrethroid-like biological activity, including their actions on crayfish axons. Like allethrin these other synthetic compounds are slowly or poorly reversible after washing with normal saline solutions.

The interpretation of these various findings using allethrin and other pyrethroids presupposes an acceptance of the ionic hypothesis of nerve action as outlined by Hodgkin and his co-workers over a period of many years (Hodgkin, 1964). This hypothesis states that the action potential of a nerve (and certain other excitable tissues also) is caused by ionic movements across the axon membrane as permeability changes occur. Although this mechanism is not the only approach to excitation that has been proposed (e.g., Tasaki, 1968), it offers a useful working hypothesis and is widely accepted. Under normal conditions the extracellular fluid surrounding nerve cells is high in sodium ion and low in potassium ion, whereas the reverse is true for the inside axoplasm of nerve cells. The rising phase of the spike of an action potential is caused by sodium influx and this is referred to as the *early transient current*. The early transient current is a sign of increase in sodium conductance (sodium activation). The falling phase of the action potential is caused by a turning off of the sodium activation and an increase in potassium efflux (potassium activation). The potassium current is often referred to as the *late steady-state conductance change*. The actions of allethrin, then, may be interpreted as caused by an inhibition or block of the normal sodium and potassium currents associated with the nerve action potential. Moreover, these actions imply that allethrin somehow combines with the excitable membrane and subsequently exerts its blocking actions. More will be said about these mechanisms in Section V.

V. Concerning the Mode of Action

One cannot help but approach the subject of mode of action with a sense of humility and frustration. In spite of the numerous discussions on this complex subject in various articles and reviews, it is difficult to gain a satisfactory understanding of the subject. Although this chapter may not cast much additional light on the subject, it will at least attempt to put the subject of the mode of action of pyrethrins into more critical perspective.

Part of our difficulty lies in our hesitation to recognize that pyrethrins have several actions. For purposes of this discussion, let us list these actions as (*1*) knockdown, (*2*) paralysis, and (*3*) lethality. These may be partially or wholly independent actions as suggested by other reviews (Yamamoto, 1970a,b). Alternatively, the knockdown and kill may be stages in the same process occurring at the same site in the CNS, rather than separate phenomena, one of which occurs very rapidly at a peripheral nervous site (Elliott and Janes, Chapter 4 in this volume). One thing to recognize is that knockdown, paralysis, and lethality are phenomena that are observed in intact organisms. In contrast, the phenomena of excitation and block that were discussed earlier can be best analyzed at the tissue and cellular levels, and our reductionist efforts take us closer and closer to understanding the molecular basis of block, for example. In that sense, we can speak of the mode of action of excitation block in arthropod nerve with pyrethrins. However, this may not permit us to conclude that this cellular mode of action applies to the important and highly relevant knockdown, paralysis, and lethality observed with intact insects.

Mainly because of studies with allethrin, we now know that the mode of action of pyrethroids on invertebrate nerve is to interfere with the changes in sodium and potassium currents (Narahashi, 1971). As early as 1947, Welsh and Gordon suggested that pyrethrum and other insecticides exert their actions by a physical interference with the lipid surface of the nerve axon. As mentioned in Section IV, we know that allethrin can act on both sides of the giant axon of the squid, although its actions on potassium currents are more effective from the inside (Narahashi and Anderson, 1967). Furthermore, pyrethrins are highly lipid soluble and they are not electrically charged (as are local anesthetics, for example). Therefore, we may conclude that pyrethroid molecules are bound or dissolved in the lipid layer of the nerve cell membrane and thus exert their blocking effects. This dissolution in the lipid compartment of the nerve cell membrane probably accounts for the poor reversibility after washing. If the number of molecules in the nerve membrane is sufficiently large, interference with cationic conductances finally leads to nerve block. It is this nerve block that ultimately leads to the symptoms of paralysis observed in intact insects.

Several threads of evidence are combining to give us some lead into understanding knockdown. Various studies have demonstrated repetitive discharge in arthropod nerve following exposure to pyrethroids (e.g., Welsh and Gordon, 1947; Narahashi, 1962a). This occurs in peripheral nerve as well as in nerve cells of the CNS. The CNS of both insects (Burt and Goodchild, 1971a) and crayfish (Camougis and Davis, 1971) show high frequency discharges, indicating excitation, before block. Moreover, these excitatory actions in ganglia can appear with much lower concentrations than those required to block axonal conduction (Burt and Goodchild, 1971a). Finally, in *in vivo* studies with crayfish Camougis and Davis (1971) observed rapid excitation and convulsions after administration of pyrethrins. Paralysis and death occurred later. Stimulation, loss of coordination, and eventual paralysis are all symptoms of pyrethrum poisoning in insects (Shepard, 1951). An insect knocked down by pyrethrum may show intense activity and may not subsequently undergo paralysis and die. It is tempting, therefore, to attribute knockdown to the stimulatory phases of pyrethrins. The flying insect has its exoskeleton and associated sensory fibers initially exposed to pyrethrins. Before any block occurs, the unstabilizing actions leading to trains of repetitive discharges in sensory fibers occurs. As the poison is absorbed systemically, other peripheral nerves, including motor fibers, are affected. Finally, the highly sensitive ganglia are excited. It is difficult to envision how a flying insect could maintain precise neuromotor control under these circumstances. Knockdown would occur immediately and long before paralysis and block became manifest.

How drugs and poisons kill animals is not a simple matter, and indeed a whole discipline is devoted to its study. Generalizations are difficult, especially when insects and mammals are compared (O'Brien, 1967). Some lethal mechanisms are fairly well understood, i.e., organophosphorus poisoning in mammals. However, critical review of our knowledge of pyrethrins should lead us to be cautious about our conclusions on fatal lesions in insects. It may be true that the fatal lesions of pyrethrin I may be within the ganglia as suggested by Burt and Goodchild (1971a). However, we need more information before we can feel more confident on this important point. The fatal lesions from pyrethroids

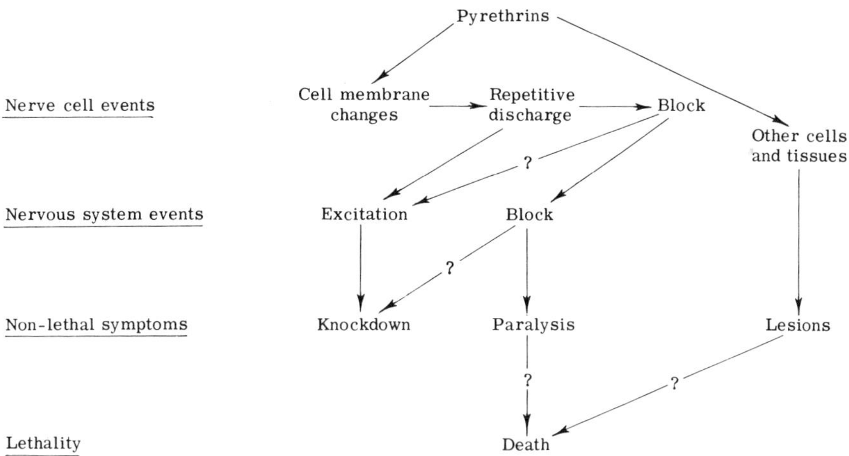

Fig. 1. Diagram of proposed relationships among various events in the nervous systems of arthropods following exposure to pyrethrins.

in arthropods will probably be found to include a combination of cells, tissues, and organ systems.

Figure 1 is an attempt to tie together the various experimental findings with the phenomena of knockdown, paralysis, and death. Pyrethrins combine with the nerve cell membrane, causing changes in excitability properties. Repetitive discharge and block follow. At the nervous system level, these events may account for excitation and block. Note that it is proposed that nerve cell block may cause nervous system excitation. This is because differential block of inhibitory fibers may unmask excitation. Proceeding to knockdown and paralysis, excitation and block are proposed to underlie these two symptoms. Whereas the proposed hypothesis for knockdown implicates excitation as the main mechanism, block of part of the nervous system may also be involved in knockdown. Finally, speculations on the basis of lethality focus on paralysis and unknown lesions. Paralysis itself may not be fatal; however, prolonged paralysis may cause major systemic pathology, directly or indirectly. At any rate, the events outlined in Fig. 1 are an attempt to diagram various component actions of pyrethroids on the arthropod nervous system. Future discussions on mode of action will be more meaningful if they focus more critically on specific components.

VI. Some Interesting Problems

Before this review is concluded, it will be useful to discuss some problems that appear both interesting and relevant to the actions of pyrethrins on nerve tissue. Some of the discussions may be useful in further research.

Of foremost interest is the observation that pyrethrins are relatively inactive on isolated vertebrate nerve. This difference cannot be explained solely by penetration. First, pyrethrins are highly apolar and hence should penetrate cell membranes. Second, the desheathed nerve experiments presumably removed a potential penetration barrier. These experiments should be repeated by another laboratory. Such findings are unexpected because pyrethrins and certain synthetic pyrethroids can be highly toxic to mammals on intravenous administration even though they are of low oral and dermal toxicity.

Insects recover completely from sublethal doses, if they have not passed beyond the initial stages of paralysis (Yamamoto, 1970a). This suggests that pyrethrins are detoxified, and it is easy to picture the pyrethrin molecules in tissue generally undergoing metabolism. What is more difficult to envision is how pyrethroid molecules leach out of the nerve cell membrane so that recovery from paralytic block can occur. Perhaps they are metabolized *in situ* in the nerve. Isolated nerve experiments suggest very poor recovery.

It is curious that synergists by themselves have not been tested on isolated nerve preparations. It is known that synergists reduce the concentrations of pyrethrins required for both knockdown and lethal actions. Assuming that synergists act as alternative substrates for metabolism, sparing pyrethrins from detoxication (Casida *et al.*, 1966), this action must be fast indeed in the case of knockdown. Metabolism of synergists and pyrethroids must occur in significant amounts within minutes, otherwise the synergists would not act in potentiating the actions of pyrethrins in knockdown. However, it is possible that synergists may have neuroexcitatory actions of their own. This should be investigated.

VII. Summary

The mode of action of pyrethrum on the arthropod nervous system is assumed to result from induction of toxic lesions at this site. Actions of pyrethrins on the arthropod nervous system include both excitation and block. Little is known about the direct actions of pyrethrins on the nervous systems of other species. Recent evidence suggests that vertebrate nerve is relatively insensitive to pyrethrins applied directly. Over the past decade studies with synthetic pyrethroids at the cellular level have provided us with considerable insight into the mechanism of excitation block. Pyrethroids probably combine with the lipid layer of the nerve cell membrane and interfere with cationic conductances. However, discussions on the mode of action should also consider the phenomena of knockdown, paralysis, and lethality. Knockdown may be caused by excitation. Nerve block at the cellular and finally at the systemic levels is responsible for paralysis. Lethality is not understood, and lesions in tissues other than nerve are probably also involved. Because of the complexity of the subject, discussions of the mode of action of pyrethrins should focus on specific events and symptoms.

References

Berteau, P. E., Casida, J. E., and Narahashi, T. (1968). Pyrethroid-like biological activity of compounds lacking cyclopropane and ester groupings. *Science* **161**, 1151–1153.

Burt, P. E., and Goodchild, R. E. (1971a). The site of action of pyrethrin I in the nervous system of the cockroach *Periplaneta americana* L. *Entomol. Exp. Appl.* **14**, 179–189.

Burt, P. E., and Goodchild, R. E. (1971b). The spread of topically-applied pyrethrin I from the cuticle to the central nervous system of the cockroach *Periplaneta americana. Entomol. Exp. Appl.* **14**, 255–269.

Camougis, G., and Davis, W. M. (1971). A comparative study of the neuropharmacological basis of action of pyrethrins. *Pyrethrum Post* **11** (1), 7–14.

Casida, J. E., Engel, J. L., Esaac, E. G., Kamienski, F. X., and Kuwatsuka, S. (1966). Methylene-C^{14}-dioxyphenyl compounds: Metabolism in relation to their synergistic action. *Science* **153**, 1130–1132.

Ellis, C. H., Thienes, C. H., and Wiersma, C. A. G. (1942). The influence of certain drugs on the crustacean nerve-muscle system. *Biol. Bull.* **83**, 334–352.

Hayes, W. J., Jr. (1965). Insecticides, rodenticides, and other economic poisons. *In* "Drill's Pharmacology in Medicine" (J. R. Di Palma, ed.), 3rd ed., pp. 989–1004. McGraw-Hill, New York.

Hodgkin, A. L. (1964). "The Conduction of the Nerve Impulse." Liverpool Univ. Press, Liverpool.

Lowenstein, O. (1942). A method of physiological assay of pyrethrum extract. *Nature (London)* **150**, 760–762.

Metcalf, R. L. (1955). "Organic Insecticides. Their Chemistry and Mode of Action." Wiley, New York.

Narahashi, T. (1962a). Effect of the insecticide allethrin on membrane potentials of cockroach giant axons. *J. Cell. Comp. Physiol.* **59**, 61–65.

Narahashi, T. (1962b). Nature of the negative after-potential increased by the insecticide allethrin in cockroach giant axons. *J. Cell. Comp. Physiol.* **59**, 67–76.

Narahashi, T. (1971). Mode of action of pyrethroids. *Bull. W.H.O.* **44**, 337–345.

Narahashi, T., and Anderson, N. C. (1967). Mechanism of excitation block by the insecticide allethrin applied externally and internally to squid giant axons. *Toxicol. Appl. Pharmacol.* **10**, 529–547.

Negherbon, W. O. (1959). "Handbook of Toxicology," Vol. III. Saunders, Philadelphia, Pennsylvania.

O'Brien, R. D. (1967). "Insecticides. Action and Metabolism." Academic Press, New York.

Roeder, K. D. (1953). "Insect Physiology." Wiley, New York.

Schallek, W., and Wiersma, C. A. G. (1948). The influence of various drugs on a crustacean synapse. *J. Cell. Comp. Physiol.* **31**, 35–47.

Shepard, H. H. (1951). "The Chemistry and Action of Insecticides." McGraw-Hill, New York.

Takman, B. H., and Camougis, G. (1970). Local anesthetics. *In* "Medicinal Chemistry" (A. Burger, ed.), 3rd ed., Part II, pp. 1607–1632. Wiley (Interscience), New York.

Tasaki, I. (1968). "Nerve Excitation: A Macromolecular Approach." Thomas, Springfield, Illinois.

Welsh, J. H., and Gordon, H. T. (1947). The mode of action of certain insecticides on the arthropod nerve axon. *J. Cell. Comp. Physiol.* **30**, 147–172.

Yamamoto, I. (1970a). Problems in mode of action of pyrethroids. *In* "Biochemical Toxicology of Insecticides" (R. D. O'Brien and I. Yamamoto, eds.), pp. 193–200. Academic Press, New York.

Yamamoto, I. (1970b). Mode of action of pyrethroids, nicotinoids, and rotenoids. *Annu. Rev. Entomol.* **15**, 257–272.

Part IV

USE AREAS

Chapter 12

Pyrethrum for Control of Insects Affecting Man and Animals

CARROLL N. SMITH

Few insecticides equal pyrethrum in spanning the entire period of scientific chemical control of insect pests. As early as 1880 pyrethrum powder was used to control adult mosquitos (Shepard, 1951), and today pyrethrum aerosols are used in the disinsection of the swiftest transworld aircraft. Pyrethrum formulations are used to protect dwellings and other structures and to control pests in outdoor areas, as well as for direct application to livestock and to man himself. The details of this long history of successful use are reviewed by McLaughlin (this volume, Chapter 1). This chapter will be concerned only with recent research, generally with developments within the last 10 years. Research with useful or promising synthetic pyrethroids will be included only incidentally. Use of pyrethrum for the control of cockroaches and other household pests, except for flies and mosquitos, is considered by Gillenwater and Burden (this volume, Chapter 13).

I. Insect Control in Homes and Other Structures

A. Aerosols and Aircraft Disinsection

Pyrethrum sprays of one type or another have been a perennial favorite for the control of mosquitos, houseflies, and other flying insects in homes, from the earliest kerosene solutions applied with hand atomizers to the modern pressurized aerosols. Synergists conferred increased mortality, and other insecticides were sometimes added to extend the range of effectiveness of the aerosols. One recent development of special significance in medical entomology is the use of aerosols to kill insects in aircraft, particularly mosquitos and other disease vectors on aircraft making international flights.

Sullivan *et al.* (1964) compared two experimental aerosols containing pyrethrum and DDT for use in the disinsection of aircraft at "blocks away" (World Health Organization, 1971) in tropical areas. Earlier studies (e.g., Sullivan *et al.*, 1962) had shown that the method was suitable as a standard procedure for aircraft disinsection for international quarantine purposes. The essential element of the method consists of the discharge of single-use aerosol containers within the aircraft passenger cabins after the doors have been closed (and, incidentally, the wheel blocks taken away) but prior to take off, when air circulation within the cabin is at a minimum. To meet quarantine requirements the method requires suitable administrative safeguards to assure that an adequate dosage has actually been dispensed. The flight deck is treated prior to occupancy by the flight crew, and cargo holds and wheel wells are sprayed shortly before take off. One of the formulations used by Sullivan *et al.* (1964) contained 1.6% of pyrethrum extract (25% pyrethrins) and 3% DDT, and was designated as the World Health Organization Standard Reference Aerosol (SRA), whereas the other contained 6% of pyrethrum extract (20% pyrethrins) and 2% DDT and was designated G-1492. Both formulations contained solvents and propellents but no synergists. The optimum dosage was considered to be 10 gm per 1000 ft^3 (28.3 m^3), and the dosages actually applied ranged from about 8.6 to about 11.3 gm per 1000 ft^3 (31–40 gm per 100 m^3), depending on the volume of the cabin and the size of the single-use containers. The tests were made on flights in the Philippines and between the Fiji Islands and New Zealand. Female mosquitos were exposed in cages in various parts of the cabin; strains of *Aedes aegypti* and *A. albopictus* were used which showed two- to three-fold resistance to DDT; some of the strains of *Culex fatigans* (= *Culex pipiens quinquefasciatus*) that were used were apparently resistant and others were susceptible. Formulation G-1492 gave 100% kills of mosquitos in four tests and 84–100% in one, whereas the SRA formulation gave kills of 100% in five tests, 96–100% in two, and 40–100% in two. Both formulations gave 100% kills of nonresistant mosquitos. The G-1492 formulation gave faster knockdown than the SRA aerosol.

McGuire *et al.* (1966) compared the SRA and the G-1492 formulations with a synergized pyrethrum aerosol in laboratory tests. The third formulation, designated PBK, contained 1.7% of pyrethrum extract (23.5% pyrethrins), 2% sulfoxide, and 2.4% isopropyl myristate plus solvent and propellents. The tests were made in a 1000 ft^3 chamber. The insects used were male and female houseflies and mosquitos (*A. aegypti*) from DDT-resistant colonies. Various dosages were used, and LD_{50}'s were computed in the tests with houseflies. The SRA was practically ineffective against the resistant houseflies, and the LD_{50}'s were 6.71 and 4.07 gm per 1000 ft^3 for the G-1492 and PBK aerosols, respectively. The PBK aerosol was also superior in 30-min knockdown. In the tests against mosquitos no regressions were calculated, but the PBK formula was slightly more effective than G-1492, and both were superior to SRA.

The present International Health Regulations (World Health Organization, 1971) recommend the use of aerosol formulations that have been demonstrated to be equal or superior to the SRA for use in the "blocks away" disinsection procedure.

Pyrethrum aerosols remain a mainstay for control of flying insects in homes. J. Lee (1969), in a review of aerosol developments, reported comparable results with Tropital and piperonyl butoxide as synergists for pyrethrins in laboratory tests with aerosols against houseflies.

Fales *et al.* (1970) conducted laboratory tests with a variety of pyrethrum aerosols against several species of insects affecting man and animals. Their standards were the Official Test Aerosol (OTA), containing 0.4% pyrethrins and 2% DDT, and G-1357, containing 0.6% pyrethrins and 1.4% piperonyl butoxide. All tests were made in a Peet-Grady chamber. The dosages used approximated either 3 or 7 gm per 1000 ft^3. In tests against DDT-resistant houseflies aerosols containing 0.4% or 0.5% of pyrethrins and 4% of Tropital were equal to G-1357 in kill and only slightly less effective in knockdown, whereas aerosols with lower concentrations of Tropital were less effective. Both formulations were also about equal to G-1357 in tests against stable flies and three species of mosquitos (*C. p. pipiens, A. aegypti,* and *Anopheles quadrimaculatus.*) Formulations containing 0.45% of pyrethrins in combination with 2.7% of Tropital or sulfoxide, or 0.5% of pyrethrins in combination with 3% of either synergist, were equal to or slightly better than G-1357 in tests against resistant houseflies, stable flies, face flies, *A. aegypti, A. atropalpis, A. triseriatus, Anopheles quadrimaculatus,* and *C. p. pipiens.*

B. Smoke Devices

Smoke has been used by one method or another to repel mosquitos from early times, and the burning of pyrethrum powder for this purpose was mentioned by Howard in 1911. Within recent years there has been a resurgence of interest in the burning of such formulations as impregnated candles or mosquito coils for protection from mosquitos in rooms and some outdoor areas. Recent research has included attention to the preparation and evaluation of mosquito coils containing pyrethrins and synthetic pyrethrin-like compounds (e.g., Nagasawa *et al.,* 1951; Paulini and Ricciardi, 1954; Kogan, 1961).

Maciver (1963) reviewed the formulation and manufacture of pyrethrum mosquito coils. These became popular in eastern Asia, especially Japan (Nagasawa *et al.,* 1951, 1952), and apparently were derived from the traditional incense or joss sticks. Maciver reports that in 1962 more than 2×10^6 kg of mosquito coils were exported from Japan to various countries on every continent. The mosquito coils were designed to burn for about 8 hr in order to provide protection to sleepers from biting mosquitos throughout the night. The basic ingredients of the coils, in addition to the pyrethrum, are an organic filler, capable of smoldering for a long time, a binder, and such auxiliary materials as fungistats and dyes. The pyrethrum is often present in the form of powder from ground flowers, although coils may sometimes be impregnated with a solution of pyrethrum extract. The organic filler may be ground leaves specially selected for burning qualities or aroma, wood flour, coconut shell flour, or sometimes pyrethrum marc, which is the residue from the pyrethrum flowers left after the extraction process has been completed. The binder may be tabu powder, which is the ground leaves or bark of a tree, *Machilus thunbergii,* and which contains a polysaccharide gum. These ingredients are made into a paste, then either extruded as a filament that is formed into a coil or as a flat sheet from which coils are stamped.

The pyrethrum in the smoke from the coils may act in several related ways to provide relief from annoyance: (*1*) by repelling mosquitos, i.e., preventing their entry into treated areas; (*2*) by expelling mosquitos already present in the areas when the coils are lit; (*3*) by inhibiting the feeding activity of the mosquitos; or (*4*) by causing their knockdown or death (Maciver, 1963).

Maciver (1964) also conducted experiments on the knockdown, mortality, and repellency to mosquitos caused by smoke from coils containing pyrethrum with and without piperonyl butoxide. His tests were conducted in a room with a volume of 1400 ft^3 (40 m^3) at 27°C and 39–44% relative humidity. In the knockdown and mortality tests the mosquitos were exposed for 1 hr in glass cages with gauze-covered ends. Coils containing different concentrations of pyrethrins were used, and the times required for 50% and 90% knockdown were determined. In tests with *A. aegypti* the time for 90% knockdown (KD_{90}) decreased from 105 min at 0.1% pyrethrins to 20 min at 0.5% pyrethrins, but showed no further decrease at higher concentrations. Mortality of 100% (after 24 hr) was obtained with concentrations of 0.4% or higher. In repellency (or expellency) tests the insects were caged in a glass bottle, the coils were burned for 1 hr, and then the bottle was unstoppered and the opening covered with gauze. As the smoke gradually penetrated the bottle the mosquitos tended to move away from the open end.

Maciver (1964) obtained no increase in knockdown effect by adding piperonyl butoxide to the pyrethrins in the coils at ratios of 1:1, 5:1, or 10:1; he commented that this was in conformity with the results of Nagasawa *et al.* (1952) in tests with *C. p. pallens,* but differed somewhat from those of Kogan (1961), who found no synergism at a ratio of 5:1, but found some effect at a ratio of 8:1, in tests with *C. p. fatigans.* Maciver also found pyrethrins to be about twice as effective in knockdown as allethrin.

Smith and Obudho (1967) used a specially constructed verandah-trap hut to evaluate the protection against *Anopheles gambiae* provided by mosquito coils under natural conditions. The volume of the hut was about 17.6 m^3 (approximately 500 ft^3). Two men slept in the hut and one pyrethrum coil containing 0.23–0.24% pyrethrins was burned throughout the night, about 8 hr. It would be difficult to measure the repellency to mosquitos seeking to enter the hut, but the expellent effect on mosquitos already in the hut was computed from the numbers of mosquitos in the traps on the following morning and the numbers remaining in the hut. The treatment had a pronounced expellent effect: 86–92% of the mosquitos left the treated hut as opposed to an average of 39% leaving an untreated hut. The treatment also inhibited the feeding activity of the mosquitos; only 30% of the mosquitos leaving untreated huts were unfed, whereas 75–81% of the mosquitos left the treated huts without feeding. The subsequent mortality of mosquitos taken from the traps was 8% from untreated huts and 19–27% from treated huts.

Fales *et al.* (1968) tested coils containing 0.5% pyrethrins and 0.5% allethrin against *C. p. pipiens* in a 1000-ft^3 room. In knockdown tests the mosquitos were free flying. When 8 gm of coil were burned within the room, 100% knockdown required 60 min in one test with pyrethrins and 120 min in another, and 120 min in one test with allethrin. Repellency (or expellency) was determined from the percent of mosquitos that left a cage at the wall of the treated room and flew through a glass passageway to a covered cage outside the room. The maximum repellency obtained was 58% in one test with 1 gm of coil containing 0.5% allethrin, followed by 42% in three tests with 0.25 gm of coil containing 0.5% of pyrethrins.

Winney (1969) obtained results similar to those of Maciver (1964) in the knockdown and kill of *A. aegypti*. He also found that the knockdown with pyrethrins was not increased by the addition of piperonyl butoxide and, like Fales *et al.* (1968), he found pyrethrins to be about three times as effective as allethrin in repellency. Exposure of the mosquitos for 3 min in rooms treated with pyrethrum coils reduced the biting rate from 60–100% to 0–10%.

Winney and Webley (1969) reported that the biological activity of various pyrethrum clones incorporated into mosquito coils varied with the ratio of pyrethrin I to pyrethrin II, and with the pyrethrin I and cinerin content. Winney (1971) also found the activity of mosquito coils was enhanced by the inclusion of material from a clone (4331) with a high ratio of pyrethrin I to pyrethrin II, with respect to knockdown of *A. aegypti*.

Chadwick (1970) tested coils containing various active ingredients against *A. aegypti* in a Peet-Grady chamber. Coils containing 0.2 or 0.3% pyrethrins caused significantly faster knockdown than coils with the same amounts of pyrethrins plus five times as much piperonyl butoxide. The KD_{50} was significantly shorter (5.5 min) with coils containing 0.5% of *d-trans*-allethrin (bioallethrin) than with coils containing 0.5% pyrethrins, allethrin, resmethrin, or *d-trans*-resmethrin (bioresmethrin), which gave KD_{50}'s of 8.8 to 9.6 min. Other tests at a range of concentrations confirmed this approximate ratio of effectiveness (about 0.6 to 1). Whereas 85% of unexposed mosquitos bit readily, and 61% of the mosquitos fed after exposure to smoke from blank coils, only 2% fed after exposure for 5 min. to smoke from coils containing 0.5% pyrethrins or bioallethrin. Chadwick also commented on his previous work (Smith and Chadwick, 1964) in which fewer mosquitos were caught in a hut treated with pyrethrum emulsion, either because they were repelled from entering or expelled after entry.

Fales *et al.* (1971) used the method previously mentioned (Fales *et al.*, 1968) to compare the repellency of coils containing pyrethrins and synthetic pyrethroids. At concentrations of 0.5%, and dosages of 0.25 gm of coil, pyrethrins caused 36% repellency; bioresmethrin caused 36%; resmethrin, 27%; bioallethrin, 18%; and tetramethrin, 10%.

Nishizawa (1971) reported quicker knockdown and kills of *C. p. pallens* and houseflies with allethrin and 5-propargyl-2-furylmethyl chrysanthemate (prothrin) than with pyrethrins in mosquito coils, in contrast to the reports mentioned above.

Bremer (1964) reported that burning a powder consisting of 50% ground pyrethrum flowers, 6% piperonyl butoxide, and 44% of a composition containing 3,7-dinitroso-1,3,5,7-tetraazabicyclo[3.3.1]nonane gave quicker knockdown and higher kills of houseflies than burning a standard coil, when tested at comparable total dosages of pyrethrins.

C. Toxicity of Pyrethrins to the Brown Dog Tick

Gladney *et al.* (1972) found pyrethrins plus piperonyl butoxide (1:10) to be the most effective of 26 widely used insecticides in laboratory tests against nymphs of the brown dog tick (*Rhipicephalus sanguineus.*) The tests were made by enclosing engorged larvae in envelopes made of teabag paper, allowing the larvae to molt, dipping the envelopes containing the unfed nymphs in emulsions of the insecticides, and observing the mortality after 24 hr. Each insecticide was tested at a range of concentrations and the LC_{50} and LC_{90} values were computed from the mortality data. Pyrethrum plus piperonyl butoxide had an LC_{50} of 0.000028% and LC_{90} of 0.000059%, in comparison to LC_{50}'s of, for example, 0.00016% for Dursban and resmethrin, 0.00075% for DDT, 0.0052% for carbaryl, 0.0057% for ronnel, 0.023% for malathion and 0.38% for chlordane.

D. Fly Control in Barns

Adkins *et al.* (1971) compared pressurized formulations containing 0.2% pyrethrins plus 1% piperonyl butoxide with formulations containing 0.25 or 0.2% of tetramethrin plus 0.12 or 0.05%, respectively, of resmethrin for the control of houseflies in barns. All three treatments were about equally effective, reducing fly counts from an average of about 8 per grill to 0.2–0.5 per grill after 15 to 30 min, but the synthetic materials were less irritating. The aerosols were applied at 6 gm per 1000 ft^3.

E. Control of Poultry Mites

Chadwick (1960) obtained good control of the red mite, *Dermanyssus gallinae*, in poultry houses with two treatments, at a 7-day interval, with a spray containing 0.025% pyrethrins, and Westermarck (1967) reported control with an emulsion containing 0.01% pyrethrins and 0.04% piperonyl butoxide.

II. Insect Control in Outdoor Areas

A. Ultralow Volume Applications

The use of ultralow volume (ULV) applications of insecticides, particularly as aerial sprays, permits the rapid treatment of large areas with formulations containing minimum quantities of inert carriers. Several recent studies indicate that pyrethrum formulations may be competitive with synthetic insecticides for this type of application against tsetse flies, tabanids, and mosquitos.

The toxicity of pyrethrins to tsetse flies is illustrated by the work of Burnett (1961) who compared pyrethrins with various other insecticides for toxicity to *Glossina morsitans* by topical applications with a microburet. Pyrethrins were about as toxic as dieldrin (LD_{50}, 0.002 μg per insect), and pyrethrins synergized with piperonyl butoxide at a ratio of 15:1 were about twice as toxic.

Lee *et al.* (1968) applied an aerial spray containing 0.4% pyrethrins with 2% piperonyl butoxide to 1920 acres of forest and grassland area in Tanzania for the control of tsetse flies, mainly *Glossina pallidipes* concentrated in dense thicket. Six applications were made at a dosage rate of 0.016 gal/acre at intervals of 3 weeks. Fly abundance was determined by fly-round counts; reductions of 79, 87, and 95% were obtained after the first, second, and third applications, and the authors concluded the results were comparable to those ordinarily obtained with synthetic insecticides. However, Tarimo *et al.* (1970) used similar applications in a similar area and found that, although reductions of *G. pallidipes* were indicated to be 64, 36, and 29% after three applications when measured by fly-round counts as used by Lee *et al.* (1968), they were only 11, 7, and 0% when measured by Langridge-trap collections. They concluded the trap collections were more indicative of the effect of insecticide treatments on this particular species, and that estimates from fly rounds may give a false assessment.

C. W. Lee (1969) reviewed the use of aerial sprays for control of tsetse flies by the Tropical Pesticides Research Institute at Arusha, Tanzania between 1948 and 1967, including treatments with DDT, BHC, dieldrin, isobenzan, endosulfan, fenthion, and synergized pyrethrum, and recommended additional research with pyrethrum and improved methods of application.

Tarimo *et al.* (1971) obtained 55–79% control of *Glossina swynnertoni* and 87–94% control of *G. pallidipes* in three ULV aircraft applications of synergized pyrethrins. The sprays were oil solutions containing 1% pyrethrins and 4% piperonyl butoxide and were applied at an average rate of 148 ml/hectare (about 0.6 gm actual pyrethrins per acre). The sprays were applied to an area of about 878 hectares (about 2170 acres) of savannah woodland, protected from reinfestation by a barrier treated with residual applications of dieldrin. The results were as good as those obtained by the same investigators with a formulation containing 0.5% pyrethrins, 2% piperonyl butoxide, and 5% DDT. The authors concluded that pyrethrum was competitive with synthetic insecticides in cost as well as effectiveness.

Ezell (1971) tested ULV applications of several pyrethrum formulations for the control of adult tabanids in a park bordered by a salt marsh in Charleston, South Carolina. The most abundant species were *Chrysops atlanticus* and *Tabanus nigrovittatus.* One formulation contained piperonyl butoxide as a synergist, one contained an ultraviolet (UV) inhibitor but no synergist, and one contained both the synergist and the UV inhibitor. The treated areas ranged from about 175 acres when ground applications were made to about 640 acres with aerial applications. The synergist was used at five times the concentration of pyrethrins. Application rates varied from 3.7 oz/acre of a spray containing 2.5% synergized pyrethrins to 7.3 oz/acre of a 5% synergized spray. All formulations gave good immediate reductions (about 80–95%) in adult tabanid populations, but infestations in the treated areas usually began to build up after 2 to 4 days. Although comparisons between tests at different times were only indicative, the pyrethrum formulations appeared to be about as effective as ULV applications of naled at 1 to 2 oz/acre or malathion at 3 to 4 oz/acre, but were more expensive. There were only slight differences in favor of the synergized formulations and the heavier dosages.

Mount and Pierce (1972) tested synergized pyrethrins and other insecticides as ULV cold aerosols dispersed with ground equipment to determine their toxicity to caged female mosquitos (*Aedes taeniorhynchus.*) The concentrated insecticide formulations were dispersed with a Leco ULV cold aerosol generator. The tests were conducted in an open field between 6 and 9 PM under favorable climatic conditions. Female mosquitos were exposed in cages, suspended on stakes 150 and 300 ft downwind from the line of travel of the aerosol generator. Each insecticide was applied at a range of dosages, mortality of the mosquitos was recorded after 18 hr, and an LD_{90} for each insecticide was estimated. The LD_{90} for pyrethrins was 0.008 lb/acre with piperonyl butoxide as a synergist and 0.015 lb/acre with Tropital as a synergist; both synergists were used at five times the dosage of the pyrethrins. The LD_{90}'s for two standard insecticides, naled and malathion, were 0.025 and 0.06 lb/acre, respectively.

A. W. Buzicky (personal communication, 1972) reports good results with ULV applications of synergized pyrethrum dispersed with ground equipment for the practical control of pest mosquitos (*Aedes vexans* and other species) by the Metropolitan Mosquito Control District (St. Paul, Minn.). Piperonyl butoxide was used as the synergist with the pyrethrins at a ratio of 5:1. Approximately 30,000 acres were treated in 1972 at the rate of 0.004 lb actual pyrethrins per acre for control of adult mosquitos. Sprays were dispersed with a truck-mounted cold aerosol generator and a ULV back pack sprayer. Resmethrin, *s*-bioallethrin, and bioresmethrin gave field results approximately equivalent to synergized pyrethrins at rates of 0.003 to 0.005 lb/acre. A striking action of all these compounds was the rapid knockdown and cessation of biting shortly after ULV applications were made. Preliminary tests with synergized pyrethrins diluted in fuel oil and applied with a hand-carried pulse jet thermal fogger gave promising results at 0.004 lb/acre.

B. Fogs for Housefly Control

Davies (1966) obtained about 49% control of houseflies in a congested residential area by fogging with a kerosene solution containing 0.05% pyrethrins and 0.4% piperonyl butoxide and 75% control with an emulsion containing 0.05% pyrethrins and 0.5% piperonyl butoxide, applied with a TIFA fog generator. The reduction 6 days after fogging was also high, 65 and 78%, respectively.

C. Larviciding for Gnat Control

Hitchcock and Anderson (1968) tested various insecticides for the control of larvae of *Chironomus atrella* in a tidal cove. Open-ended drums placed in the bottom of the cove were used to isolate small plots for tests under field conditions. When the larvicides were applied as granules, to carry the insecticide to the bottom mud, 93% mortality was obtained with a dosage of 2 lb/acre of pyrethrum plus 20 lb/acre of piperonyl butoxide. Naled was equally effective at 2 lb, malathion at 0.5 lb, and Abate at 0.1 lb. Liquid formulations were unsatisfactory.

III. Direct Application to Animals and Man

A. Livestock Sprays

Pyrethrum sprays have been among the popular methods for protecting livestock from biting flies for a number of years. Among the earlier investigations that established the effectiveness of pyrethrins, usually in combination with piperonyl butoxide or other materials, for this purpose may be cited the work of MacCreary and Goddin (1937), Howell (1949), Bruce and Decker (1951, 1955), Goodwin *et al.* (1952), Granett *et al.* (1955), Cutkomp and Harvey (1958), and Cheng (1958).

Roberts *et al.* (1963) studied the effectiveness of pyrethrum sprays containing various additives for the control of stable flies (*Stomoxys calcitrans*) and horn flies (*Haematobia irritans*) on cattle. Individual animals were mist sprayed with 2 fl. oz of insecticide solution and exposed in large cages to starved flies from laboratory colonies. Control was determined by the percentage of stable flies that fed and the percentage mortality after 24 hr, and by the time required for complete elimination of horn flies from the animal. Formulations containing 0.02% pyrethrins with various combinations of 0.08% R-11 [2,3,4,5-bis(Δ^2-butylene)tetrahydrofurfural], 0.08% R-326 (di-*n*-propyl isocinchomeronate), 0.62–1% MGK 264 [*N*-(2-ethylhexyl)-5-norbornene-2,3-dicarboximide], or 0.38–1% piperonyl butoxide were less effective against both species than sprays containing 0.1% pyrethrins without additives, which in turn were less effective than 0.1% pyrethrins plus 1% piperonyl butoxide. In another series of tests, formulations containing 0.1% pyrethrins synergized with either 1% piperonyl butoxide, 1% MGK 264, or 0.38% piperonyl butoxide plus 0.62% MGK 264 did not differ significantly in effectiveness; complete elimination of horn flies occurred in 0.8 to 2.75 hr and control of stable flies ranged from 42 to 54% when the tests were begun 8 hr after treatment, but all treatments were considerably less effective after 24 hr. In topical applications against stable flies, however, piperonyl butoxide produced marked synergism whereas MGK 264 did not.

In field trials Wilson *et al.* (1963) obtained 77% control of horseflies on cattle during the peak period of activity by daily applications of a spray containing 0.06% pyrethrins, 0.12% piperonyl butoxide, 0.2% MGK 264 and 0.4% MGK R-1207 (3-chloropropyl octyl sulfoxide) in oil. A treated herd showed a lower rate of infection with anaplasmosis than an untreated herd (19% versus 33%); the species of tabanids present were not mentioned.

Roberts (1962) reported that radiation in the range of 2600 to 4000 Å was principally responsible for the loss in toxicity to biting flies of the pyrethrins in sprays applied to cattle. He found antioxidants to be more effective than UV absorbers in retarding breakdown; lanolin and diamyl hydroquinone were the best protectants in retarding photodecomposition.

B. Control of Ear Ticks

Drummond *et al.* (1967) tested 26 insecticides for the control of the ear tick (*Otobius megnini*) in the ears of naturally-infested cattle. Various formulations were applied to the infested ears, and the cattle were examined after 1 week and again after 1 month. Emulsion containing 0.1% pyrethrins plus 1% piperonyl butoxide gave 95–100% control of larval and nymphal ear ticks after 1 week, as did 24 of the other 25 insecticides. The control after 1 month was quite variable with all insecticides; it was less than 50% with pyrethrins but better than 70% in some tests with, for example, 5% dust or 4% solution of coumaphos, 0.1% emulsion of Dursban, 0.3% emulsion of Ciodrin, and 1% solution of trichlorfon.

C. Louse Powders

During the first years of World War II pyrethrum louse powders played an important part in preventing the spread of lice and typhus and in the suppression of outbreaks when they occurred. Pyrethrum powders were generally superseded by DDT and, when resistance to DDT developed, by lindane and, later, malathion. However, some research with pyrethrins has continued (e.g., Cole *et al.*, 1958; Clark and Cole, 1960), as has evaluation of the synthetic pyrethroids, and pyrethrum powders are still mentioned among the alternative methods of louse control (e.g. 0.2% pyrethrins plus 2% piperonyl butoxide; World Health Organization, 1970). In general, synergized pyrethrins have been shown to be highly effective against lice when first applied but they do not give as much residual protection in the clothing as DDT or malathion.

D. Mosquito Repellents

Weaving and Sylvester (1967) tested various pyrethrum formulations in the laboratory as repellents against *A. aegypti* when applied to the skin of human arms. All formulations also contained piperonyl butoxide at a 1:1 ratio to the pyrethrins. They found that oil-based formulations gave better protection than water- or alcohol-based formulations and that the protection period increased with the concentration. They obtained complete protection for 0 to 2 hr with oil-based formulations containing 1% pyrethrins and for 4-5 hr with 2% pyrethrins.

IV. Development of Resistance to Pyrethrins

Fine reviewed the status of resistance to pyrethroid insecticides in 1963. At that time he reported resistance in field strains of cockroaches in the United States (*Blattella germanica,* 8- to 13-fold), bedbugs in Kenya (*Cimex hemipterus,* 10-fold), houseflies in Sweden (11- to 13-fold), and cattle ticks in South Africa (*Boophilus decoloratus,* 18-fold) and low level resistance in body lice in France. The resistant houseflies showed even higher levels of resistance (25- to 200-fold) to synthetic pyrethroids, such as barthrin, allethrin, cyclethrin, and dimethrin. Resistance had also been developed experimentally in laboratory colonies of houseflies and body lice (*Pediculus h. humanus*). In houseflies, selection for resistance to knockdown with DDT produced resistance to pyrethrins (Busvine, 1953) and in body lice, selection with synergized pyrethrins induced resistance to DDT (Cole and Clark, 1961). Plapp and Hoyer (1967) also found low levels of resistance to synergized pyrethrins in colonies of houseflies selected with DDT or Dilan and DDT.

Farnham (1971) subjected strains of houseflies that were already resistant to DDT and synergized pyrethrins to additional laboratory selection; selection with unsynergized pyrethrins increased the resistance to pyrethrins, resmethrin, and other synthetic pyrethroids to higher levels than did additional selection with resmethrin.

V. Summary

This chapter is concerned with the current status and recent developments in the use of pyrethrum insecticide; research with synthetic pyrethroids is included only incidentally.

Aerosols containing pyrethrum and a synergist, such as piperonyl butoxide, are still a favored method for controlling mosquitos, houseflies, and other flying insects in homes. Formerly these formulations sometimes contained DDT, but new formulations without DDT but with increased concentrations of pyrethrins and synergists are equally effective against nonresistant and DDT-resistant strains. The present International Health Regulations recommend aerosols of this type for use in the disinsection of aircraft on international flights.

Interest is growing in the use of burning pyrethrum coils for protection from mosquitos in rooms and some outdoor areas. Mosquito coils became popular in eastern Asia, and in 1962 more than 2×10^6 kg were exported from Japan. The smoke acts by repelling or expelling flying mosquitos, by inhibiting their avidity to bite, or by causing their knockdown or death. Synergists generally cause increases in the mortality but do not shorten the time to knockdown or increase repellency.

Pyrethrum plus piperonyl butoxide was the most effective of 26 widely used insecticides in laboratory tests against unfed nymphs of the brown dog tick.

Pressurized spray formulations containing pyrethrins and piperonyl butoxide gave good control of houseflies in barns, and emulsions controlled the red mite in poultry houses.

Ultralow volume applications of pyrethrum with piperonyl butoxide or other synergists have been effective for the control of tsetse flies and tabanids when applied from the air at dosages as low as 0.6 gm/acre, and against mosquitos when applied with ground equipment at dosages as low as 0.004 lb/acre. Thermal fogs have also been effective against houseflies in open areas.

Granules containing synergized pyrethrins killed the chironomid larvae in the mud in isolated plots in a tidal cove, but a heavy dosage (2 lb pyrethrins per acre) was required.

Pyrethrum sprays applied directly to the skin of livestock have for several years been a popular method for protecting the animals from attack by biting flies; piperonyl butoxide and other additives usually increase the period of effectiveness, which is about 8 to 24 hr. Pyrethrum formulations applied in the ears of cattle have also been effective against infestations of the ear tick.

During the first years of World War II pyrethrum louse powders played an important part in preventing the spread of lice and typhus. Later, pyrethrum was largely replaced by DDT and other insecticides, but resistance to these newer insecticides continues to spread, and pyrethrum powders are still listed as alternative methods in recommendations by health agencies on vector control.

In laboratory studies with mosquito repellents, formulations containing 1–2% pyrethrins applied to the skin have been effective for several hours against *Aedes aegypti.*

Resistance to pyrethrins or to synergized pyrethrins has been reported in cockroaches, bedbugs, houseflies, cattle ticks and body lice. However, all these species have also developed high levels of resistance to organochlorine, organophosphorus, and carbamate insecticides.

References

Adkins, T. R., Jr., Kissam, J. B., and Krebs, W. F., III. (1971). Field evaluation of natural and synthetic pyrethroid water-based pressurized formulations against the house fly. *J. Econ. Entomol.* **64**, 459–461.

Bremer, H. G. (1964). Pyrethrum reaction aerosol. *J. Econ. Entomol.* **57**, 62–67.

Bruce, W. N., and Decker, G. C. (1951). Tabanid control on dairy and beef cattle with synergized pyrethrins. *J. Econ. Entomol.* **44**, 154–159.

Bruce, W. N., and Decker, G. C. (1955). Factors affecting the performance of treadle sprayers, *J. Econ. Entomol.* **48**, 167–169.

Burnett, G. F. (1961). The susceptibility of tsetse flies to topical applications of insecticides. II. Young adults of *Glossina morsitans* Westw. and organophosphorus compounds, pyrethrins and Sevin. *Bull. Entomol. Res.* **52**, 763–768.

Busvine, J. R. (1953). Forms of insecticide resistance in house flies and lice. *Nature (London)* **171**, 118.

Chadwick, P. R. (1960). Pyrethrum for the control of red mite in poultry houses. *Pyrethrum Post* **5**(3), 3–4.

Chadwick, P. R. (1970). The activity of *dl*-allethrolone *d-trans* chrysanthemate and other pyrethroids in mosquito coils. *Mosquito News* **30**, 162–170.

Cheng, T. (1958). The effect of biting fly control on weight gain in beef cattle. *J. Econ. Entomol.* **51**, 275–278.

Clark, P. H., and Cole, M. M. (1960). Evaluation of synergists for pyrethrum and allethrin against the body louse. *U.S. Dep. Agr., ARS* **ARS 33-62**, 1–63.

Cole, M. M., and Clark, P. H. (1961). Development of resistance to synergized pyrethrins in body lice and cross resistance to DDT. *J. Econ. Entomol.* **54**, 649–651.

Cole, M. M., Clark, P. H., and Weidhaas, D. E. (1958). Sleeve tests with malathion powders against DDT-resistant body lice. *J. Econ. Entomol.* **51**, 741–742.

Cutkomp, L. K., and Harvey, A. L. (1958). The weight responses of beef cattle in relation to control of horn and stable flies. *J. Econ. Entomol.* **51**, 72–75.

Davies, M. S. (1966). Field trials with synergised pyrethrins in the Todd Insecticidal Fog Applicator (TIFA). *Pyrethrum Post* **8**(4), 18–22.

Drummond, R. O., Whetstone, T. M., and Ernst, S. E. (1967). Insecticidal control of the ear tick in the ears of cattle. *J. Econ. Entomol.* **60**, 1021–1025.

Ezell, W. B., Jr. (1971). Biology and control of Tabanidae at the Charles Towne Landing Park, Charleston, South Carolina. Ph.D. Thesis, Clemson University, Clemson, South Carolina.

Fales, J. H., Mills, G. D., Jr., and Durbin, C. G., Jr. (1968). Evaluation of smoke from insecticidal coils against mosquitos. *Mosquito News* **28**, 547–553.

Fales, J. H., Bodenstein, O. F., Waters, R. M., and Fields, E. S. (1970). Development of synergized pyrethrins aerosols for use against public health insects on aircraft and ships. *Aerosol Age*, May 1970, 41–60.

Fales, J. H., Durbin, C. G., Jr., and Mills, G. D., Jr. (1971). Evaluation of smoke from insecticidal coils containing synthetic pyrethroids as mosquito repellents. *Mosquito News* **31**, 35–36.

Farnham, A. W. (1971). Changes in cross-resistance patterns of house flies selected with natural pyrethrins or resmethrin. *Pestic. Sci.* **2**, 138–143.

Fine, B. C. (1963). The present status of resistance to pyrethroid insecticides. *Pyrethrum Post* **7**(2), 18–21 and 27.

Gladney, W. J., Dawkins, C. C., and Drummond, R. O. (1972). Insecticides tested for control of nymphal brown dog ticks by the "tea-bag" technique. *J. Econ. Entomol.* **65**, 174–176.

Goodwin, W. J., Sloan, M. J., and Schwardt, H. H. (1952). Repellency test for horse flies and horn flies in New York State. *J. Econ. Entomol.* **45**, 121–122.

Granett, P., Hansens, E. J., and O'Connor, C. T. (1955). Automatic cattle sprayers for fly control in New Jersey. *J. Econ. Entomol.* **48**, 386–389.

Hitchcock, S. W., and Anderson, J. F. (1968). Field-plot tests with insecticides for control of *Chironomus atrella*. *J. Econ. Entomol.* **61**, 16–19.

Howard, L. O. (1911). Remedies and preventives against mosquitoes. *U.S., Der. Agr., Farmers' Bull.* **444**, 1–15.

Howell, D. E. (1949). Piperonyl formulations as horse fly repellents. *J. Econ. Entomol.* **42**, 401–402.

Kogan, M., (1961). Teste para á determinaçao do grau de repelencia do piretro. *Rev. Brasil. Biol.* **21**, 17–36.

Lee, C. W. (1969). Aerial applications of insecticides for tsetse fly control in East Africa, *Bull. WHO* **41**, 261–268.

Lee, C. W., Irving, N. S., and Parker, J. D. (1968). Aerial applications of pyrethrum aerosol to control tsetse fly. *Pyrethrum Post* **9**(4), 37–40.

Lee, J. (1969). Aerosol insecticide formulations—review of recent developments. *Pyrethrum Post* **10**(2), 9–13.

MacCreary, D. and Goddin, A. H. (1937). Tests of fly sprays by the "one half cow" method. *J. Econ. Entomol.* **30**, 478–482.

McGuire, J. U., Jr., Sullivan, W. N., and Starkweather, R. (1966). Analysis of the effectiveness of aircraft-disinsection aerosols against resistant houseflies and mosquitoes. *Bull. WHO* **34**, 151–154.

Maciver, D. R. (1963). Mosquito coils. Part I. General description of coils, their formulations and manufacture. *Pyrethrum Post* **7**(2), 22–27.

Maciver, D. R. (1964). Mosquito coils. Part II. Studies on the action of mosquito coil smoke on mosquitoes. *Pyrethrum Post* **7**(3), 7–14.

Mount, G. A., and Pierce, N. W. (1972). Adult mosquito kill and droplet size of ultralow volume ground aerosols of insecticides. *Mosquito News* **32**, 354–357.

Nagasawa, S., Katsuda, Y., Okamoto, A., and Ohno, M. (1951). On the knockdown effect of so-called pyrethrins and allethrin coating mosquitocide incense to adults of the common house mosquito (*Culex pipiens* var. *pallens* Coquillett). Studies on the biological assay of insecticides. XVII. *Botyu-Kagaku* **16**, 176–181.

Nagasawa, S., Ohno, M., and Katsuda, Y. (1952). Bioassay of insecticides. XX. Effectiveness of pyrethrum mosquitocidal incense used with piperonyl butoxide. *Botyu-Kagaku* **17**, 47–56.

Nishizawa, Y. (1971). Development of new synthetic pyrethroids. *Bull. WHO* **44**, 325–336.

Paulini, E., and Ricciardi, I. (1954). The repellent effect of pyrethrin vapour to *Culex fatigans*. *Trans. Roy. Soc. Trop. Med. Hyg.* **48**, 276–277.

Plapp, F. W., Jr., and Hoyer, R. F. (1967). Insecticide resistance in the house fly: Resistance spectra and preliminary genetics of resistance in eight strains. *J. Econ. Entomol.* **60**, 768–774.

Roberts, R. H. (1962). The effect of radiation to pyrethrins applied to cattle. *J. Econ. Entomol.* **55,** 851–853.

Roberts, R. H., Harris, R. L., and Graham, O. H. (1963). Effects of additives on the toxicity of pyrethrins to stable flies and horn flies. *J. Econ. Entomol.* **56,** 699–702.

Shepard, H. (1951). "The Chemistry and Action of Insecticides." McGraw-Hill, New York.

Smith, A. and Chadwick, P. R. (1964). A reassessment of pyrethrum against *A. gambiae* with particular reference to its action as a resident repellent. *East Afr. Med. J.* **41,** 137–141.

Smith, A. and Obudho, W. O. (1967). Trials with pyrethrum coils against *A. gambiae* entering a veranda-trap hut. *Pyrethrum Post* **9**(2), 15–17.

Sullivan, W. N., Keiding, J., and Wright, J. W. (1962). Studies on aircraft disinsection at "blocks away." *Bull. WHO* **27,** 263–273.

Sullivan, W. N., Azurin, J. C., Wright, J. W., and Gratz, N. G. (1964). Studies on aircraft disinsection at "blocks away" in tropical areas. *Bull. WHO* **30,** 113–118.

Tarimo, C. S., Lee, C. W., Parker, J. D., and Matechi, H. T. (1970). Aircraft applications of insecticides in East Africa. XIX. A comparison of two sampling techniques for assessing the effectiveness of pyrethrum applications on *Glossina pallidipes* Aust. *Bull. Entomol. Res.* **60,** 221–223.

Tarimo, C. S., Parker, J. D., and Kahumbura, J. M. (1971). Aircraft applications of insecticides in East Africa. XX and XXI. *Pyrethrum Post* **II**(1), 18–23.

Weaving, A. J. S., and Sylvester, N. K. (1967). Pyrethrum as an insect repellent. Part II. A laboratory technique for its evaluation as a mosquito repellent, and the influence of formulation on persistence. *Pyrethrum Post* **9**(1), 31–35.

Westermarck, H. (1967). The use of pyrethrins against red mite in poultry in Finland. *Pyrethrum Post* **9**(1), 10–11.

Wilson, B. H., Burns, E. C., Oglesby, W. T., Myers, R. B., Rogers, T. E., and Wimblery, J. (1963). The effect of horse fly control on rate of infection of bovine anaplasmosis under field conditions in Louisiana. *J. Econ. Entomol.* **56,** 578–579.

Winney, R. (1969). The biological activity of mosquito coils based on pyrethrum and coils based on other active ingredients. *Pyrethrum Post* **10**(1), 3–6.

Winney, R. (1971). The biological activity of mosquito coils with high pyrethrin I content. *Pyrethrum Post* **11**(2), 55–57 and 71.

Winney, R., and Webley, D. J. (1969). The biological activity of mosquito coils of different "pyrethrins" composition. *Pyrethrum Post* **10**(1), 44–48.

World Health Organization. (1970). Insecticide resistance and vector control. Seventeenth report of the WHO expert committee on insecticides. *World Health Organ., Tech. Rep. Ser.* **443,** 1–279.

World Health Organization. (1971). "International Health Regulations (1969)," pp. 79–81. World Health Organ., Geneva.

Chapter 13

Pyrethrum for Control of Household and Stored-Product Insects[1]

H. B. GILLENWATER and G. S. BURDEN

I. Introduction

The use of pyrethrum flowers as an insecticide apparently originated in Persia centuries ago, but the early developers and users were secretive about their product so the exact date of the discovery of the insecticidal activity cannot be established. Pyrethrum powder was introduced into Europe early in the nineteenth century and into the United States about 1860 (Gnadinger, 1936). Today an extract of the pyrethrum flowers, referred to as pyrethrins, is the basic pyrethrum product used as an insecticide.

[1] This paper reports the results of research only. Mention of a pesticide or a commercial or proprietary product in this paper does not constitute a recommendation or an endorsement of this product by the United States Department of Agriculture.

In the United States we probably have more approved[2] methods of utilizing pyrethrins against household and stored-product insects than we have for any other insecticide. Moreover, most of these have been in effect for many years, and some are among the oldest still in effect in the United States. As a result, the efficacy and safety of pyrethrins are time proven. However, some important refinements in methods of application have been developed recently, and promising new concepts of usage have been conceived. These will be discussed in this chapter.

The authors cite only recent literature pertaining to the control of household pests and stored-product insects and make no attempt to include a complete literature review. For specific information about formulations and the many approved methods of using pyrethrins, the reader is referred to the technical bulletins and sample labels that are available.

II. Household Insects

The discussion in this section does not include the dipteran insects commonly associated with households: the use of pyrethrins in their control is discussed in the chapter on insects affecting man and animals (Smith, Chapter 12 in this volume). Also the intent is not to mention specific formulations per se but rather to discuss the known activity of pyrethrins for the control or possible control of insects associated with households.

A. Control Methods

The methods of controlling household insects include the use of approved and recommended contact sprays, residual sprays, dusts, aerosols, fogs, mists, and baits. Many household insects can be controlled by a single type of treatment; however, some are best controlled by combination treatments. For example, in cockroach control the combination of residual sprays and dusts is more effective than either type of treatment alone (Burden and Smittle, 1961). Also, because residues of pyrethrins are not persistent in their toxicity to insects, a combination of a contact spray, aerosol, fog, or mist of pyrethrins or synergized pyrethrins and an insecticide with residual toxicity is desirable in many insect control programs.

[2] Until December 1970 all proposed uses of pesticides in the United States were registered with and subject to approval by the Pesticide Regulation Division of the United States Department of Agriculture. This responsibility is now with the Environmental Protection Agency.

B. Toxicity and Use of Pyrethrins

Before World War II, oil-base sprays with pyrethrins as the primary toxicant were sold widely as contact insecticides for the control of many insects (Mallis, 1969). More recently, laboratory studies using topical applications have demonstrated that pyrethrins are more toxic than allethrin against both resistant (organochlorine) and susceptible strains of the German cockroach, *Blattella germanica* (L.), and against susceptible strains of the American cockroach, *Periplaneta americana* (L.) (Mitlin and Babers, 1955). Also, the laboratory research of Lofgren *et al.* (1958) indicated residues of pyrethrins (0.4%) plus piperonyl butoxide (4.0%) were effective against susceptible common bedbugs, *Cimex lectularius* L. Additionally, sprays containing pyrethrins (0.1%) plus piperonyl butoxide (1.0%) applied against natural populations of bedbugs in buildings controlled all mobile stages of the bugs within 2 days although the rooms were reinfested with newly hatched nymphs after 1 week.

1. Synergists

Many compounds have been considered and evaluated as synergists for pyrethrins. However, in a recent study by Broadbent and Grubber (1971) in which 15 esters of 6-*n*-propylpiperonylic acid were evaluated as synergists, none of the candidates possessed the biological activity of a mixture of pyrethrins–piperonyl butoxide (1:4) against the German cockroach. Also, in laboratory studies with nymphal stages of the brown dog tick, *Rhipicephalus sanguineus* (Latreille), Gladney *et al.* (1972) showed that a formulation of pyrethrins–piperonyl butoxide (1:10) had the lowest LC_{50} (0.000028%) of 26 insecticides tested as emulsion dips.

2. Dusts and Aerosols

Because formulations of micronized dust applied in confined areas may be an efficient and practical means of controlling many insect pests associated with households, a formulation of pyrethrins (2.8%) at a dose of 2 gm per 1000 ft^3 was tested in truck vans. The formulation was effective against susceptible strains of the common bedbug and the oriental flea, *Xenopsylla cheopis* (Rothschild), but was not satisfactory against the German and American cockroaches (Burden, 1972). Burden also reported that aerosol formulations containing 0.45% pyrethrins plus 2.7% sulfoxide or 2.7% Tropital (piperonal bis[2-(2-butoxyethoxy)-ethyl]acetal) dispersed at doses of 30 gm per 1000 ft^3 were highly effective against the common bedbug, the American cockroach, and the oriental rat flea but ineffective against the German cockroach. (These formulations were not effective at doses of 10 gm per 1000 ft^3.)

3. *Thermal Fogs*

Single applications of insecticidal aerosols, fogs, and mists are often made for the control of various household insects. G. S. Burden (unpublished data) therefore used an electric (thermal) fog dispenser to apply oil-base formulations containing either pyrethrins (0.4%) plus piperonyl butoxide (2.0%), malathion (5.0%), or dichlorvos (3.0%) against natural infestations of the German cockroach in individual rooms of infested homes. The formulations were applied at the rate of about 6 fl. oz per 1000 ft^3. The doors of the treated rooms were kept closed, and each house was closed for a period of 2 to 3 hr after treatment. Each treatment was replicated twice, using separate buildings. Before the tests, thorough inspections were made of the infested homes to locate the harborages and to determine the magnitude of the populations. Thus, the effectiveness of the insecticide, expressed as the percentage of reduction in the number of live cockroaches, could be determined by comparing pretreatment and posttreatment counts.

The fog of synergized pyrethrins did not satisfactorily reduce the number of live cockroaches (Table I). The fogs of malathion and dichlorvos produced greater reductions than the synergized pyrethrins, but these reductions were less than 100% 1 day after treatment, and

TABLE I

Evaluation of Insecticidal Fogs Against Natural Infestations of *Blattella germanica*

Formulation	Replicate (no.)	Pretreatment count	Reduction in cockroaches after indicated days (%)		
			1	7	14
Pyrethrins (0.4%) + piperonyl butoxide (2.0%)	1	1042	43	22	35
	2	699	0	0	0
	Average	870	22	11	18
Malathion (5.0%)	1	430	80	55	24
	2	229	16	36	31
	Average	330	48	45	28
Dichlorvos (3.0%)	1	100	76	85	74
	2	92	90	66	64
	Average	96	83	76	69

the surviving populations began to increase within 7 or 14 days. These results indicate that a single treatment with an insecticidal fog alone is not satisfactory for controlling infestations of the German cockroach.

4. Guidelines for Use

Guidelines to the use of pyrethrins or synergized pyrethrins against bedbugs, cockroaches, and fleas in households may be found in such publications as Burden and Smittle (1965), World Health Organization (1970), and United States Department of Agriculture (1971). These publications advise the user of pyrethrins (*1*) that frequent retreatment may be necessary for successful control of bedbugs or fleas in buildings and (2) that the addition of pyrethrins to formulations of insecticides with residual toxicity or the use of a mist spray or aerosol containing pyrethrins will stimulate bedbugs or cockroaches to leave their harborages and contact the surfaces that have been treated with a residual spray or dust. The labels of the various pyrethrins formulations contain instructions for use against other assorted insects that may be occasional or frequent invaders of households.

C. Repellency of Pyrethrins

Consideration must also be given to the fact that residues of pyrethrins are sometimes repellent to various species of insects. For example, Ebeling *et al.* (1967, 1968) showed that pyrethrins (1.0%) were effective against the oriental [*Blatta orientalis* (L)], German, and American cockroaches when continuously exposed to residues. Moreover, the residues were repellent to the cockroaches when they were exposed in choice-box tests. Smittle *et al.* (1968) also showed that a formulation of pyrethrins (0.1%) plus piperonyl butoxide (1.0%) was repellent to the German cockroach, which probably contributed to the ineffectiveness of the compound as a residual toxicant.

D. Resistance

Many household insect pests have developed resistance or show cross-resistance to the organochlorine, organophosphorus, or carbamate insecticides. Species in this category include: the oriental cockroach; the German cockroach; the brown cockroach, *Periplaneta burnnea* (Burmeister); the tropical bedbug, *Cimex hemipterus* (F.); the common bedbug; the dog flea, *Ctenocephalides canis* (Curtis); the cat flea, *Ctenocephalides felis* (Bouché); the human flea, *Pulex irritans* (L.); the brown dog tick;

the rat flea, *Xenopsylla astia* (Rothschild); and the oriental rat flea (World Health Organization, 1970; Brown and Pal, 1971). However, resistance among household insect pests to pyrethrins or synergized pyrethrins has been confirmed in only three species: the German cockroach and the tropical and common bedbugs. Strains of the German cockroach collected in the United States and tested for susceptibility to contact sprays and residues of pyrethrins were shown to have slight to high levels of resistance (Keller *et al.*, 1956). Also, an African strain of bedbugs highly resistant to DDT and dieldrin also showed a ten-fold cross-resistance to pyrethrins, and strains from Israel and Hong Kong showed lower levels of resistance to pyrethrins (Busvine, 1958).

III. Stored-Product Insects

Numerous species of insects, mostly beetles and moths, are known pests of stored foods and feeds. We maintain colonies of about 30 species for research purposes at the Stored-Product Insects Research and Development Laboratory (SPIRDL), Savannah, Georgia. Most of these species are cosmopolitan in distribution, and at least 25 are economically important pests of stored products in the United States.

A. Control Methods

Four general approaches to using pyrethrins to control or prevent infestation of agricultural commodities by stored-product insects are: (*1*) space treatments in which formulations are dispensed into an enclosed area as very small particles that will remain suspended for several hours, (*2*) bulk commodity treatments in which formulations are sprayed on or otherwise admixed with certain raw agricultural commodities as they are placed into storage, (*3*) surface treatments in which formulations are applied to surfaces such as walls and floors of a facility to produce toxic deposits, and (*4*) repellent treatments in which formulations are applied on the exterior surfaces of food packages to minimize penetration.

Many formulations of pyrethrins are suitable for use against stored-product insects. However, the formulations most widely accepted for all four treatments contain pyrethrins synergized with piperonyl butoxide in a 1:10 ratio. This ratio is used in most formulations labeled for post-harvest application to grains and in cereal-product milling and storage facilities. Therefore, all subsequent reference to synergized pyrethrins in this section will be limited to this ratio.

1. Space Treatments

Pyrethrins used as space treatments, referred to collectively as aerosols, consist of such forms as smoke, fog, and mist sprays and are applied by either thermal or mechanical generators. Aerosols are used to control stored-product insects by being applied into the air space of enclosed facilities, such as warehouses, mills, and transport vehicles. Such treatments are directed against flying insects and against any that may be resting or crawling on exposed surfaces. The well-known rapid knockdown and toxicity of synergized pyrethrins to a broad spectrum of insect species and the low toxicity to mammals have made it one of the most widely used insecticides for space treatments. These treatments have been used effectively to prevent or protect against insect infestations but they have afforded little or no control of established infestations.

The two basic formulations of synergized pyrethrins that produce useful aerosols for thermal and mechanical application have the following ingredients (percent by weight): 0.2% pyrethrins, 2.0% piperonyl butoxide, 50.0% tetrachloroethylene, and 47.8% deodorized kerosene for the thermal aerosol and 0.5% pyrethrins, 5.0% piperonyl butoxide, 50.0% tetrachloroethylene, and 44.5% deodorized kerosene for the mechanical aerosol. Each formulation is applied in sufficient volume to achieve a dose of pyrethrins of about 0.3 gm per 1000 ft^3 of space. This dose is effective against many species of stored-product insects, although commercially packaged formulations contain greater or lesser amounts of pyrethrins.

Commercial formulations usually do not contain a flame retardant, such as tetrachloroethylene. Instead, a highly refined petroleum distillate, such as deodorized kerosene, is used as the carrier. When these formulations are applied according to the directions on the label, the concentration of deodorized kerosene produced in the air is well below the flash point, and there is little or no danger of causing a fire. (The kerosene usually provides a more economical formulation than a mixture containing a flame retardant.)

A matter of concern with some types of thermal generators is the high temperature to which the formulation is exposed. Such temperatures have been reported to partially destroy the effectiveness of pyrethrins. For example, Whiting and Bartlett (1963) found as much as a 57% reduction in the effectiveness of thermal aerosols compared with mechanical aerosols against cockroaches. Unpublished research at this Laboratory (SPIRDL) also showed a difference in effectiveness between the two methods of application (Table II). A much larger dose of the

TABLE II

Doses of Aerosols of Synergized Pyrethrins that Killed 90 to 100% of Exposed Insects

Species tested	Dose (gm per 1000 ft³)[a]	
	Mechanical	Thermal
Adults		
Plodia interpunctella	<0.15	<0.15
Rhyzopertha dominica	<0.15	<0.15
Sitophilus zeamais	0.15	0.15
Lasioderma serricorne	0.15	0.15
Tribolium confusum	0.50	0.75
Larvae		
Plodia interpunctella	0.75	0.75
Attagenus megatoma	0.35	0.50
Trogoderma inclusum	0.50	0.50

[a] Aerosols contained 0.5% pyrethrins + 5.0% piperonyl butoxide mixed with methylene chloride and deodorized kerosene (1:1 v/v).

synergized pyrethrins was needed to kill adults of the confused flour beetle, *Tribolium confusum* (Jacquelin duVal), (50%) and larvae of the black carpet beetle, *Attagenus megatoma* (F.), (43%) with thermal than mechanical application. The tests reported in Table II were not intended to establish LD values for the species tested. However, the data show that only 0.15 gm per 1000 ft^3 was an adequate dose for adults of several of the test species. To kill adults of the confused flour beetles, however, a dose of thermal aerosol of 0.75 gm per 1000 ft^3 was needed. Studies by Yerington (1971) with thermal aerosols showed that a dose of pyrethrins of about 0.07 gm per 1000 ft^3 gave satisfactory control of *Drosophila melanogaster* (Meigen) (vinegar flies) in wine cellars. Obviously, the susceptibility of species of insects to synergized pyrethrins varies greatly. Therefore, as flour beetles are a common pest of stored products, their low susceptibility to synergized pyrethrins may explain some reports of ineffectiveness with synergized pyrethrins space treatments. Moreover, most infestations of stored products involve a complex of insect species so the dose applied as a space treatment must be directed against the least susceptible species if the treatment is to be effective.

The ultralow volume (ULV) application of insecticide is a new, promising approach to the use of synergized pyrethrins as a space treatment. These applicators are capable of producing small droplets of a formulation within a narrow range of sizes, so highly efficient distribution and greater efficacy can be achieved with less insecticide. The utility of the method has been adequately demonstrated with other insecticides in the field, and the National Pest Control Association (1972) has a preliminary report on the method for indoor use. This report does not specifically recommend the ULV method for use against stored-product insects; however, it appears that the method would be highly effective in such space treatments.

2. *Bulk Commodity Treatments*

Bulk commodity treatments are applied to certain nonperishable raw commodities, such as stored grains and oil seeds, to protect them from insects. The formulations may consist of either oil- or water-base sprays or dusts, but the water-base sprays are most common. Whether it is a spray or dust, it is usually applied to the stream of grain as it moves into storage on a conveyor system.

For many years, synergized pyrethrins were the only insecticide approved for postharvest application to or admixing with any nonperishable raw agricultural product. However, during the past 10 to 12 years, malathion has been used rather extensively. Now, with the development of resistance to malathion in several species of insects (Parkin, 1965; Speirs *et al.*, 1967), a return to more extensive use of synergized pyrethrins appears imminent.

The formulations of pyrethrins and the doses that are effective for bulk treatment of commodities depend largely on the product and insects involved; the amounts of pyrethrins applied to most products do not exceed 3 ppm. In the United States, the tolerances for residues of synergized pyrethrins on any commodity do not exceed 3 ppm for the pyrethrins and 20 ppm for the piperonyl butoxide.

Dove and Schroeder (1955) showed the value of water-miscible emulsions of synergized pyrethrins for the protection of stored grains in laboratory tests. Walkden and Nelson (1959) demonstrated that both dust and spray formulations of synergized pyrethrins applied at rates ranging from 1.5 to 2.5 ppm were effective protectants for wheat and corn in bulk storage. La Hue (1966) reaffirmed that a water-base spray of synergized pyrethrins applied at 2.75 ppm was an effective protectant for shelled corn. Also, most researchers agree that the toxicity of synergized pyrethrins applied to grain is of short duration and that the extended period of protection (9 to 12 months) is rendered by its repellent action.

Several United States Department of Agriculture publications give detailed information about bulk treatment of specific commodities. For example, *Marketing Bulletin No. 20* (Agricultural Marketing Service, 1962a) describes methods of application and equipment (which are basically the same for most products) for treatment of grain; *Agricultural Handbook No. 129* (Agricultural Research Service, 1971) gives complete information about the control of insects in stored rice; and *AMS-453* (Agricultural Marketing Service, 1961) describes methods of controlling insects in farmers stock peanuts. Although the formulations and doses vary, depending on the specific product, there is little danger of producing excessive residues if applications conform to label instructions. However, only a formulation that is properly labeled for the intended use should be applied.

3. *Surface Treatments*

Surface treatments with pyrethrins can be used effectively to protect insect-free commodities against infestation or to supplement other treatments in the control of insect infestations. For example, during the preparation of grain bins, elevators, and warehouses for storage, all exposed and accessible surfaces of the facilities may be sprayed to eliminate insects not removed by routine sanitation procedures. Similar treatments in transit vehicles as railroad boxcars, truck vans, and ships' holds before a commodity is loaded will likewise provide protection against infestation in transit.

Also, sprays or dusts can be applied to the surface of stored grains as a supplement to other control methods, a procedure that is particularly beneficial when it is used as a "top dressing" after a bulk treatment of a commodity or is used to deter insects entering a product after fumigation. However, surface treatments of a commodity will not control infestations already established below the surface.

In the same way, sprays of synergized pyrethrins may be applied to the exterior surface of bags containing farmers stock peanuts and dried citrus pulp animal feed after these items have been bagged in burlap and stacked in storage. The treatment will aid in preventing insect infestation of these products.

Finally, surface treatments of synergized pyrethrins may be used in or on certain processing equipment and facilities. For example, sprays of an oil solution may be applied to milling machinery, and either oil- or water-base sprays may be applied to certain other surfaces in flour mills. These applications are most useful as a followup to routine sanitation procedures. Also, surface treatments with sprays and dusts may be used in plants where dry milk is processed but not in the machinery. Care must be exercised to avoid any residues in the product.

The formulations of synergized pyrethrins used for surface treatment should be applied to produce a deposit of about 5 mg of pyrethrins and 50 mg of piperonyl butoxide per square foot of surface. Close adherence to this rate is not critical unless there is a possibility of depositing excessive residues on a commodity. Although the formulation and the rate and frequency of application will vary for different commodities, if surface treatments are made according to label instructions, there is little or no possibility of depositing excessive residues on a commodity.

Methods and techniques for utilizing surface treatments are described in *USDA Leaflet No. 345* (United States Department of Agriculture, 1966) for farm-stored wheat, in the revised *AMS-302* (Agricultural Marketing Service, 1962b) for dry milk processing plants, by Laudani *et al.* (1959) for dried citrus pulp animal feed, and by Chapman and Schmitt (1959) for stored grains. Most publications that describe methods of controlling stored-product insects for a specific commodity will also include information, if applicable, about surface treatments.

4. Repellent Treatments

The repellency of synergized pyrethrins to insects has been recognized for many years and is a major factor in the success of pyrethrins in protecting stored commodities from insect attack. Laudani and Davis (1955) showed the value of repellency versus toxicity for protecting flour in paper bags treated with synergized pyrethrins. Further studies by Davis and Laudani (1956) and Highland *et al.* (1964, 1966) led to the development of an insect-resistant treatment (IRT) for multiwall paper bags. As a result, the United States Armed Forces now use IRT multiwall paper bags for packaging flour and other cereal products intended for long-term storage and ocean shipment. Also, large quantities of IRT bags are used to protect dry cereal products for foreign aid programs from insect infestation when they are shipped overseas and stored.

However, the commercial use of IRT bags for domestic packaging has been relatively limited because of the greater costs and the limitation of the treatment to 50-lb-size bags. The cost of the IRT is from 3 to 5¢ per bag more than for untreated bags, which is an important factor in the competitive marketing of products (H. A. Highland, SPIRDL, personal communication). Although the treatment is safe and effective, researchers have continued efforts to develop barriers (Highland *et al.*, 1970) and formulations (Langbridge, 1970) that will prevent migration of the IRT into the packaged product and extend its period of effectiveness.

Until 1971, the IRT could not be used on bags containing less than 50 lb of a product. However, studies by Yerington (1972) showed that the IRT was safe and effective for prunes packaged in 2-lb units and a use label for this treatment has been approved. Therefore, the IRT has potential for further development as a method of protecting many other types of products packaged in small containers.

The effectiveness and further potential of the IRT bag for protecting commodities against insects were emphasized by the results of studies conducted at the request of the United States Armed Forces Pest Control Board and completed in 1970 at the SPIRDL. Because of insect infestations, the Armed Forces in Southeast Asia were losing large quantities of flour and other cereal products that had been shipped and stored in plywood overpacks. The overpacks were cubicles of about 80 ft^3 and each one contained 48 bags of flour in 50-lb IRT bags. The overpacks were tightly constructed to minimize pilferage of the flour and physical damage to the bags; therefore, inspections to determine the extent of insect infestation could not be made without destroying their integrity. The attempts to find a method of preventing insect infestation of the flour included both the use of residual insecticidal sprays on the exterior surface and internal treatments of the overpacks with insecticidal vapor or inert dust. The effectiveness of each treatment was assessed at 3-month intervals for 12 months. Regardless of the treatment, insects entered the overpacks and infested several bags of flour in each overpack. However, none of the bags, whether infested or not, were actually penetrated by insects, and each infested bag was physically damaged or had been imperfectly sealed. Disregarding the effects of the various treatments (except the IRT), we therefore made the following analysis of data from inspections of 396 bags of flour in 22 overpacks stored 12 months in an infested warehouse:

1. Number of insect penetrations	0
2. Percentage of bags infested	15.2
3. Percentage of bags with imperfect seals	13.4
4. Percentage of bags physically damaged	3.3

We thus demonstrated the value of the IRT bag as a food container and the importance of properly sealing the bag to exclude invading insects. A sound multiwall IRT bag filled with an insect-free product that is properly sealed will remain insect free during 9 to 12 months of storage.

During the same 12-month study, the degradation of the IRT on the bags and the migration of the treatment from the bags into the flour stored both inside the overpacks and on open pallets were also assessed. From the analyses for piperonyl butoxide on the bags, storage in the plywood overpacks slowed the degradation of the IRT (Table III),

TABLE III

Residues of Piperonyl Butoxide from the Insect-Resistant Treatment (IRT) of Flour Bags[a]

Months in storage	Sample location in stack	Residue on			
		Paper bags (mg/ft^2)		Flour (ppm)	
		Inside overpack	Outside overpack	Inside overpack	Outside overpack
0	Random[b]	27.3	—	0.25	—
3	Top	23.9	27.7	0.19	0.94
	Middle	27.7	26.4	0.37	0.22
	Bottom	28.4	26.1	0.32	0.25
6	Top	30.9	15.9	0.94	0.55
	Middle	34.2	26.5	0.66	1.39
	Bottom	29.4	28.0	1.23	1.22
9	Top	23.7	8.5	2.92	1.05
	Middle	22.2	17.9	2.01	1.42
	Bottom	19.5	21.1	2.02	1.56
12	Top	21.5	7.9	1.04	0.90
	Middle	20.0	10.3	1.96	1.29
	Bottom	20.5	11.3	1.79	1.23

[a] Data based on four samples from each location; IRT applied September 1968 and random samples collected December 1968.

[b] Average of eight samples.

and, as expected, the greatest amount of degradation occurred among the top bags stored on open pallets. Also, after 12 months of storage, about twice as much piperonyl butoxide was present on the bags stored in the overpacks as on bags stored on open pallets. However, greater amounts of piperonyl butoxide migrated into the flour stored in overpacks than into that stored on open pallets although all residues in the flour were well below the established tolerance of 10 ppm. In fact, 2.92 ppm was the highest residue found in the flour.

For many years, synergized pyrethrins have been used at the SPIRDL as a standard for comparison in a continuing program of evaluation of new compounds as repellents. *Marketing Research Report No. 324* (Agricultural Marketing Service, 1959) and *Supplement I* (Agricultural Marketing Service, 1964) show that 49 compounds evaluated in this program were as repellent as or more repellent to flour beetles, *Tribolium* spp., than the standard. In addition, unpublished data collected during the last 5 years show that 65 other compounds had a similar potential. However, followup studies on this potential are incomplete. Therefore, continued high priority should be given to the development of the repellent property of pyrethrins.

B. Resistance

There are relatively few reports of resistance to pyrethrins among stored-product insects. However, by selection pressure, Lloyd and Parkin (1963) developed a strain of the granary weevil, *Sitophilus granarius* (L.), that showed a 52-fold resistance to pyrethrins in the twenty-ninth generation, but this strain showed only a 5.5-fold resistance to synergized pyrethrins. Also, Speirs and Zettler (1969) reported a 13.3-fold resistance to pyrethrins in a field-collected strain of red flour beetles, *Tribolium castaneum* (Herbst), and J. Larry Zettler and L. L. McDonald (personal communication) of SPIRDL found a low level of resistance to synergized pyrethrins in the Indian meal moth, *Plodia interpunctella* (Hübner) (2.5 fold); the almond moth, *Cadra cautella* (Walker) (2.0-fold); and the merchant grain beetle, *Oryzaephilus mercator* (Fauvel) (7.1-fold). That resistance to pyrethrins among stored-product insects only occurs at low levels or not at all may be "due merely to its relatively feeble selective powers in the absence of a long residual insecticidal action," as Busvine (1960) stated. Also, the repellency of pyrethrins, which results in minimum exposure of insects to residues, undoubtedly plays a part. However, the lack of extensive research to determine resistance to pyrethrins among stored-product insects may also be a factor. A global survey of resistance to insecticides and fumigants among selected species of stored-product insects is currently being conducted under the auspices of the Food and Agriculture Organization of the United Nations (B. R. Champ, CSIRO, Canberra, Australia, personal communication), but unfortunately the survey does not specifically include studies of resistance to pyrethrins. Perhaps the need for such studies will be demonstrated during this survey.

C. Tolerances

Tolerances for residues of synergized pyrethrins have been established for a host of nonperishable, raw agricultural commodities and for several processed dried foods and animal feeds. The establishment of a tolerance for residue on a product is not a license to indiscriminately apply an insecticide, even when the application is made at levels within the tolerance. One must also comply with specific rules and regulations that are designed to promote the safe use of the insecticide. There are many regulations covering the various uses of pyrethrins and piperonyl butoxide; however, one should be thoroughly informed about the methods and techniques before applying insecticide on, in, or around food commodities.

IV. Summary

Formulations of pyrethrins or synergized pyrethrins have usefulness in control programs directed against insect pests associated with households and food commodity storage and processing facilities. Both research and many years of practical experience have shown no adverse effects caused by the use of pyrethrins as an insecticide in situations that are so unusually close to man. However, the reader is reminded that for the safe use of any insecticide, he must understand and follow label instructions. Also, for satisfactory results in controlling insects, the differences and limitations of the activity of pyrethrins (direct-contact, residual toxicity, and repellency) must be recognized. Low levels of resistance to pyrethrins have been shown to occur in several insect species, but resistance has not been a serious problem, although the possibility exists that important levels will be found in untested species, especially in geographical areas that have not been surveyed.

References

Agricultural Marketing Service. (1959). Laboratory evaluation of promising compounds as repellents to flour beetles, *Tribolium* spp. *U.S., Agr. Mkt. Serv.* [*AMS Ser.*], *Res. Rep.* **324**, 1–46.

Agricultural Marketing Service. (1961). Insect prevention and control in farmers stock peanuts. *U.S., Agr. Mkt. Serv.* [*AMS Ser.*] **453**, 1–12.

Agricultural Marketing Service. (1962a). Methods and equipment for bulk treatment of grain against insects. *U.S., Agr. Mkt. Serv.* [*AMS Ser.*], *Bull.* **20**, 1–7.

Agricultural Marketing Service. (1962b). Insect prevention and control in plants processing dry milk. *U.S., Agr. Mkt. Serv.* [*AMS Ser.*] **AMS-302**, 1–12 (revised).

Agricultural Marketing Service. (1964). Laboratory evaluation of promising compounds as repellents to flour beetles, *Tribolium* spp. *U.S., Agr. Mkt. Serv.* [*AMS Ser.*], *Res. Rep.* **324**, Suppl. I, 1–46.

Agricultural Research Service. (1971). Controlling insect pests of stored rice. *U.S., Dep. Agr., Agr. Handb.* **129**, 1–19.

Broadbent, D. J., and Grubber, M. J. (1971). The synergistic abilities of esters of 6-*n*-propylpiperonylic acid with pyrethrins and neopynamin. *Pyrethrum Post* **11**(2), 63–65 and 68.

Brown, A. W. A., and Pal, R. (1971). "Insecticide Resistance in Arthropods," 2nd ed., Monogr. Ser. No. 38. World Health Organization, Geneva.

Burden, G. S. (1972). Gas-propelled aerosols and micronized dusts for the control of insects in enclosed areas including aircraft and vans. Part 6. Insects of medical importance. *J. Econ. Entomol.* **65**, 1458–1462.

Burden, G. S., and Smittle, B. J. (1961). New and current insecticides for German cockroach control. *Pest Control* **29**(6), 30.

Burden, G. S., and Smittle, B. J. (1965). Household insects. *In* "Consumers All, The Yearbook of Agriculture," pp. 187–193. U.S. Dept. Agr., Washington, D.C.

Busvine, J. R. (1958). Insecticide-resistance in bed bugs. *Bull. WHO* **19**, 1041–1052.

Busvine, J. R. (1960). Resistance to pyrethrins. *Bull. WHO* **22**, 592–593.

Chapman, H. C., and Schmitt, J. B. (1959). Stored-grain insects and their control in the Northeastern States. *N.J. Agr. Bull.* **791**, 1–16.

Davis, D. F., and Laudani, H. (1956). Long-term insecticide tests. *Mod. Packag.* **29**, 236–240, 332, 334, 337, and 338.

Dove, W. E., and Schroeder, H. O. (1955). Protection of stored grain with sprays of pyrethrins-piperonyl butoxide emulsion. *J. Agr. Food Chem.* **3**, 932–936.

Ebeling, W., Reierson, D. A., and Wagner, R. E. (1967). Influence of repellency on the efficacy of Blatticides. II. Laboratory experiments with German cockroaches. *J. Econ. Entomol.* **60**, 1375–1390.

Ebeling, W., Reierson, D. A., and Wagner, R. E. (1968). Influence of repellency on the efficacy of Blatticides. IV. Comparison of four cockroach species. *J. Econ. Entomol.* **61**, 1213–1219.

Gladney, W. J., Dawkins, C. C., and Drummond, R. O. (1972). Insecticides tested for control of nymphal brown dog ticks by the "tea-bag" technique, *J. Econ. Entomol.* **65**, 174–176.

Gnadinger, C. B. (1936). "Pyrethrum Flowers," 2nd ed. McLaughlin Gormley King, Minneapolis, Minnesota.

Highland, H. A., Davis, D. F., and Marzke, F. O. (1964). Insect-proofing multiwall bags. *Mod. Packag.* **37**, 133, 134, 136–138, and 195.

Highland, H. A., Jay, E. G., Phillips, M., and Davis, D. F. (1966). The migration of piperonyl butoxide from treated multiwall kraft bags into four commodities. *J. Econ. Entomol.* **59**, 543–545.

Highland, H. A., Secreast, M., and Merritt, P. H. (1970). Packaging materials as barriers to piperonyl butoxide migration. *J. Econ. Entomol.* **63**, 7–10.

Keller, J. C., Clark, P. H., and Lofgren, C. S. (1956). Susceptibility of insecticide-resistant cockroaches to pyrethrins. *Pest Contr.* **24**(11), 14–15, and 30.

La Hue, D. W. (1966). Evaluation of malathion, synergized pyrethrum, and diatomaceous earth on shelled corn as protectants against insects—in small bins. *U.S., Agr. Mkt. Serv. [AMS Ser.], Res. Rep.* **768**, 1–10.

Langbridge, D. M. (1970). Treatment of paper to protect packaged food from insect attack. *Appita* **24**, 45–51.

Laudani, H., and Davis, D. F. (1955). The status of federal research on the development of insect-resistant packages. *Tappi* **38**, 322–326.

Laudani, H., Gillenwater, H. B., Kantack, B. H., and Phillips, M. (1959). Protection of citrus pulp against insect infestation with surface applications of pyrethrum-piperonyl butoxide wettable powder. *J. Econ. Entomol.* **52**, 224–227.

Lloyd, C. J., and Parkin, E. A. (1963). Further studies on a pyrethrum-resistant strain of the granary weevil, *Sitophilus granarius* (L.). *J. Sci. Food Agr.* **9**, 655–663.

Lofgren, C. S., Keller, J. C., and Burden, G. S. (1958). Resistance tests with the bed bug and evaluation of insecticides for its control. *J. Econ. Entomol.* **51**, 241–244.

Mallis, A. (1969). "Handbook of Pest Control." McNair-Dorland Co., New York.

Mitlin, N., and Babers, F. H. (1955). Relative toxicity of topically applied allethrin and pyrethrins to house flies and cockroaches. *J. Econ. Entomol.* **48**, 747–748.

National Pest Control Association. (1972). "ULV Insecticide Application Indoors," Tech. Release No. 10–72. Nat. Pest Contr. Ass., 1–6.

Parkin, E. A. (1965). The onset of insecticide resistance among field populations of stored-product insects. *J. Stored Prod. Res.* **1**, 3–8.

Smittle, B. J., Burden, G. S., and Banks, W. A. (1968). Cockroach insecticides—how repellent are they? *Pest Contr.* **36**(11), 9–10.

Speirs, R. D., and Zettler, J. L. (1969). Toxicity of three organophosphorus compounds and pyrethrins to malathion-resistant *Tribolium castaneum* (Herbst) (Coleoptera, Tenebrionidae). *J. Stored Prod. Res.* **4**, 279–283.

Speirs, R. D., Redlinger, L. M., and Boles, H. P. (1967). Malathion resistance in the red flour beetle. *J. Econ. Entomol.* **60**, 1373–1374.

U.S. Department of Agriculture. (1966). Insects in farm-stored wheat; how to control them. *U.S. Dep. Agr. Leafl.* **345**, 1–8.

U.S. Department of Agriculture. (1971). Controlling household pests. *U.S., Dep. Agr., Home Gard. Bull.* **96**, 1–32.

Walkden, H. H., and Nelson, H. D. (1959). Evaluation of synergized pyrethrum for the protection of stored wheat and shelled corn from insect attack. *U.S., Agr. Mkt. Serv.* [*AMS Ser.*], *Res. Rep.* **322**, 1–48.

Whiting, R. W., and Bartlett, P. G. (1963). High temperatures cut effectiveness. *Food Process.* August, 1–3.

World Health Organization. (1970). Insecticide resistance and vector control. *World Health Organ., Tech. Rep. Ser.* **443**, 1–280.

Yerington, A. P. (1971). Evaluation of pyrethrins thermal aerosols for control of vinegar flies and dried fruit beetles in wine cellars. *J. Econ. Entomol.* **64**, 986.

Yerington, A. P. (1972). New studies on insect-resistant packages. *Proc. Dried Fruit Tree Nut Ind. Res. Conf., 1972*, Fresno, California.

Chapter 14

Pyrethrum for Control of Agricultural Insects

W. HARRY LANGE and NORMAN B. AKESSON

I. Introduction

With the present controversy over the use of large quantities of persistent or "hard" insecticides with resultant problems of toxic residues, resistance, resurgence of new insect pests, contamination of the agroecosystem, possible contamination of air, water, and soil, and many other problems, it appears that one of our oldest insecticides, pyrethrum, can make a comeback in the agricultural field. Perhaps the greatest possibility of its expanded use is in the fields of vegetable insect control and for the control of forest pests. Some of these newer uses have been pointed out by Bogaard (1969) and new developments in forest insect problems have been outlined by Moore (1969) and Parker (1972).

In an attempt to replace the more persistent types of insecticides, there are two lines of research being carried out at the present time: (*1*) that concerned with developing new types of materials, such as hormone analogs or pyrethrin analogs, and (2) research aimed at getting better control out of existing types of insecticides, particularly in regard to formulation and method of application. Hagley (1959) has pointed out that the physical and chemical properties of pyrethrum, together with its biological characteristics, have resulted in its widespread use in certain fields of applied entomology. Certainly its low mammalian toxicity, its nonpersistence, and its wide spectrum of activity on many insects make it a material that can be used with safety in many situations.

In recent years pyrethrum has not been used widely in the agricultural field, probably because of its high cost and the fact that frequent applications have to be made because of its nonpersistent characteristics. It has been more widely used for the protection of stored food and food products and in the control of flies, mosquitos, and other insects of public health importance.

II. Historical Review

Perhaps one of the first advocates of the agricultural and horticultural use of pyrethrum in this country was C. V. Riley (1881, 1881–1882, 1885), who reported on the agricultural and horticultural possibilities of this product. In 1881, for example, he wrote a paper on "Safe Remedies for Cabbage Worms and Potato Beetles." Although pyrethrum was grown on a large scale in 1828, Riley says that the first man to point out its agricultural possibilities was a Mr. C. Willemot in 1857. It is known that a Mr. G. N. Milco, a native of Dalmatia, cultivated *Chrysanthemum cinerariaefolium* near Stockton, California.

During the 1930's, up until the advent of DDT, pyrethrum was used as a dust or spray for the control of many insects and impregnated Pyrocide dust (McLaughlin Gormley King Company) was widely used for insects on alfalfa, artichokes, asparagus, beans, beets, cabbage, cantaloupes, cauliflower, celery, corn, cranberries, cucumbers, grapes, melons, peppers, potatoes, spinach, squash, and sugar beets, and it was even recommended for ants, box elder bugs, and gypsy moths. Essig and Hoskins (1934, rev. 1944) recommended pyrethrum preparations for tulip bulb mite, iris root aphid, black citrus aphid, pea aphid, mealy plum aphid, potato aphid, grape leafhopper, leafhoppers on cotton, western plant bug, cotton dauber, squash bugs on melons, glacial whitefly on bush fruits, grape thrips, bean thrips, pear thrips on plum and prune, thrips on several other crops, artichoke plume moth, false plume moth on egg plant, cabbage worms, tomato fruit worm on corn and other crops, grape leafroller, alfalfa semilooper on lettuce, western parsley caterpillar, red-humped caterpillar, western spotted cucumber beetles, and many other insects.

With the advent of World War II and a shortage in the supply of pyrethrum, plus the fact that DDT and more potent insecticides became available, the use of pyrethrum on agricultural and horticultural crops became more limited. In recent years, we have been interested in ultralow volume (ULV) types of application and also better types of formulations. Ultralow volume applications are presently being carried out by the authors on artichokes, lettuce, and grapes in cooperation with Norman F. McCalley, Farm Advisor at Salinas, California.

Investigations by Miskus and Andrews (1971, 1972) have resulted in the stabilization of thin films of pyrethrins using a combination of an antioxidant and an ultraviolet screening agent in a mineral oil formulation. The stabilization of pyrethroid compositions makes it advisable to reexamine pyrethrins for agricultural uses with the hope of getting longer-lasting effects than are presently possible.

III. Effects on Bees and the Natural Enemies of Harmful Insects

In any integrated control program it is necessary that a material be safe on bees and also not disrupt the natural and introduced parasite and predator complex. The investigations of Anderson and Atkins (1958) indicate that pyrethrins are relatively nontoxic to bees. Ginsburg (1930) found that honeybees are not affected by the volatile material emanating from pyrethrum. In some laboratory studies, Bartlett (1958) found that pyrethrins at higher rates killed adult ladybird beetles but did not affect the larvae. This would indicate that although pyrethrins do have adverse effects on certain natural enemies the effects are probably fairly selective on different stages, and being nonpersistent, the effects would not be long lasting.

IV. Insects Controlled

A. Vegetable and Field Crops

One of the first investigators to point out the value of pyrethrins for tomato insect control was Thomas (1936a,b), who reported on the control of the tomato pinworm in Pennsylvania. *Drosophila* spp., often called vinegar flies, frequently pose a serious problem for the tomato canning industry. The advent of mechanical harvest types of tomatoes in some areas has reduced the threat of this particular group of insects, but pyrethrins are still commonly used in most areas. The initial work by Collins (1956) was followed by Bickley *et al.* (1956) and then an excellent publication by Stombler *et al.* (1957). This research combined with investigations by the National Canners Association resulted in the use of a 0.11% stabilized pyrethrins or 0.08% stabilized pyrethrins plus 1.1% piperonyl butoxide as a postharvest control measure. This treatment is still widely used. Ebeling (1958) pointed out that vinegar flies have been steadily increasing in importance as pests in California fruit crops, and in addition to tomatos, figs and Thompson seedless grapes for raisins are also affected. A summary of the use of pyrethrum in the tomato canning industry in the United States is given by Watts (1962).

Pyrethrins have long been effective for the control of leafhoppers on potatos and an interesting note was published by Mader and Udey (1937) who found that pyrethrins dust acted as a stimulant to the growth of potatos which apparently was not correlated with insect control. The work of DeLong (1934) demonstrated the value of pyrethrum in reducing populations of the potato leafhopper, *Empoasca fabae* (Harris). Allen, in 1936, demonstrated that pyrethrum was quite specific in its toxicity as a kerosene steepate to leafhoppers whereas rotenone was specific to flea beetles. The work of Skaptason (1938) demonstrated the value of a pyrethrum–rotenone–sulfur combination dust. Prior to the introduction of DDT, a pyrethrum-in-oil spray containing 0.08% pyrethrins was used to control the beet leafhopper (Hills *et al.*, 1948). Michelbacher and his associates (1955) were among the first to point out the fact that DDT increases spider mites and the melon aphid on melons and therefore suggested the use of pyrethrum as an integrated control concept.

Pyrethrum has been used for many years in cabbage insect control as it is particularly potent on the cabbage looper. As early as 1934, Walker and Anderson showed that pyrethrum was very effective on the cabbage looper. Huckett (1936, 1937) demonstrated the value of pyrethrum and also combinations with rotenone and nicotine for the control of cabbage worms in New York. However, not all cabbage insects are controlled with pyrethrum. The value of auxiliary materials in rotenone and pyrethrum insecticides for the control of cabbage caterpillars was evaluated by Reid and Cuthbert (1952), who found that auxiliary materials were most useful in pyrethrum insecticides in the control of imported cabbage worm and diamondback moth. The impregnated pyrethrum dust usually proved more effective against the caterpillars than ordinary dust mixtures of the same pyrethrins content, but they were more difficult to blend and apply. The auxiliary materials that proved of significant value were mineral oil and sulfur, piperonyl cyclonene, and an organic thiocyanate.

Celery leaf tier is controlled by pyrethrum as shown by the work of Wisecup (1936) and White (1945). Wisecup, using the celery leaf tier as a test insect, found that undiluted pyrethrum would be effective when it was stored as long as 3 years in closed containers, but definite deterioration could be expected when stored in open containers.

On artichokes, Lange (1941) found that pyrethrins were not effective in the control of the artichoke plume moth as contrasted with rotenone, which gave some degree of control. However, pyrethrins were very effective in the control of the artichoke aphid. The work of Walton and Howell (1954) indicated that pyrethrins were ineffective for the control of aphids on spinach and commercial greens. In 1942, Smith recommended pyrethrum for the control of Say's stink bug and for the control of the western twelve-spotted cucumber beetle, when used with 1% aliphatic thiocyanate. The work of Pepper (1941) demonstrated the value of a pyrethrum-oil mixture in the control of the corn earworm on sweet corn. The materials were applied to the silks by means of oilers. The virus disease known as lettuce yellows is transmitted by the six-spotted leafhopper. Pepper and Haenseler (1939) demonstrated the value of pyrethrum dusts in the control of this particular insect.

B. Forage Crops

On alfalfa, Sorenson (1939, 1942) demonstrated that although pyrethrum was satisfactory in cage tests in killing *Lygus* bugs, it was not effective under field conditions.

C. Fruits and Grapes

The investigations of Madsen and McNelly (1961) showed that pyrethrum dusts gave good control of the western twelve-spotted cucumber beetle on apricots. The beetles attack the maturing fruit so persistent insecticides cannot be used at this particular time. On avocados, Ebeling and Pence in 1953 demonstrated the value of oil and pyrethrum in the control of greenhouse whitefly. For the control of pear psylla, Madsen and Westigard in 1963 found that the addition of pyrethrins to oil offered little extra effect. Several species of Lygus damaging to peaches in the Pacific Northwest were adequately controlled with Pyrocide dust. In the work three applications of 0.2% pyrethrins were used and the resultant crop was 98.3% free of *Lygus* injury.

Gnadinger *et al.* (1940) carried out a series of experiments with pyrethrum for the control of codling moth including the use of a pyrethrins–oil spray for controlling overwintering larvae and the use of pyrethrum dust in conjunction with oil sprays for the control of adult moths, eggs, newly hatched larvae, and mature larvae. The efficacy of sprays to kill overwintering codling moth larvae was also tested in some work by Yothers *et al.* (1942), who demonstrated that pyrethrum would kill the hibernating codling moth larvae in cocoons on the trees. Pyrethrum in this instance was shown to be a supplemental control measure.

Bailey (1938) successfully used pyrethrum to control several species of thrips on many fruit crops, including grape thrips, flower thrips, and citrus thrips. The western twelve-spotted cucumber beetle in orchards was controlled in tests carried out by Michelbacher *et al.* (1941), who found that the most effective dusts used contained pyrethrum and, particularly, a material like Lethane. They found that the dusts should be applied at the rate of 50 lb/acre at temperatures not higher than 63°F and that two dustings were usually necessary. The grape leafhopper was controlled on grapes by Lamiman in 1933. The cranberry fruit worm on blueberries was controlled by pyrethrum in some work carried on by Beckwith in 1941.

V. Formulations

The field experiments of Warner (1963) demonstrated the value of piperonyl butoxide in extending the period of effectiveness of aqueous pyrethrum formulations.

The importance of suitable pyrethrum formulations is demonstrated by some of the recent work in forest insect control (Miskus and Lyon, this volume, Chapter 15) and by the stabilized pyrethrum formulation developed by Miskus and Andrews (1971, 1972).

From the standpoint of agricultural use, the addition of various synergists is still somewhat of a matter of debate inasmuch as there are many environmental conditions that also influence the effectiveness of pyrethrum. For example, Harries *et al.* (1945) discussed some of the factors affecting the insecticidal action of pyrethrum extracts on the beet leafhopper. Their tests indicated a marked increase in insecticidal effect at lower temperatures and a similar effect in the case of sprays. From a practical standpoint, they found that the best results in control would be obtained by treatments made in the late afternoon, which would naturally be followed by lower temperatures at night. Mortality was higher among the insects held at the lower percentages of relative humidity for the observation period but the effect of humidity was much less pronounced than that of temperature. They found greater efficiency was obtained with a lighter oil, such as kerosene. A pyrethrum–water spray showed a greater efficiency at the lower holding temperature with no effect of humidity. Variations in humidity caused no significant changes in the efficiency of the dusts. As far back as 1932, Hartzell and Wilcoxon (1932) demonstrated the loss of pyrethrins content in pyrethrum flowers caused by heat, sunlight, ultraviolet light, and natural aging. A paper by Maciver (1962) described an emulsion concentrate formulation, useful in preparing an aqueous emulsion, in which the pyrethrins were shown to have a high degree of stability even after 1 year shelf storage.

Various synergists are commonly used in pyrethrum formulations and the literature is rather extensive along this particular line. Certain preparations are on the market that contain piperonyl butoxide, such as Pyrenone (Niagara Chemical Division of FMC Corporation). Smith as far back as 1936 discussed the relation between the degree of fineness of pyrethrum powders and deterioration by light and air, demonstrating that the finer the pyrethrum powder is ground, the more rapid the rate of kill and also the more rapid the rate of deterioration of the toxic principles upon exposure to light. They found that titanium dioxide acted as a protectant in that this preparation held its toxicity slightly longer than one containing tannic acid. Trentini (1962) discussed field trials with water emulsions of synergized pyrethrum against horticultural pests, in which piperonyl butoxide, Panasol AN-2, and Triton X171 were utilized. He obtained good results in field trials with water-based emulsions of synergized pyrethrum against many insects.

Pyrethrins can also be combined with many other insecticides. Mistric (1964) compared the effectiveness of pyrethrins when mixed with various other chemicals and Stafford and Jensen (1957) combined pyrethrins and parathion. In the literature can be found all kinds of combinations with pyrethrins, some to the advantage of the product, and some to the disadvantage.

VI. Equipment for Application of Pyrethrins as Aerosols

Considerable research on effectiveness of small liquid drops from submicron to coarse aerosols (greater than 100 μm) has been conducted over the years, beginning in the early 1930's and reaching a peak of activity in the late 1940's (Schechter *et al.*, 1960). Whereas the early work dealt with such materials as Paris Green and various petroleum spray oils, the work in the late 1940's received a tremendous stimulus from the advent of the organochlorine synthetic insecticides. During this period, work with pyrethrins and other botanical type materials continued.

With some of the highly toxic insecticides, such as parathion and other organophosphates, the liquid spray drop size can be related to a specific insect lethal dose. For example, Weidhaas *et al.* (1970) have shown that the LD_{100} for an adult *Aedes taeniorhynchus* mosquito was contained in one 20 μm drop of fenthion. However, many less toxic materials including pyrethrins would require a much larger drop, which would not be very efficient as an airborne aerosol. In part, then, it is necessary to produce very fine aerosol size drops (less than 25 μm) in order to obtain maximum insect contact efficiency and impaction of drops, particularly by flying insects. Several observers have arrived at a drop size in the 10–15 μm range as being most effective for aerosoling of adult flies and mosquitos out of doors, and aerosols of 2–5 μm as being equally effective for aerosoling indoors.

Table I shows the tremendous gain in coverage to be obtained with small particle size aerosols. The 10-μm drops settle very slowly (about 1/100th the rate of a 100-μm drop), which keeps them suspended in

TABLE I

Terminal Velocities of Water Drops and Numbers per Given Volume in Relation to Unit Area and Air Volume

Drop diameter (μm)	Terminal or steady-state velocity		Number of drops at applied rate of 10 oz/acre (730 ml/hectare)		
	ft/sec	m/sec	per in^2	per cm^2	In air to depth of 32.8 ft (10 m) per cm^3 air
1	0.01×10^{-2}	0.033×10^{-3}	90×10^6	14×10^6	14,000
5	0.25×10^{-2}	0.76×10^{-3}	0.716×10^6	0.11×10^6	110.0
10	1×10^{-2}	3.0×10^{-3}	0.09×10^6	0.014×10^6	14.0
20	4×10^{-2}	12×10^{-3}	0.011×10^6	0.0017×10^6	1.7
50	0.25	0.076	716	110	0.11
70	0.4	0.12	265	41	0.042
100	0.9	0.27	90	14	0.014
150	1.5	0.46	27.1	4.2	
200	2.5	0.76	11.3	1.75	
500	7.0	2.13	0.7	0.11	
1500	13	4.0	0.088	0.0136	
2000	21	6.4	0.011	0.0017	

TABLE II

Spray Drop Size Range, Use, and Approximate Recoveries[a]

General description	Drop size (μm vmd)	Estimated recovery within 1000 ft downwind (%)	General use
Fine aerosols; two-fluid atomizers	<50	15	adulticiding by ground equipment for vector control
Coarse aerosols; high pressure hydraulic or high-speed rotary atomizers	50–100	15–40	aircraft aerosoling, low dosage and toxicity materials
Fine sprays; cone, fan, and rotary atomizers	100–250	40–60	large area programs, low dosage and toxicity materials

[a] Temperature gradient, neutral; wind velocity, 3–5 mph; release height, <10 ft.

the air for greater flying insect exposure time. At 10 oz/acre there are 1000 times more 10-μm drops than 100-μm drops available per unit of flat surface; also, it follows that there would be 1000 times more 10-μm drops per cubic centimeter of air than 100-μm drops.

However, this shows only one side of the problem; the other side, loss of these drops to wind drift, may prevent buildup of a sufficient quantity to provide impaction on flying insects or fallout on those resting on plant foilage and thus poor overall control. Table II gives a brief description of the fine and coarse aerosols and fine spray ranges along with the general pest control use for each drop size category. Although Table I gives fall rates for drops to 1 μm, in field practice drops up to 50 μm can remain airborne for a half mile or more on a light wind of 3–5 mph. Table II shows that aerosols below 50 μm vmd[1] are not practical when released by aircraft at some distance above ground because the small drops will not fall below the release height and may actually be carried higher by the turbulent wake of the aircraft. Even with ground equipment aerosols below 50 μm vmd must be applied in very light winds and with warm overhead air (temperature inversion) to hold the aerosol close to the ground for effective insect contact.

As the vmd of the released liquid drops increases the amount of re-

[1] vmd = volume mean diameter.

TABLE III

Estimated Aircraft Spray Fallout of Various Drop Size Ranges and Approximate Applied Swath Widths[a]

Spray particle size range (μm vmd)	Estimated fallout within 1000 ft distance when released from 10-ft height (%)	Approximate swath width (ft) at designated release heights (ft)		
		10	25	100
100–250	40–60	250	350	650
50–100	15–40	1000	3000	5280
20– 50	15 or less	3000	5280	10,560

[a] Neutral to small temperature inversion; wind velocity, 3–5 mph.

covered fallout of the spray goes up. However, even with fine sprays of 100–250 μm vmd, the recovery in 1000 ft downwind seldom exceeds 50–60% of that released.

Table III indicates liquid drop size range, relative to the estimated fallout recovery within 1000 ft. as percent of that released, when liquid is released from a 10-ft height, and the swath or downwind distance that will receive significant amounts of fallout when the aircraft is operated at 10-, 25-, and 50-ft elevations. As can be seen, a fine aerosol (under 50 μm vmd) will spread downwind to over one-half mile when released at a 10-ft height, and to over 2 miles when released at a 100-ft height in light wind and slight temperature inversion (warm air overhead) weather.

Table IV shows estimated ground recovery (fallout on plants and ground) resulting from application with a ground rig aerosol generator

TABLE IV

Estimated Ground Rig Aerosol Fallout of Various Drop Sizes at Several Distances Downwind[a]

Spray particle size (μm vmd)	Percent recovery (cumulative) downwind at:				
	50 ft	100 ft	325 ft	500 ft	1000 ft
10	0	1	3	3.3	3.5
25	1.0	1.5	5	10	13
50	10	25	30	35	38
100	25	50	70	75	80

[a] Neutral to small temperature inversion; 3–5 mph wind velocity; release height, 3 ft.

in low wind and slight inversion to neutral temperature gradient. The 10 μm vmd drop size range does not deposit and less than 3.5% is likely to collect on plants and ground within 1000 ft downwind. However, the 100 μm vmd coarse aerosol will deposit to 80% within 1000 ft under these conditions.

Table V shows some typical drop size distributions obtained by closed room settling techniques for drop size collection. The fine aerosols (cold fogger) may have 5% of drop volume in the 1–5 μm class, the 50% point or vmd is seen to be just above the 5–10 μm class and actually can be found to be about 11 μm. Similarly, the coarse aerosol shows a vmd at 86 μm using a two-fluid air shear atomizer and the spinning gauze atomizer produces a 130 μm vmd of a fine spray category. All atomizers produce a range of drop sizes as shown; the aerosol types appear to narrow this range somewhat, but not significantly.

Types of Atomizers. As noted in Tables II and V, there are a number of atomizer types available to produce drops down to about 50 μm vmd. To go below this, however, the energy required is so high that two-fluid types of atomizers (using air energy) are the only practical

TABLE V

Drop Size Distribution from Three Atomizers, Showing Cumulative Percent by Volume with Increasing Drop Size Class[a]

	Cumulative percent by volume at size class on left		
Drop size classes (μm)	Fine aerosols, cold foggers	Mists, coarse aerosols, two-fluid nozzles	Fine sprays, spinning gauze
1–5	5	0.1	
5–10	45	0.4	0.001
(11) vmd	(50)		
10–15	77	2.0	
15–20	97		0.01
20–40	100	12.0	
40–60		35.0	0.3
60–80			
(86) vmd		(50)	
80–100		59.0	15.8
100–120			
120–140 (130) vmd			(50)
140–180		100	
180–200			81.0
200–300			100

[a] Note that vmd is found at the 50% level.

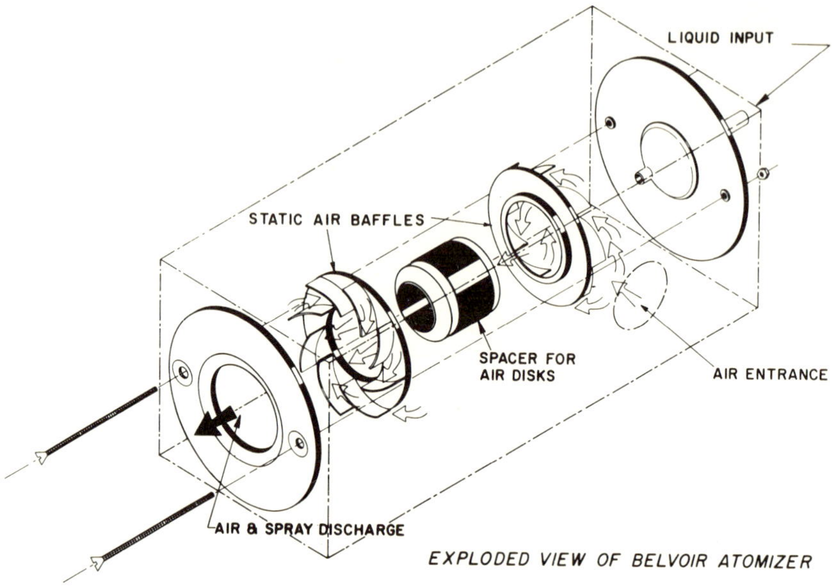

Fig. 1. Cold fogger Belvoir nozzle.

means for the job. The cold foggers use a form of air shear nozzle, as shown in Fig. 1. Here the air pressure is low, not over 5 psig, and volume is high, at around 150 ft^3/min for atomizing at about 135 ml/min flow rate.

Figure 2 shows two types of fluid atomizers, at the left an internal mix type and at the right an external mix or paint sprayer type. These are both capable of atomizing to very fine aerosols and can be easily installed on ground equipment. Air flow is low at 1–2 ft^3/min per nozzle, and air pressure is high at 75 to 100 psig to produce fine aerosols. Liquid pressure must equal air pressure in the nozzle at the left but need only be at 1–5 psig for the external mix type on the right. These atomizers can be mounted on an aircraft, but the air compressor will be large and the double system for both air and liquid will cause added problems.

Drop size will ordinarily stay in the 75–100 μm vmd range for aircraft use and the usual cone- or fan-type hydraulic pressure nozzles are satisfactory. Figure 3 shows two such nozzles. At the right is a hollow cone-type, showing the interchangeable tips and whirl plates to control drop size. Tips are listed by Spraying Systems Company as D1, 2, etc., and whirl plates as #13, 23, etc. The D1 is a $\frac{1}{64}$ in diameter orifice and 13, 23, etc., apply to increased size whirl plates for coarse drop size sprays. Also shown at the left is a fan nozzle identified, for example,

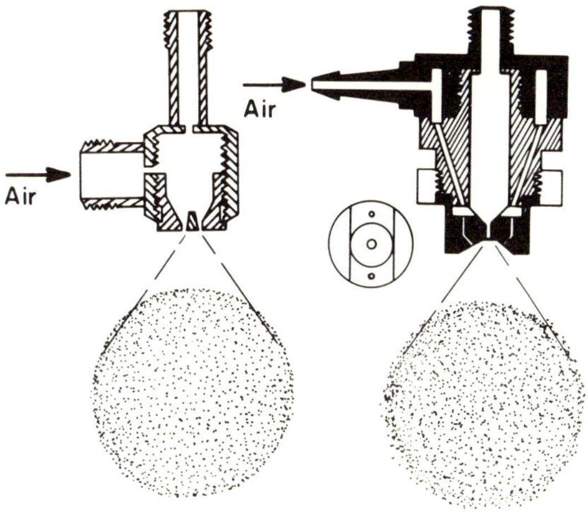

Fig. 2. Two types of fluid atomizers: left, internal mix type; right, external or paint sprayer type.

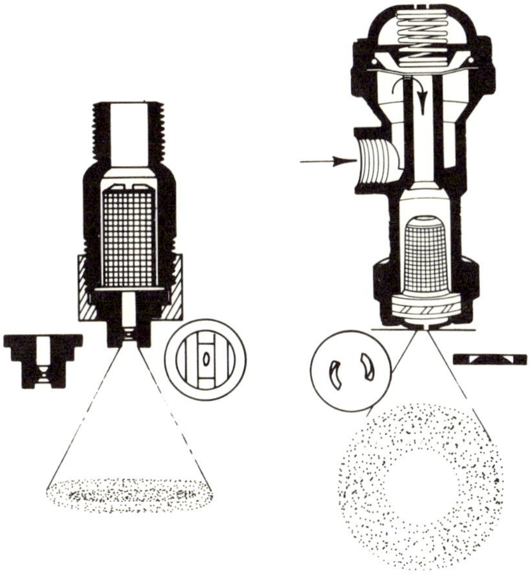

Fig. 3. Hydraulic pressure nozzles: left, fan nozzle; right, hollow cone nozzle with check valve.

by the numbers 80005. The first two digits (80) are the included fan angle of the discharged spray and the next three digits (or zeros) give the liquid flow rate in gallons/minute at 40 psig, or in this case, 0.05

gpm. These nozzles produce drops of 125 to 200 μm vmd when operated at 40–50 psig and directed 90° to an air stream (aircraft) of 100 mph. In order to reduce the drop size below 100 μm, it is necessary to increase liquid pressure to 250–300 psig, which can be done with either windmill, direct power take off, or electric or hydraulic motor drive on fixed wing aircraft. On helicopters the lower flight speed (usually under 60 mph) produces less atomization shear and so higher liquid pressure is used. Also, the helicopter is not practical for a windmill drive so either direct mechanical coupling or electric or hydraulic motor drives are used.

Pyrethrins appear to be most effective when applied as fine to coarse aerosols, the former by ground equipment and the latter by aircraft. A variety of atomizing equipment is available and, with proper design (adequate pump size and power source) and installation on the aircraft or ground rig, application of the aerosol drop size that is desired may be readily obtained.

VII. Recent Tests

In 1972 trials in cooperation with N. F. McCalley and J. Root, ULV applications of pyrethrins applied by helicopter for artichoke plume moth, *Platyptilia carduidactyla* (Riley), control did not yield significant results. Eight applications of 1 qt/acre using 4–9 cone nozzles at 0.1 lb/acre of pyrethrins (20 ft boom and 27 ft swath) yielded no data because of the lack of a significant infestation.

Trials by N. F. McCalley (unpublished) on lettuce from April 20 to April 27, 1972 using the same nozzle setup by helicopter yielded poor control of the alfalfa looper.

Tests by W. H. Lange and J. Kishiyama in 1971 and 1972 have shown little value of pyrethrins alone and in combination with other materials, such as rotenone, malathion, and *Bacillus thuringiensis,* using hand-applied full coverage sprays for the control of tomato caterpillars. Tables VI and VII indicate the poor control obtained with pyrethrins in the 1971 trials. Similar 1972 trials by the same authors (unpublished) were also relatively ineffective in the control of the tomato fruitworm and beet armyworm on tomatos.

It appears from recent tests that pyrethrins have more of a place in the agricultural field in reducing adult populations than in controlling worms. Whether ULV applications of pyrethrum aerosols will be more effective in controlling worms under agricultural conditions will depend on further experimentation as the present tests are inconclusive. Certainly more persistent and effective formulations should be tried. In addition, the quick knockdown properties of pyrethrins and their relative

TABLE VI

Results of Worm Control Experiment on Tomatoes, Del Monte Ranch, Yolo, California, 1971[a]

Treatment[b]	Formulation	Formulation per acre	Average worm damage (%)[c] Green fruit	Average worm damage (%)[c] Ripe fruit
Resmethrin	2 lb/gal EC[e]	946 ml	7.8 abcd[d]	13.0 bc[d]
NIA 18739	1 lb/gal EC	379 ml	12.5 d	17.8 c
NIA 26021	1 lb/gal EC	379 ml	8.8 bcd	11.8 abc
NIA 24110	1 lb/gal EC	379 ml	8.8 bcd	13.0 bc
Pyrocide + Rotenone	0.21% + 1.06% EC	946 ml	8.8 bcd	14.8 bc
Pyrocide + Rotenone + QP 4400	0.21% + 1.06% EC + 5 lb/100 gal	946 ml 40 gal	10.5 cd	12.3 abc
Pyrenone + Thiodan	1 lb + 2 lb/gal EC	946 ml	5.5 abc	7.5 ab
Carbaryl	4 lb/gal flowable	1892 ml	3.3 a	4.5 a
Dipel (*Bacillus thuringiensis*)	3.2% WP[f]	454 g	5.0 ab	10.3 abc
Dipel	2% Dust	40 lb.	9.8 bcd	11.0 abc
Thuricide (*Bacillus thuringiensis*)	6×10^6 viable spores per milligram	946 ml	7.0 abc	7.5 abc
Check	—	—	9.8 bcd	12.0 abc

[a] Randomized complete block design, plots 4-rows wide (20 ft), 50 ft long; treated 9/3, 9/15, and 9/29/71, except Dipel dust treatment on 9/8, 9/16, and 9/29/71, using 60 psi CO_2 sprayer, 3 #8004 T-jet nozzles per row, 40 gal water per acre.

[b] The chemical names for the test compounds are: NIA 18739 = (5-benzyl-3-furyl)-methyl (+)-*trans*-2,2-dimethyl-3-(2-methyl-1-propenyl)cyclopropane-1-carboxylate; NIA 26021 = (5-benzyl-3-furyl)methyl (+)-*cis*-2,2-dimethyl-3-(2-methyl-1-propenyl)cyclopropane-1-carboxylate; NIA 24110 = (5-benzyl-3-furyl)methyl 1*R*,2*R*-2[(cyclopentylidene)methyl]-3,3-dimethylcyclopropane-l-carboxylate; QP 4400 = hydroxyethylcellulose, 5 lb actual per 100 gal water, pH 6.8.

[c] Rated 10/5/71, examining 100 green and 100 ripe fruit per plot (800 total fruit per treatment). Worm damage combination of *Heliothis zea* and *Spodoptera exigua*.

[d] Green fruit counts transformed using $\sqrt{x + 0.5}$ for analysis of variance; ripe fruit counts not transformed. Means associated with the same letter are not significantly different at the 5% level, based on Duncan's New Multiple Range Test. All means associated with the same letter or the same combination of letters are in the same class or grouping relative to effectiveness of treatment as determined by worm damage.

[e] EC = Emulsifiable concentrate.

[f] WP = Wettable powder.

TABLE VII

Results of Worm Control Experiment on Tomatoes, Jiminez Ranch, Madison, California, 1971[a]

Treatment	Formulation	Formulation per acre	Average worm damage (%)[b]	
			Green fruit	Ripe fruit
Pyrethrins	Pyrocide,[c] 1.4% pyrethrins	946 ml	1.2 ab[d]	4.0 abc[d]
Pyrethrins + rotenone	0.21% +1.06% EC[e]	3785 ml	2.2 b	6.4 c
Pyrethrins + *Bacillus thuringiensis*	Pyrocide + Dipel	946 ml 454 gm	1.7 ab	2.5 ab
Carbaryl	4 lb/gal flowable	1892 ml	0.2 a	1.1 a
Check	—	—	3.6 b	5.4 bc

[a] Latin square design, plots 5-rows wide (25 ft), 200 ft long; treated 7/29, 8/11, and 8/25/71, using 60 psi CO_2 sprayer, 3 #8004 T-jet nozzles per row, 40 gal water per acre.

[b] Rated 9/10/71, examining 200 green and 200 ripe fruit per plot (2000 total fruit per treatment). Worm damage combination of *Heliothis zea* and *Spodoptera exigua*.

[c] McLaughlin Gormley King Company.

[d] Means associated with the same letter are not significantly different at the 5% level, based on Duncan's New Multiple Range Test. See Table VI.

[e] EC = Emulsifiable concentrate.

safety to many biological control agents make it possible to combine them with reduced rates of materials having more pronounced adverse effects on natural enemies. They can also be combined with materials having more persistent or selective properties or with microbial control agents. It also appears that pyrethrins are more effective as aerosols applied either by ground equipment or by aircraft and additional investigations are indicated to determine the practical uses for aerosol applications under agricultural situations.

VIII. Summary

Prior to World War II pyrethrum was used extensively for the control of agricultural and horticultural insect pests. It has been used less frequently in recent years because other effective compounds afford the advantages of greater persistence and lower cost. Research on formulations and methods of application is opening up new avenues for possible utilization of pyrethrum, particularly as fine to coarse aerosols in vegetable insect control.

References

Allen, T. C. (1936). Toxicity of kerosene steepates of derris and pyrethrum to some potato insects. *J. Econ. Entomol.* **29**, 742–743.

Anderson, L. D., and Atkins, E. L., Jr. (1958). Effects of pesticides on bees. *Calif. Agr.* **12**, 3–4.

Bailey, S. F. (1938). Thrips of economic importance in California. *Calif., Agr. Exp. Sta., Circ.* **346**, 10, 19, 32, 38, 39, 44, 61, 64, and 69.

Bartlett, B. R. (1958). Laboratory studies on selective aphicides favoring natural enemies of the spotted alfalfa aphid. *J. Econ. Entomol.* **51**, 374–378.

Beckwith, C. S. (1941). Control of cranberry fruit worm on blueberries. *J. Econ. Entomol.* **34**, 169–171.

Bickley, W. E., Harrison, F. P., and Ditman, L. P. (1956). *Drosophila* as a pest of canning tomatoes. *J. Econ. Entomol.* **49**, 417–418.

Bogaard, T. (1969). A comeback for pyrethrum? *Agr. Chem.* **24**(12), 23.

Collins, W. E. (1956). On the biology and control of *Drosophila* on tomatoes for processing. *J. Econ. Entomol.* **49**, 607–610.

DeLong, D. M. (1934). The relative value of Bordeaux mixture, sulphur and pyrethrum products in reducing populations of the potato leafhopper (*Empoasca fabae* Harris). *J. Econ. Entomol.* **27**, 525–533.

Ebeling, W. (1958). Vinegar fly control treatments. *Calif. Agr.* **12**(6), 12–14.

Ebeling, W., and Pence, R. J. (1953). Avocado pests. *Calif., Agr. Exp. Sta., Circ.* **428**, 15.

Essig, E. O., and Hoskins, W. M. (1934, rev. 1944). Insects and other pests attacking agricultural crops. *Calif., Agr. Ext. Serv., Circ.* **87**, 1–197.

Ginsburg, J. M. (1930). Test to determine toxicity of pyrethrum vapors to honeybees. *J. Agr. Res.* **40**, 1053–1057.

Gnadinger, C. B., Moore, J. B., and Coulter, R. W. (1940). Experiments with pyrethrum for the control of codling moth (*Carpocapsa pomonella* L.). *J. Econ. Entomol.* **33**, 143–152.

Hagley, E. A. C. (1959). Pyrethrum—a note on its horticultural application. *Pyrethrum Post* **5**(1), 23–24.

Harries, F. H., De Courscy, J. D., and Hofmaster, R. N. (1945). Some factors affecting the insecticidal action of pyrethrum extracts on the sugar beet leafhopper. *J. Agr. Res.* **71**, 553–565.

Hartzell, A., and Wilcoxon, F. (1932). Some factors affecting the efficiency of contact insecticides. II. Chemical and toxicological studies of pyrethrum. *Contrib. Boyce Thompson Inst.* **4**, 107–117.

Hills, O. A., Mast, A. A., and Wood, R. C. (1948). Field insecticide tests against the beet leafhopper in sugar beets grown for seed. *Amer. Soc. Sugar Beet Technol. Proc., 5th Gen. Meet.* pp. 480–486.

Huckett, H. C. (1936). Some tests of pyrethrum, derris and nicotine mixtures against cabbage worms. *J. Econ. Entomol.* **29**, 575–580.

Huckett, H. C. (1937). Further tests with pyrethrum as an insecticide for cabbage worms. *J. Econ. Entomol.* **30**, 323–328.

Lamiman, J. F. (1933). Control of grape leafhopper in California. *Calif., Agr. Ext. Serv., Circ.* **72**, 12–14.

Lange, W. H., Jr. (1941). The artichoke plume moth and other pests injurious to the globe artichoke. *Calif., Agr. Exp. Sta., Bull.* **563**, 1–71.

Maciver, D. R. (1962). Preliminary experiments on the stability of pyrethrins in aqueous emulsion. *Pyrethrum Post* **6**(4), 20–21.

Mader, E. O., and Udey, E. C. (1937). Blooming of potatoes as influenced by pyrethrum dust. *Phytopathology* **27**, 112–113.

Madsen, H. F., and McNelly, L. B. (1961). Important pests of apricots. *Calif., Agr. Exp. Sta., Bull.* **783**, 26.

Madsen, H. F., and Westigard, P. H. (1963). Control of pear psylla with oils and oil-pyrethrins. *Calif. Agr.* **17**(1), 9–10.

Michelbacher, A. E., MacLeod, G. F., and Smith, R. F. (1941). A preliminary report on control of the western twelve-spotted cucumber beetle in orchards. *J. Econ. Entomol.* **34**, 709–716.

Michelbacher, A. E., Middlekauff, W. W., and Bacon, O. G. (1955). Controlling melon insects and spider mites. *Calif., Agr. Exp. Sta., Bull.* **749**, 18.

Miskus, R. P., and Andrews, T. L. (1971). Stabilization of pyrethroid compositions. U.S. Patent 3,560,613.

Miskus, R. P., and Andrews, T. L. (1972). Stabilization of thin films of pyrethrins and allethrin. *J. Agr. Food Chem.* **20**, 313–315.

Mistric, W. J., Jr. (1964). Effectiveness of rotenone and pyrethrins when mixed with other chemicals for boll weevil control. *J. Econ. Entomol.* **57**, 765–766.

Moore, A. D. (1969). New developments in forest insect problems. *Trans. N. Amer. Wildl. Natur. Resourc. Conf.* **34**, 155–159.

Parker, L. (1972). New work for an ancient insecticide. *Agrichem. Age January* pp. 6–10.

Pepper, B. B. (1941). The corn earworm and its control on sweet corn. *N.J., Agr. Exp. Sta., Circ.* **413**, 1–13.

Pepper, B. B., and Haenseler, C. M. (1939). Pyrethrum and derris as a control for the six-spotted leafhopper, a vector of lettuce yellows. *J. Econ. Entomol.* **32**, 291–296.

Reid, W. J., Jr., and Cuthbert, F. P., Jr. (1952). Value of auxiliary materials in rotenone and pyrethrum insecticides for control of cabbage caterpillars. *U.S. Dep. Agr., Bur. Entomol.* **834**, 1–24.

Riley, C. V. (1881). Experiments with pyrethrum: Safe remedies for cabbage worms and potato beetles. *Amer. Natur.* **15**, 145–147.

Riley, C. V. (1881–1882). Pyrethrum: Its use as an insecticide. Report of the Entomologist. *Rep. U.S. Comm. Agr.* pp. 76–87.

Riley, C. V. (1885). Vegetable insecticides. *Rep. U.S. Entomol. Comm.* **4**, 164–180.

Schechter, M. S., Sullivan, W. N., and Yeomans, A. H. (1960). Studies on fine-particle, coarse-particle, and concentrated aerosols. *J. Econ. Entomol.* **53**, 908–914.

Skaptason, J. B. (1938). Some comparisons of dusts for potato leafhopper control on Long Island. *Amer. Potato J.* **15**, 271–277.

Smith, C. L. (1936). The relation between the degree of fineness of pyrethrum powder produced by different periods of grinding to toxicity to insects and to deterioration by light and air. *J. N.Y. Entomol. Soc.* **44**, 317–333.

Smith, G. L. (1942). California cotton insects. *Calif., Agr. Exp. Sta., Bull.* **660**, 1–50.

Sorenson, C. J. (1939). Lygus bugs in relation to alfalfa seed production *Utah, Agr. Exp. Sta., Bull.* **284**, 51–57.

Sorenson, C. J. (1942). Insecticidal tests for field control of *Lygus* bugs in seed alfalfa. *J. Econ. Entomol.* **35**, 884–886.

Stafford, E. M., and Jensen, F. L. (1957). Grape leaf-folder. *Calif. Agr.* **11**(6), 4, 10, and 15.

Stombler, V., Pelekassis, C. D., and Doyle, E. S. (1957). *Drosophila* control on harvested tomatoes for processing in California—1956. *J. Econ. Entomol.* **50**, 476–480.

Thomas, C. A. (1936a). The tomato pin worm. *Pa., Agr. Exp. Sta., Bull.* **337**, 12.

Thomas, C. A. (1936b). Status of the tomato pin worm *Gnorimoschema lycopersicella* Busck in Pennsylvania. *J. Econ. Entomol.* **29**, 313–317.

Trentini, R. (1962). Field trials with water emulsions of synergised pyrethrum against certain horticultural pests. *Pyrethrum Post* **6**(3), 13–19.

Walker, H. G., and Anderson, L. D. (1934). Notes on the use of derris and pyrethrum dusts for the control of certain insects attacking cruciferous crops. *J. Econ. Entomol.* **27**, 388–392.

Walton, R. R., and Howell, D. E. (1954). Control of aphids on commercial green crops by insecticides. *J. Econ. Entomol.* **47**, 780–785.

Warner, J. L. (1963). Field experiments on the persistence of aqueous formulations of synergised pyrethrum for use in horticulture. *Pyrethrum Post* **7**(1), 34–42.

Watts, C. N. (1962). The use of pyrethrum powders in the tomato canning industry in the U.S.A. *Pyrethrum Post* **6**(6), 3–6.

Weidhaas, D. E., Bowman, M. C., Mount, G. A., Lofgren, C. S., and Ford, H. R. (1970). Relationship of minimum lethal dose to the optimum size of droplets of an insecticide for mosquito control. *Mosquito News* **30**, 195–200.

White, W. H. (1945). Control of celery leaf tier through pyrethrum powder. *U.S. Dep. Agr., ARS, R.A.S.* **26**(E), 1.

Wisecup, C. B. (1936). Decrease in effectiveness of stored pyrethrum dusts as shown by biological tests with the celery leaf tier in the laboratory. *J. Econ. Entomol.* **29**, 1000–1003.

Yothers, M. A., Carlson, F. W., and Cassil, C. C. (1942). Sprays to kill overwintering codling moth larvae. *J. Econ. Entomol.* **35**, 450–451.

Chapter 15

Pyrethrum for Control of Forest Insect Pests

RAYMOND P. MISKUS and ROBERT L. LYON

I. Introduction

Only since the early 1960's have pyrethrins been seriously considered for forest insect control. This interest stems from two developments in the United States Forest Service.

In 1964, the Forest Service set up an Insecticide Evaluation Research Unit at its Pacific Southwest Forest and Range Experiment Station, headquartered in Berkeley, California. The unit was assigned the task of developing, evaluating, and testing suitable substitutes for DDT and other chlorinated hydrocarbon insecticides that had been banned from use on the National Forests. Their replacements had to be safe, effective, nonpersistent, and highly toxic to insect pests. In addition, they had to produce minimal impact on the forest environment.

In 1966, a stabilized formulation of pyrethrins was first field tested on the Bitterroot National Forest in Montana. A United States patent on the formulation was granted in 1971 (Miskus and Andrews, 1972).

This review summarizes work done mainly by the United States Forest Service and by the Canadian Chemical Control Research Institute in developing, evaluating, and testing pyrethrins for the control of forest insect pests.

II. Stabilization of Pyrethrins

In laboratory tests, we found that pyrethrins were usually ten to more than 100 times more toxic than DDT to lepidopterous larvae (Table I). The pyrethrins also have a low order of toxicity to vertebrates (Negherbon, 1959). However, they are highly unstable in the field and can quickly convert to nontoxic products (Crowe *et al.,* 1961; Chen and Casida, 1969). Consequently, small amounts of pyrethrins could not be successfully sprayed on forested areas. It was necessary to develop a formulation that would remain stable after aerial application.

We tried several formulations developed by others, but none performed adequately. Many combinations of antioxidants, ultraviolet (UV) absorbents, and saturated paraffin oils provided about 4 hr of stability when exposed to sunlight as thin films on glass plates. After 3 years of effort, the Pacific Southwest Station's Insecticide Evaluation Research Unit developed a stabilized formulation for pyrethrins that persisted for at least 4 hr in sunlight (Miskus and Andrews, 1972).

The formulation that has proved most effective to date consists of an oil-soluble antioxidant, 2,6-dioctadecyl-*p*-cresol (DOPC), and the UV screening agent, benzyl cinnamate. The stabilizing adjuvants are used at 1% (w/v) in the light mineral oil, Klearol (Witco Chemical Company, Sonneborn Division). The UV screening agents we tested have a characteristic absorption spectrum that parallels the sunburn spectrum of sunlight (DasGupta, 1967). This characteristic accounts for the increased stability of pyrethrins in greenhouses and homes. Window or soft glass transmits radiation only above 300 nm; hard or borosilicate glass transmits radiation beginning at 290 nm. We found a much greater destruction of pyrethrins in sunlight when borosilicate glass was used as a filter than when soft glass was used.

III. Spruce Budworm

To date, field tests of pyrethrins have been conducted on two species of spruce budworm: the western form, *Choristoneura occidentalis* (Freeman), and the eastern form, *C. fumiferana* (Clemens). They were first tested against *C. occidentalis* on the Salmon National Forest in Idaho

TABLE I

Comparative Toxicity of Pyrethrins and DDT to Selected Forest Insect Pests

Insect	LD_{50} (μg/gm)		Toxicity ratio at LD_{50} (DDT/pyrethrins)	References
	Pyrethrins	DDT		
Elm spanworm (*Ennomos subsignarius*)	0.055	1.6	29.0	Robertson and Lyon (1973)
Forest tent caterpillar (*Malacosoma disstria*)	0.12	6.5	54.0	Lyon *et al.* (1972)
Boxelder tussock moth (*Hemerocampa* n. sp.)	0.14	7.9	56.0	Flake and Lyon (1967)
Western tussock moth (*Hemerocampa vetusta*)	0.14	12.0	86.0	Page and Lyon (1973b)
Western tent caterpillar (*Malacosoma californicum lutescens*)	0.26	36.0	138.0	Page and Lyon (1973a)
Douglas-fir tussock moth (*Hemerocampa pseudotsugata*)	0.28	4.1	15.0	Lyon *et al.* (1970)
Pandora moth (*Coloradia pandora*)	0.34	63.0	185.0	Lyon (1971)
Western hemlock looper (*Lambdina fiscellaria lugubrosa*)	0.36	95.0	264.0	R. L. Lyon *et al.* (unpublished)
Western (California) tent caterpillar (*Malacosoma californicum californicum*)	0.61	0.34	0.56	Robertson and Gillette (1973)
Spruce budworm (*Choristoneura fumiferana*)	0.91	56.0	62.0	R. L. Lyon *et al.* (unpublished)
Red-humped oakworm (*Symmerista canicosta*)	1.0	1.1	1.1	Robertson *et al.* (1972)
Western spruce budworm (*Choristoneura occidentalis*)	1.0	24.0	24.0	R. L. Lyon *et al.* (unpublished)
California oakworm (*Phryganidia californica*)	1.3	9.3	7.2	Schwartz and Lyon (1971)
Pine tussock moth (*Dasychira* sp. near/= *grisefacta*)	1.4	15.0	11.0	Lyon *et al.* (1967)
Fall cankerworm (*Alsophila pometaria*)	2.2	39.0	18.0	Lyon and Brown (1970)
Green spruce budworm (*Choristoneura viridis*)	2.9	50.0	17.0	R. L. Lyon *et al.* (unpublished)
Jack pine budworm (*Choristoneura pinus pinus*)	3.1	16.0	5.2	R. L. Lyon *et al.* (unpublished)

in 1964 before a stabilized formulation had been developed. Two tests were made, one with 0.01 lb of pyrethrins in 1 gal of diesel oil per acre applied by helicopter, the other with 0.03 lb pyrethrins in 2 gal per acre as an invert emulsion (Instemul of Stull Chemical Company) applied by fixed-wing aircraft. Mortality was low in both tests, ranging from 0 to 19% with the invert emulsion and from 11 to 48% with the diesel oil solution.

In 1965, formulations of pyrethrins in Instemul were tested against *C. occidentalis* in New Mexico (Buffam *et al.*, 1966). The formulation included dyes in an attempt to stabilize pyrethrins against photochemical degradation (Warner, 1963). Pyrethrins were applied by helicopter at 0.02 lb/acre against larvae in the fourth and fifth instars. Results were very poor; population reduction ranged from 0 to 19%.

In 1966, a stabilized formulation (Miskus and Andrews, 1972) was evaluated for its activity against a population of *C. occidentalis*, mainly in the fourth, fifth, and sixth instars, on the Bitterroot National Forest, Montana (Buffam *et al.*, 1967). The formulation consisted of 0.1 lb of pyrethrins in 1 gal of a saturated isoparaffin oil, Isopar M (Humble Oil Company). The formulation contained DOPC and the UV absorber, Permasorb MA [2-hydroxy-4-(2-hydroxy-3-methacrylyloxy)propoxybenzophenone]. Both stabilized and unstabilized pyrethrins were applied by helicopter equipped with size 80015 nozzles (Spraying Systems Company, Bellwood, Ill.) The stabilized formulation reduced the insect population on the average by 73%; the nonstabilized formulation reduced the population by 65%. This difference was not considered satisfactory, and later work on the formulation indicated that substantial degradation had occurred in both applications. This degradation was caused by the solvent. In subsequent work, we found that mineral oil was a satisfactory replacement.

In 1971, the United States Forest Service planned a field test of pyrethrins on *C. fumiferana* in Minnesota. The formulation contained mineral oil. Benzyl cinnamate was used in place of Permasorb MA because Permasorb did not stay in solution under some field conditions. The plan for application included three dosage levels of 0.025, 0.05, and 0.1 lb pyrethrins per acre. A heavy rainfall occurred shortly after two plots had been sprayed. The third plot was sprayed a day later. The effect of the rain and the occurrence of extremely low populations of budworm made an analysis of efficacy highly unreliable.

The Canadian Chemical Control Research Institute has also tested pyrethrins against *C. fumiferana*. This pest is a serious problem in New Brunswick and Quebec. In trials in 1970 and 1971, the effectiveness of pyrethrins seemed to be related to the stage of development of the larvae, i.e., less effective when applied to the second and third instars then when applied to the more exposed older larvae (Randall *et al.*, 1970, 1971).

In 1970, the tests were conducted at Shediac, New Brunswick, using the MGK formulation #7014. This formulation contained 14.3% by weight of 20% pyrethrins extract, 1.2% DOPC, 1.72% Permasorb MA 70, 4.28% Dowanol EB, and 78.5% Klearol mineral oil. This provided a solution containing 0.2 lb pyrethrins per United States gallon.

Two 400-acre plots in heavily infested spruce–balsam fir forest were treated at an emitted dose of 0.5 oz (0.03 lb) a.i.[1] per acre. One plot was treated at the peak of the third instar and the other at the peak of the fifth instar. The formulation was applied to each plot by using a Stearman aircraft fitted with four AU-3000 Micronair atomization units calibrated to provide a droplet spectrum of 175 μm mmd.[2] Pyrethrins were not effective (10–20% population reduction) against the early instar larvae but were highly effective (70–90% population reduction) against the more exposed fifth instar larvae. A beneficial side effect of the spray was a drastic reduction in populations of biting insects, such as mosquitos, black flies, and biting midges.

Trials were repeated in 1971, using the MGK formulation #7063 and the same aircraft, spray system, and dose rate. These trials were directed at the second and third instar larvae during the needle and bud mining stages. Again, these early budworm populations were reduced only slightly—even at the higher deposit levels (>20 spray droplets[2] per square centimeter). Laboratory bioassay using field-collected foliage from the heavier deposit areas suggested no toxic effects on fourth instar larvae. The investigators concluded that 0.5 oz a.i. per acre had little effect on second and third instars but that pyrethrins at this dosage were highly effective against older larvae in the fifth and sixth instars.

IV. Western Hemlock Looper

In 1969, stabilized pyrethrins were tested on a population of western hemlock looper, *Lambdina fiscellaria lugubrosa* (Hulst), in the state of Washington. This formulation contained DOPC and Permasorb MA in mineral oil. Application was made by helicopter, using size 80015 tee jet nozzles at a rate of 0.1 lb per 0.5 gal per acre. This test consisted of two parts.

In the first part, small-scale tests were conducted to determine if the dosage of 0.1 lb/acre would be toxic enough to justify a full-scale field test to follow. Larvae in the second and third instars were placed on

[1] a.i. = Active ingredient.

[2] mmd = Mass median diameter.

individual trees and also on wooden platforms containing hemlock foliage supported off the surface on wire screening. The stabilized pyrethrins application at the 0.1 lb/acre dose was high enough to kill or paralyze 94 to 99% of the larvae (Mason, 1970).

In the second part of the test, pyrethrins were applied to a natural population of hemlock looper in a 145-acre heavily timbered area near Mt. Baker. The larvae were mainly third and fourth instars at the time of spraying. Results were not as favorable as those in the preliminary test; population reduction averaged about 65% (range, 20 to 80%) (Mounts *et al.*, 1970). The lower mortality may have been caused by much denser foliage providing cover for the insects.

Observations during the field test showed that many aquatic insects were killed. Subsequent information on the effects of pyrethrins on non-target organisms indicated that a dose of 0.1 lb or less per acre created no hazard to fish, birds, small game, or aquatic insects but that doses as high as 0.20 lb/acre could kill many aquatic insects (Shea, 1972). No other harmful effects were observed on wildlife populations.

V. Douglas-Fir Tussock Moth

A test involving pyrethrins was conducted against the Douglas-fir tussock moth, *Hemerocampa pseudotsugata* (McDunnough), in 1972 in Washington. Stabilized pyrethrins containing DOPC and benzyl cinnamate in mineral oil were applied by a helicopter equipped with size 80015 tee jet nozzles. The larvae were mainly third and fourth instars. Two applications were made 7 days apart at the rate of 0.05 lb per 0.5 gal per acre for each application. The total effect of the pyrethrins spray was a reduction of about 78% in the larval population (Shea, 1972). This is a preliminary estimate because a full statistical analysis of the data has not been completed. Spray assessment cards indicated that the deposit of insecticide was low (Neisess *et al.*, 1972) which may help explain the low mortality.

VI. Forest Tent Caterpillar

A single field test was made in 1965 (Buffam *et al.*) against the forest tent caterpillar, *Malacosoma disstria* (Hubner), in Texas before the stabilized formulation of Miskus and Andrews (1972) was developed. The test was staged using oak twigs on which larvae were placed just before spraying. Five formulations were used (Table II), three in

TABLE II

Tests of Pyrethrins Sprays against Forest Tent Caterpillar[a]

Pyrethrins formulation	Knockdown (%)	Mortality (%)
Diesel oil	83	12
Diesel oil with orange dye and piperonyl butoxide	92	13
Instemul	75	23
Instemul with orange dye	55	37
Instemul with yellow dye	88	45
Check (unsprayed)	0	0

[a] Source: Buffam *et al.*, 1965.

Instemul and two in diesel oil. Dyes were used as stabilizers in three formulations. All were applied by fixed-wing aircraft at 0.02 lb pyrethrins in 1 gal/acre.

Knockdown was high (55 to 92%) but mortality was low, ranging from 12 to 45%—depending on the formulation (Table II).

VII. Gypsy Moth and Elm Spanworm

A field test was conducted in 1972 by Dunbar and Doane against the gypsy moth, *Porthetria dispar* (Linnaeus), and the elm spanworm, *Ennomos subsignarius* (Hubner). A stabilized pyrethrins formulation (MGK Pyrocide 2083) was applied to infested apple trees at the rate of 1.5–2 qt per tree with a Kieben knapsack mistblower. An unstabilized formulation (Niagara Pyrenone) containing the synergist piperonyl butoxide at 10 times the concentration of pyrethrins was also tested. The formulations were emulsified in water with 2 oz DuPont Spreader Sticker per 10 gal.

In one test, the stabilized pyrethrins were applied at a concentration of 0.056 lb a.i. per 10 gal (dose of 1.27 gm a.i. at 2 qt per tree) and the unstabilized pyrethrins were applied at a concentration of 0.014 lb a.i. per 10 gal (dose of 0.32 gm a.i. at 2 qt per tree). At the time of spray application, the gypsy moth larvae were predominantly in the fourth instar and the elm spanworm larvae were mainly in the fifth instar.

Both stabilized and unstabilized pyrethrins were very effective in reducing the number of gypsy moth and elm spanworm larvae. There was rapid and almost complete knockdown within the first 20 min after treatment by both formulations. Forty-eight hours after treatment, there were 96 to 100% fewer larvae of both species on trees treated with stabilized pyrethrins compared to the number of larvae on untreated trees and 82 to 92% fewer larvae on trees treated with the unstabilized pyrethrins.

A second test against gypsy moth larvae in the fifth and sixth instars was conducted with sprays at the same concentrations as mentioned above and also at one-half these concentrations. The reduction in number of larvae 48 hr after treatment by the stabilized pyrethrins varied from 81 to 99% when compared to the number of larvae on untreated trees. The unstabilized pyrethrins reduced the larvae numbers by 21 to 80%. A large proportion of the larvae knocked down by the unstabilized pyrethrins recovered and returned to the trees.

VIII. Spray Application Technology

With such short-lived insecticides as pyrethrins, which are highly active by contact and must be applied in small amounts, it is crucial to develop methods that insure impingement on the insect. Some of the field testing has shown erratic results. Such results, in our estimation, do not deny the efficacy of the insecticide but rather reflect on the state of the art in spray application technology. Effectiveness of contact sprays will depend in large part on the ability to deliver the insecticide to the target insect because only a little if any residual action can be expected.

The most effective spray drop size appears to be 50 μm or less in diameter (Himel and Moore, 1967). Such small droplets are more apt to impinge on small structures, and the geometry of the target strongly influences deposition (Hadaway and Barlow, 1965; Pieper, 1972).

To date, spray droplets in this aerosol range have had only limited use because our understanding of aerosol delivery and behavior in mountainous terrain is so sketchy. Much more research is needed before the potential of sprays with droplet sizes in the aerosol range can be exploited.

A spray application that makes the most efficient use of the prescribed dose can minimize the quantity of insecticide needed and offer environmental safety and reduced cost. Perhaps some insects can be managed by a knockdown dose only; if so, the applied dose could even be smaller than a dose intended to kill them. This would be a distinct advantage where the knockdown dose is enough to do the job. Pyrethrins have a strong knockdown effect on most insects.

IX. Summary

Stabilized pyrethrins have proved effective in suppressing certain forest insect defoliators that are highly susceptible and that are exposed in the tree crown. Kills as high as 90% have been obtained with doses as low as 0.5 oz (0.03 lb) per acre. Field trials so far seem to show that a dosage of 0.03 to 0.1 lb a.i. per acre is needed to protect trees from serious defoliation. Considering all their advantages, pyrethrins show considerable promise for specialized use in forest insect control and warrant further research and field study.

References

Buffam, P. E., Larson, J. E., Miskus, R. P., Downing, G. L., and Meyer, H. (1965). Results of April 1965 pyrethrins field tests at San Antonio, Texas, and ultraviolet exposure tests at Berkeley. Unpublished report, U.S. Forest Serv., Berkeley, California.

Buffam, P. E., Downing, G. L., Moore, A. D., Larson, J. E., Meyer, H. E., and Tiernam, C. F. (1966). Results of aerial spray trials with five insecticides against the spruce budworm in New Mexico, June 1965. Unpublished report, U.S. Forest Serv., Berkeley, California.

Buffam, P. E., Meyer, H. E., and Miskus, R. P. (1967). Smallscale trials of five insecticides . . . sprayed on spruce budworm in Montana. *U.S. Forest Serv., Res. Note* PSW-**159**, 1–4.

Chen, Y. L., and Casida, J. E. (1969). Photodecomposition of pyrethrin I, allethrin, phthalthrin, and dimethrin. Modifications in the acid moiety. *J. Agr. Food Chem.* **17**, 208–215.

Crowe, T. J., Glynne Jones, G. D., and Williamson, R. (1961). The use of pyrethrum formulations to control *Antestiopsis* on coffee in East Africa. *Bull. Entomol. Res.* **52**, 31–41.

DasGupta, V. J. (1967). Evaluation of sunscreen agents. *J. Soc. Cosmet. Chem.* **18**, 143–147.

Dunbar, D. M., and Doane, C. C. (1972). Field evaluation of pyrethrin and synthetic pyrethroids for control of the gypsy moth and elm spanworm. *Pap., Nat. Meet. Entomol. Soc. Amer., 1972.*

Flake, H. W., Jr., and Lyon, R. L. (1967). Insecticide tests against larvae of *Hemerocampa,* new species, a tussock moth that defoliates boxelder in New Mexico. *J. Econ. Entomol.* **60**, 607–608.

Hadaway, A. O., and Barlow, F. (1965). Studies on the deposition of oil drops. *Ann. Appl. Biol.* **55**, 267–275.

Himel, C. M., and Moore, A. D. (1967). Spruce budworm mortality as a function of aerial spray droplet size. *Science* **156**, 1250–1251.

Lyon, R. L. (1971). Contact toxicity of 40 insecticides tested on pandora moth larvae. *U.S., Dep. Agr., Forest Serv. Res. Note* **PSW-235**, 1–4.

Lyon, R. L., and Brown, S. J. (1970). Contact toxicity of insecticides applied to fall cankerworm reared on artificial diet. *J. Econ. Entomol.* **63**, 1970–1971.

Lyon, R. L., Brown, S. J., and Richmond, C. E. (1967). Insecticides tested on new tussock moth . . . defoliator found in Montana. *U.S., Forest Serv., Res. Note* **PSW-161**, 1–4.

Lyon, R. L., Flake, H. W., Jr., and Ball, L. (1970). Laboratory tests of 55 insecticides on Douglas-fir tussock moth larvae. *J. Econ. Entomol.* **63**, 513–518.

Lyon, R. L., Brown, S. J., and Robertson, J. L. (1972). Contact toxicity of 16 insecticides applied to forest tent caterpillar reared on artificial diet. *J. Econ. Entomol.* **65**, 928–930.

Mason, R. R. (1970). Controlled field test of stabilized pyrethrins against the western hemlock looper. *U.S., Forest Serv., Res. Note* **PNW-120**, 1–9.

Miskus, R. P., and Andrews, T. L. (1972). Stabilization of thin films of pyrethrins and allethrin. *J. Agr. Food Chem.* **20**, 313–315.

Mounts, J., Pettinger, L. F., and Phillips, R. (1970). A field test of stabilized pyrethrins against the western hemlock looper, Mt. Baker National Forest, 1969. Unpublished report, U.S. Forest Serv., Portland, Oregon.

Negherbon, W. O. (1959). "Handbook of Toxicology," Vol. III. Saunders, Philadelphia, Pennsylvania.

Neisess, J., Maksymiuk, B., Waite, R. A., and Haskett, M. J. (1972). Personal communication. U.S. Forest Serv., Pacific Northwest Forest and Range Exp. Sta., Corvallis, Oregon.

Page, M., and Lyon, R. L. (1973a). Toxicity of eight insecticides applied to the western tent caterpillar. *J. Econ. Entomol.* **66**, 995–997.

Page, M., and Lyon, R. L. (1973b). Insecticides applied to western tussock moth reared on artificial diet: laboratory tests. *J. Econ. Entomol.* **66**, 53–55.

Pieper, G. R. (1972). Effect of target size and air movement on drop impingement efficiency and drop size distribution. *J. Econ. Entomol.* **65**, 884–886.

Randall, A. P., Hopewell, W. W., Haliburton, W., and Zylstra, B. (1970). Field evaluation of the effectiveness of ultra-low volume applications of insecticidal sprays from aircraft for the control of the spruce budworm (*Choristoneura fumiferana*) in New Brunswick 1970. Part IV. Field evaluation of the new stabilized pyrethrins and pyrethroid formulations against spruce budworm larvae. Unpublished report to Interdep. Comm. Meet. on Forest Spraying Oper., Dep. of Environ. Ottawa, Canada.

Randall, A. P., Hopewell, W. W., Haliburton, W., and Zylstra, B. (1971). Studies on the control of spruce budworm (*Choristoneura fumiferana*) by aerial application of chemicals. Unpublished report to Interdep. Comm. Meet. on Forest Spraying Oper., Dep. of Environ., Ottawa, Canada.

Robertson, J. L., and Gillette, N. L. (1973). Western tent caterpillar: Contact toxicity of ten insecticides applied to the larvae. *J. Econ. Entomol.* **66**, 629–630.

Robertson, J. L., and Lyon, R. L. (1973). Elm spanworm: Contact toxicity of ten insecticides applied to the larvae. *J. Econ. Entomol.* **66**, 627–628.

Robertson, J. L., Lyon, R. L., Shon, F. L., and Gillette, N. L. (1972). Contact toxicity of 20 insecticides applied to *Symmerista canicosta*. *J. Econ. Entomol.* **65**, 1560–1562.

Schwartz, J. L., and Lyon, R. L. (1971). Contact toxicity of five insecticides to California oakworm reared on an artificial diet. *J. Econ. Entomol.* **64**, 146–148.

Shea, P. J. (1972). Personal communication, U.S. Forest Serv., Pacific Southwest Forest and Range Exp. Sta., Berkeley, California.

Warner, J. L. (1963). Field experiments on the persistence of aqueous formulations of synergized pyrethrum for use in horticulture. *Pyrethrum Post* **7**(1), 34–42.

Part V

SUMMARY

Chapter 16

Residue and Tolerance Considerations with Pyrethrum, Piperonyl Butoxide, and MGK 264

JOSEPH B. MOORE

I. Introduction

Pesticide residues on food may result from preharvest or from postharvest pesticide treatment. The same two categories are also used in considering pesticide tolerances. In the case of pyrethrins, there is no pesticide residue tolerance required for preharvest treatments on fruits and vegetables. This ruling is based on the fact that the pyrethrins are rapidly degraded in the presence of sunlight and air and there is no residue remaining after a period of 12 to 24 hr. This is an advantage to the use of pyrethrins for insect control in the preharvest applications. They are nonpersistent and biodegradable.

Pyrethrins are extensively used in postharvest applications in a wide variety of situations. They are used as protectants on stored products against stored products insects, as in the case of grain during the storage period; against insects attacking dried fruits, nuts, and cereal products; and in transportation facilities to protect raw or processed bulk or packaged commodities during transit. Pyrethrins are also applied in processing plants to prevent infestations that would otherwise be present in the product. This includes such established uses as in feed mills, corn mills, flour mills, bakeries, canneries, dehydrating plants, candy factories, pickle works, and wineries. As opposed to preharvest applications, these postharvest applications are not usually exposed to sunlight or other conditions conducive to rapid breakdown.

Gunther (1962) defines a residue as including, in addition to the chemical pesticide or adjuvants, metabolites and derivatives that might occur as a result of the action of climatic environment on the deposited pesticide as well as from reactions within the biological tissues of the treated material. The amounts and persistence of insecticide residues in stored products depend upon the physical and chemical properties of both the insecticide being used and the material being treated as well as upon the toxicity of the chemical to the insect being controlled. Therefore, there is a wide area in which to assess and evaluate the amount of residues in stored products.

Recent advances in analytical techniques have made it possible to determine the very small amount of pyrethrum constituents and synergists in biological materials, including treated food commodities. Moore (1970) summarizes the literature on the residues that occur in foods resulting from postharvest application of pyrethrins and synergists under various methods of application, processing, and storage based on data developed rather recently with the more sensitive chromatographic procedures. The equipment and the cleanup and analytical techniques referred to in Moore's paper are given here for the analysis of the pyrethrins and the synergists, piperonyl butoxide and MGK 264, normally used either together or singly with the pyrethrins in these applications. The potential sensitivity of these methods of analysis is 10 ppb for each of these synergists, using Bruce's (1967) detector cell, and 0.5 ppb for the pyrethrins. The practical sensitivity limit, considering cleanup problems with various types of samples, is usually found to be nearer 100 ppb for these compounds.

The existence of pyrethrins and synergists residues raises the question of regulatory measures for the safety of human beings and animals that will be consuming food commodities treated with these materials. Such legal regulations are intended to prevent the appearance of toxic quantities of the pesticide in or on the treated food and to permit the establishment of tolerances allowable for pyrethrins and synergists before they can be sold in interstate commerce. The established tolerances are summarized in this review.

II. Extraction of Pyrethrins and Synergists for Residue Analysis

1. General

The extraction involves a modification of Moffitt's method (1963) for extracting fat and fat-soluble pesticides. This method, devised for extracting pesticides from milk, is applicable with minor modifications to milk products, such as nonfat dry milk, evaporated milk, ice cream, ice cream mixes, and coffee cream, and to other materials. This extraction procedure should result in high recovery values for the pyrethrins and synergists. It may not be as quantitative as the Mills' (1961) procedure for the extraction of milk because only one rather than three extractions is employed. However, the use of a centrifuge is not required. Recovery of 92 to 95% of the pesticides added is reported.

2. Reagents

Nanograde quality ethyl ether and pentane; 99.8% ethanol (redistilled in glass); reagent grade Na_2SO_4 (anhydrous, heated at 450°C overnight to eliminate interferences); reagent grade potassium oxalate.

3. Procedure

Fluid whole milk (100 gm) or other dairy products the total fat content of which does not exceed 5 gm is weighed and transferred into a 1-liter separatory funnel using several rinses with water to aid the transfer and to make the total volume to approximately 100 ml. Aqueous potassium oxalate solution (5% w/v) (10 ml) is added and the contents of the funnel mixed. The beaker in which the sample was weighed is then rinsed with ethanol (100 ml) and the rinse added to the mixture in the separatory funnel. The funnel is shaken for 1 min, following which ethyl ether (200 ml) is added and the funnel shaken again for 2 min. Finally, pentane (100 ml) is added and the funnel is shaken for a third time for a 1-min period. The mixture is allowed to stand for 10 min, after which the aqueous (lower) layer is drawn off and discarded. After the mix has stood for an additional 1 min, the residual water, which separates, is drawn off and discarded.

The solvent (upper) layer containing the lipids and pesticides is filtered through Na_2SO_4 (about 25 gm) positioned over a glass-wool plug in a funnel. The separatory funnel is washed three times with pentane (15 ml each time) and the solvent poured through the same funnel. The combined solvent is evaporated to dryness on a steam bath (60°–80°C) under a current of air until the fatty liquid all appears clear. The sample should not be overheated, and heating should not continue after the last of the solvent has been evaporated. The analysis may be interrupted overnight at this point. If this is done, the beaker containing the fat should be covered with aluminum foil.

The procedure can also be applied to tissues and eggs, in which case the tissue (25 gm) or eggs (50 gm) are blended with ethanol (100 ml) and transferred to a separatory funnel with 200 ml of ethyl ether; water (100 ml) and pentane (100 ml) are added and the mixture is shaken and the layers allowed to separate before proceeding with the comparable step as detailed above.

III. Pyrethrins—Cleanup of Extracts and Analysis

1. General

Analysis of the pyrethrins requires the removal of fats and other interfering materials from the extracts of milk, eggs, and tissues. The procedure must not cause deterioration of pyrethrins. This is accomplished by using the acetonitrile–petroleum ether partitioning method of Mills (1961), which removes much of the extracted fat; then by using column chromatography of the petroleum ether extract on silica gel to separate the pyrethrins from most of the chlorinated hydrocarbons and interfering materials; and finally by a type of partition chromatography, using acetonitrile with microfiber strips. The sample is then sufficiently clean for gas–liquid chromatography (GLC) with an electron-capture detector, calculating the magnitude of pyrethrins residue from the amount of pyrethrin I because it is the most easily detected ester of pyrethrins.

2. Reagents, Materials, and Apparatus

Nanograde quality acetone, acetonitrile, hexane, and petroleum ether; reagent grade NaCl, heated at 450°C overnight to eliminate interferences; silica gel (grade 950, 60–200 mesh, Fisher Chemical Specialty) heated at 450°C overnight; Belman I.T.L.C. type A microfiber (alumina oxide on microfiber strips). It is advisable to further clean the type A microfibers prior to use by allowing 90% acetone to ascend to the top before storing them in a moist chamber. After 12 hr of equilibration in a moist chamber the microfiber will contain sufficient absorbed water for use.

A gas chromatograph equipped with an electron-capture detector and a column (3.2 mm × 91 cm) packed with 1% QF1 plus 1% Epon 1001 on Chromosorb W 80/100 is used under the following operating conditions: injection port temperature for on-column injection, 210°C; a column temperature of 180–190°C to give satisfactory chromatograms; detector temperature, 190°C; nitrogen gas flow, 40 ml/min.

3. Acetonitrile Partitioning

The fat extracted from 100 ml milk, 25 gm tissue, 50 gm eggs, or 5 gm body fat is quantitatively transferred to a 125-ml separator, using small portions of petroleum ether, so that the total volume of fat and petroleum ether in the separator is 25 ml. Thirty milliliters of acetonitrile (saturated with petroleum ether) is added, the separator is shaken vigorously for 1 min, and the layers are allowed to separate. It is important to make certain during this extraction that all of the lipids are in solution and do not appear at the interface and to note that an inversion of the hydrocarbon solvent and acetonitrile phases may sometimes occur. The acetonitrile is drained into a 1-liter separator, containing 700 ml of 2% NaCl solution and 100 ml of petroleum ether. The petroleum ether solution is extracted in the 125-ml separator three more times in the same manner, using 30 ml portions of petroleum-ether–saturated acetonitrile each time, and all extracts are combined in the 1-liter separator. The 1-liter separator is stoppered and inverted and the pressure is vented off; the first extraction of pyrethrins is performed by swirling the inverted separator, with caution to minimize emulsions. The layers are allowed to separate and the aqueous layer is drained into a second 1-liter separator. One hundred milliliters of petroleum ether is added and the separator is shaken vigorously 15 sec; then the layers are allowed to separate. The aqueous layer is discarded and the petroleum ether is combined with that in the original separator and washed with two 100-ml portions of water. (If emulsions form, the mix is washed with 1–2% NaCl solution, or 5 ml of saturated NaCl solution are added to the 1-liter separator.) The washings are discarded and the petroleum ether layer is drawn off through a 5 cm column of anhydrous Na_2SO_4 into a Kuderna-Danish concentrator. The separator and column is rinsed three times with about 10 ml portions of petroleum ether. The solvent is evaporated to about 10 ml for transfer to the silica gel column.

4. Preliminary Separation by Column Chromatography

The chromatographic column (2.2×32 cm) is prepared by adding, in sequence, 1.3 cm of Na_2SO_4 and 10 gm of silica gel, topping with 3.0 cm of Na_2SO_4. The column is prewet with 30 ml of hexane, the petroleum ether extract is transferred into the column, and the elution rate is adjusted to about 5 ml/min. The flask is rinsed with two successive 5-ml portions of hexane to effect a quantitative transfer to the column. The rinse is allowed to sink into the Na_2SO_4 layer, then the wall of the chromatographic column is rinsed down with 3–5 ml of hexane. It is usually necessary to pretest one or two such columns to establish the elution pattern of the column that effects the best separation of pyrethrins from the chlorinated hydrocarbons. This first fraction may be tested for chlorinated pesticides and discarded. The pyrethrins is eluted with 100 ml of ether–hexane (1:1) mixture.

5. *Partition Chromatography Cleanup*

The second fraction from the column chromatography containing the pyrethrins will also contain fatty materials that should be removed to prevent contamination of the GLC column and an unsatisfactory baseline on the chromatogram. The 100 ml of solvent are removed from this fraction with a vacuum rotary evaporator of a Kuderna-Danish type. The fatty residue is transferred into a 15-ml conical centrifuge tube with 3 ml of petroleum ether. The container is carefully rinsed with three more successive portions of petroleum ether to insure a complete transfer to the centrifuge tube. The petroleum ether is evaporated with a slow stream of dry nitrogen, with the centrifuge tube immersed in a water bath maintained at 40°C. The sides of the centrifuge tube are rinsed down with petroleum ether and the volume is adjusted to about 0.5 ml. This concentrate is transferred by means of a 500 μl Hamilton syringe in a band measuring 1.3 cm wide across the 5 cm width of the microfiber. The band should be about 3.8 cm above the bottom edge of the strip. The centrifuge tube is rinsed with 0.2 ml of petroleum ether, and this amount is transferred to the band. This procedure is repeated twice to complete the transfer. A pencil is used to mark the strip 0.5 cm above the residue band. After it has dried for 5 min, the strip is hung in the chromatographic tank with the bottom edge immersed 0.6 cm into pure acetonitrile contained in a 150-ml beaker. The humidity is elevated in the chromatographic tank by means of a separate microfiber wick immersed in water. The pyrethrins elutes up with acetonitrile and the fats remain at the application site. After the acetonitrile front has moved 10 cm above the fat band, the strip is removed from the tank, the solvent front is marked with a pencil, and the strip is allowed to dry 10 min. A clean pair of scissors is used to cut on the line above the fat band. This lower portion is discarded. The upper portion of the strip up to the solvent front is cut across into 1.3 cm strips and dropped into a 125-ml Erlenmeyer flask. Twenty milliliters of hexane–benzene (4:1) mixture is pipetted into the flask. The mixture is agitated in a swirling motion for 2 min. One milliliter of water is added to wet the microfiber and to replace the hexane–benzene solvent absorbed in the fiber. The mixture is agitated briefly and transferred to a 50-

or 100-ml volumetric flask. The flask is rinsed three times with 5 ml of hexane and the rinses combined in the volumetric flask and made up to volume with hexane.

6. *Analysis*

Inject one or more microliter into the gas chromatograph, depending upon its sensitivity. A standard curve is prepared for pyrethrins within the limits of the linear response (usually up to 2 ng as the amount of pyrethrins actually injected).

IV. Piperonyl Butoxide and MGK 264—Cleanup of Extracts and Analysis

1. *General*

A cleanup procedure has been developed to remove interfering materials from samples of eggs and tissue extracts. Samples are extracted with a mixture of ethanol, ether, and hexane. Large fat and tissue samples are treated by the procedure of O'Donnel *et al.* (1954). This method, involving saponification with alcoholic KOH, is useful for cleaning up fat samples for the analysis of alkali-stable pesticides; it is employed for samples containing aldrin, dieldrin, and endrin. The procedure is usually limited by the fact that many pesticides are destroyed entirely during alkaline digestion or are converted to derivatives through dehydrochlorination and other reactions. However, this is an advantage in the analysis of the alkali-stable synergists. The saponified extract in hexane is chromatographed through silica gel to separate the synergist from the chlorinated hydrocarbon pesticides. The final cleanup is accomplished by partition chromatography. The amount of synergist is determined by analysis with a special electron-capture detector.

2. *Saponification*

The sample extract, representing no more than 50 gm of glyceride, is placed in a 500-ml round-bottom flask equipped with an outer T joint. The solvent is evaporated at 60°–80°C under a current of air, using care not to overheat. When more than 50 gm of fat must be analyzed, it is recommended that the saponification be carried out on two separate portions. When less than 50 gm is used, the procedure can be scaled down by using smaller amounts of KOH and solvents. For oils and animal products, a 50-gm sample is weighed directly into the saponification flask.

Aqueous KOH solution (50% w/v) (50 ml) and ethanol (200 ml) are added to the flask. A reflux condenser is attached, two to four boiling chips are added, and the solution is refluxed for 60 min or until no oily particles are discernible. Next, water (200 ml) is added and the solution is transferred quantitatively to a 1-liter separatory funnel, using alternate streams of water and Skelleysolve B delivered from wash bottles. Skelleysolve B (150 ml) is added to the funnel, and it is stoppered and shaken vigorously. After the phases separate, the lower (soap) layer and the hydrocarbon (Skelleysolve B) layer are recovered in separate flasks. The soap solution is extracted twice more, each time with Skelleysolve B (100 ml), combining the recovered Skelleysolve layers with the preceding extract.

The Skelleysolve B solution containing the pesticides is transferred to a separatory funnel, water (150 ml) is added, and the mixture is swirled gently to extract residual soap into the aqueous phase. Vigorous shaking must be avoided at this stage to minimize the possibility of gel formation. The aqueous phase is then drained off and discarded. The hydrocarbon phase is extracted with 150-ml portions of water until the extract is colorless to phenolphthalein indicator. The washings are discarded. When the hydrocarbon phase is washed with water, an interfacial "cuff" may be formed during the first washings. This should be retained with the hydrocarbon phase.

The hydrocarbon extract is dried for 15 min over anhydrous Na_2SO_4 (20–50 gm), then filtered through Whatman No. 1 filter paper; the flask and filter are washed with small portions of Skelleysolve B and the washings are added to the extract; and the sample is evaporated at 60°–80°C under a current of air with care to avoid overheating.

3. *Preliminary Separation by Column Chromatography and Partition Chromatography Cleanup*

The silica gel column used in the preliminary separation of pyrethrins is also applicable to the separation of piperonyl butoxide and MGK 264 from chlorinated hydrocarbons and interfering materials. The eluate fraction containing these synergists is the same as that containing the pyrethrins, as noted above. The partition chromatography cleanup procedure used for the pyrethrins is also applicable in all aspects to piperonyl butoxide and MGK 264.

4. *Analysis Detector Cell for GLC Analysis of Synergists*

The detector cell described by Bruce (1967) is particularly useful for analyzing pyrethrins synergists. The detector is designed to fit into a Parr bomb, the inside measurements of which are 42 mm long and 27 mm in diameter. Appropriate gas-tight connections are welded to the Parr bomb. A movable electrode is incorporated into the detector cell, which makes it possible to select the most sensitive position for detection and to convert it into an electron-capture system by substituting a single electrode for the paired No. 28 Chromal electrodes. The tritium foil of 100 mCi is contained within a stainless steel jacket 13.0 mm long and 12.8 mm in diameter. A 30.5 cm stainless steel column 0.32 mm in diameter was used in Bruce's studies. The packing material used in the column consisted of Chromosorb W containing 1.0% Epon 1001, 0.3% DEGS, 0.2% Viton A, and 0.1% Dow Corning silicone fluid 710. Applied voltages were 450 V and 160 V DC in the new detection circuits and 6.75 V in the electron capture circuit. Column temperatures ranged between 140° and 172°C and nitrogen gas flow between 40 and 65 ml/min. The injection port temperature was maintained at a level about 20°C higher than the column temperature.

Unlike the electron-capture cell, the sensitivity of Bruce's cell to organophosphorus insecticides, piperonyl butoxide, and MGK 264 increases as the potentials to the cell are increased. The current changes are much greater in the new detector cell when compared to the electron-capture detector. For this reason, three to ten times the electrometer gain was used with the electron-capture circuit. The new detector possesses a remarkable sensitivity to piperonyl butoxide and MGK 264. It is 2000 to 10,000 times more sensitive than the electron capture detector for these chemicals. Tests with organophosphorus pesticides are less spectacular. Sensitivity increases in a range of 5–20 times, depending upon the compound. The chromatograms from the new detector system contain negative peaks for Bidrin, diazinon, Azodrin, and MGK 264. By reversing the polarity of the recorder, these appear as positive peaks and show a linear relationship with amounts. Approximate minimum detectable amounts of pesticides are given in Table I.

The usefulness of Bruce's detector has been demonstrated in the analysis of wheat flour for trace amounts of piperonyl butoxide and MGK 264. The method was sensitive to as little as 10 ppb of these chemicals in flour.

V. Tolerances Established

The Food and Drug Administration, United States Department of Health, Education, and Welfare, and later the Environmental Protection Agency established the tolerances for residues of pyrethrins and the synergists from postharvest applications. Tolerances and other permitted uses of pyrethrins and synergists are listed below with Code of Federal Regulation (CFR) and Federal Register (F.R.) references.*

* Reproduced by permission from *Food Drug Cosmetic Law Reports,* published and copyrighted by Commerce Clearing House, Inc., Chicago, Illinois 60646.

TABLE I

Picograms Minimum Detectable Quantities of Pesticides

Pesticide	Bruce's detector (pg)	Electron-capture detector (pg)
MGK 264	50–100	200,000–1,000,000
Piperonyl butoxide	50–100	100,000–200,000
Ciodrin[a]	30–50	200–500
Azodrin[b]	250–500	5000–10,000
Bidrin[c]	250–500	2500–5000
Diazinon[d]	50–100	500–1000
Carbaryl[e]	500	5000

[a] α-Methylbenzyl 3-hydroxycrotonate dimethyl phosphate.
[b] 3-Hydroxy-*N*-methyl-*cis*-crotonamide dimethyl phosphate.
[c] 3-Hydroxy-*N*,*N*-dimethyl-*cis*-crotonamide dimethyl phosphate.
[d] *O*,*O*-Diethyl *O*-(2-isopropyl-4-methyl-6-pyrimidyl) phosphorothioate.
[e] 1-Naphthyl methylcarbamate.

Pyrethrins; tolerances for residues

Section 180.128 Tolerances for residues of the insecticide pyrethrins (insecticidally active principles of *Chrysanthemum cinerariaefolium*) are established in or on raw agricultural commodities, as follows:

From postharvest application: 3 parts per million in or on barley, birdseed mixtures, buckwheat, corn (including popcorn), rice, rye, and wheat.

From postharvest application: 1 part per million in or on almonds, apples, beans, blackberries, blueberries (huckleberries), boysenberries, cherries, cocoa beans, copra, cottonseed, crabapples, currants, dewberries, figs, flaxseed, gooseberries, grain sorghum, grapes, guavas, loganberries, mangoes, muskmelons, oats, oranges, peaches, peanuts (determined on the nuts with shell removed), pears, peas, pineapples, plums (fresh prunes), raspberries, tomatoes, and walnuts.

0.5 part per million in milk fat reflecting negligible residues in milk.

0.2 part per million in meat, fat, and meat byproducts of poultry.

0.1 part per million (negligible residue) in eggs and meat, fat, and meat byproducts of cattle, goats, hogs, horses, and sheep.

0.05 part per million in or on potatoes from postharvest application.

(As Amended, 37 F.R. 11725, effective June 13, 1972.) 40 CFR.

Pyrethrins

Section 121.1075 The food additive pyrethrins may be safely used in accordance with the following prescribed conditions:

(a) It is used or intended for use in combination with piperonyl butoxide for control of insects:

(1) In cereal grain mills and in storage areas for milled cereal grain products, whereby the amount of pyrethrins is from 10 percent to 100 percent of the amount of piperonyl butoxide in the formulation.

(2) On the outer ply of multiwall paper bags of 50 pounds or more capacity in amounts not exceeding 6 milligrams per square foot, whereby the

amount of pyrethrins is equal to 10 percent of the amount of piperonyl butoxide in the formulation. Such treated bags are to be used only for dried foods.

(3) On cotton bags of 50 pounds or more capacity in amounts not exceeding 5.5 milligrams per square foot of cloth, whereby the amount of pyrethrins is equal to 10 percent of the amount of piperonyl butoxide in the formulation. Such treated bags are constructed with waxed paper liners and are to be used only for dried foods that contain 4 percent fat or less.

(4) In two-ply bags consisting of cellophane/polyethylene sheets bound together by an adhesive layer when it is incorporated in the adhesive. The treated sheets shall contain not more than 10 milligrams of pyrethrins per square foot. Such treated bags are to be used only for packaging dried prunes and are to be of 2 or more pounds capacity.

(5) In food processing areas and food storage areas: *Provided,* that the food is removed or covered prior to such use.

(b) It is used or intended for use in combination with piperonyl butoxide and *N*-octylbicycloheptene dicarboximide for insect control in accordance with Section 121.1205.

(c) A tolerance of one part per million is established for residues of pyrethrins in or on:

(1) Milled fractions derived from cereal grains when present as a result of its use in cereal grain mills and in storage areas for milled cereal grain products.

(2) Dried foods when present as the result of migration from its use on the outer ply of multiwall paper bags of 50 pounds or more capacity.

(3) Foods treated in accordance with Section 121.1205.

(4) Dried foods that contain 4 percent fat, or less, when present as a result of migration from its use on the cloth of cotton bags of 50 pounds or more capacity constructed with waxed paper liners.

(5) Foods treated in accordance with paragraph (a) (5) of this section.

(d) To assure safe use of the additive, its label and labeling shall conform to that registered with the U.S. Department of Agriculture, and it shall be used in accordance with such label and labeling.

(e) Where tolerances are established under sections 408 and 409 of the act on both raw agricultural commodities and processed foods made therefrom, the total residues of pyrethrins in or on the processed food shall not be greater than that permitted by the larger of the two tolerances.

(f) A tolerance of 0.3 part per million is established for residues of pyrethrins in or on dried prunes when present as a result of migration from its use in the adhesive layer of two-ply bags consisting of cellophane/polyethylene sheets and the bags are of 2 or more pounds capacity.

(As amended, 37 F.R. 9464, effective May 11, 1972.) 21 CFR.

Piperonyl butoxide; tolerances for residues

Section 180.127 Tolerances for residues of the insecticide piperonyl butoxide [(butyl carbityl)(6-propylpiperonyl) ether] are established in or on raw agricultural commodities, as follows:

20 parts per million from postharvest application in or on barley, birdseed mixtures, buckwheat, corn (including popcorn), rice, rye, and wheat.

8 parts per million from postharvest application in or on almonds, apples, beans, blackberries, blueberries (huckleberries), boysenberries, cherries, cocoa

beans, copra, cottonseed, crabapples, currants, dewberries, figs, flaxseed, gooseberries, grain sorghum, grapes, guavas, loganberries, mangoes, muskmelons, oats, oranges, peaches, peanuts (determined on the nuts with shell removed), pears, peas, pineapples, plums (fresh prunes), raspberries, tomatoes, and walnuts.

3 parts per million in meat, fat, and meat byproducts of poultry.

1 part per million in eggs.

0.25 part per million from postharvest application in or on potatoes.

0.25 part per million (reflecting negligible residues in milk) in milk fat.

0.1 part per million (negligible residue) in the meat, fat, and meat byproducts of cattle, goats, hogs, horses, and sheep.

(As amended, 37 F.R. 11724, effective June 13, 1972.) 40 CFR.

Piperonyl butoxide

Section 121.1074 The food additive piperonyl butoxide may be safely used in accordance with the following prescribed conditions:

(a) It is used or intended for use in combination with pyrethrins for control of insects:

(1) In cereal grain mills and in storage areas for milled cereal grain products, whereby the amount of piperonyl butoxide is at least equal to but not more than 10 times the amount of pyrethrins in the formulation.

(2) On the outer ply of multiwall paper bags of 50 pounds or more capacity in amounts not exceeding 60 milligrams per square foot, whereby the amount of piperonyl butoxide is equal to 10 times the amount of pyrethrins in the formulation. Such treated bags are to be used only for dried foods.

(3) On cotton bags of 50 pounds or more capacity in amounts not exceeding 55 milligrams per square foot of cloth, whereby the amount of piperonyl butoxide is equal to 10 times the amount of pyrethrins in the formulation. Such treated bags are constructed with waxed paper liners and are to be used only for dried foods that contain 4 percent fat or less.

(4) In two-ply bags consisting of cellophane/polyethylene sheets bound together by an adhesive layer when it is incorporated in the adhesive. The treated sheets shall contain not more than 50 milligrams of piperonyl butoxide per square foot. Such treated bags are to be used only for packaging dried prunes and are to be of 2 or more pounds capacity.

(5) In food processing and food storage areas: *Provided,* that the food is removed or covered prior to such use.

(b) It is used or intended for use in combination with pyrethrins and *N*-octylbicycloheptene dicarboximide for insect control in accordance with Section 121.1205.

(c) A tolerance of 10 parts per million is established for residues of piperonyl butoxide in or on:

(1) Milled fractions derived from cereal grains when present therein as a result of its use in cereal grain mills and in storage areas for milled cereal grain products.

(2) Dried foods when present as a result of migration from its use on the outer ply of multiwall paper bags of 50 pounds or more capacity.

(3) Foods treated in accordance with Section 121.1205.

(4) Dried foods that contain 4 percent fat, or less, when present as a result of migration from its use on the cloth of cotton bags of 50 pounds or more capacity constructed with waxed paper liners.

(5) Foods treated in accordance with paragraph (a) (5) of this section.

(d) To assure safe use of the additive, its label and labeling shall conform to that registered with the U.S. Department of Agriculture, and it shall be used in accordance with such label and labeling.

(e) Where tolerances are established under sections 408 and 409 of the act on both raw agricultural commodities and processed foods made therefrom, the total residues of piperonyl butoxide in or on the processed food shall not be greater than that permitted by the larger of the two tolerances.

(f) A tolerance of 3 parts per million is established for residues of piperonyl butoxide in or on dried prunes when present as a result of migration from its use in the adhesive layer of two-ply bags consisting of cellophane/polyethylene sheets and the bags are of 2 or more pounds capacity.

(As amended, 37 F.R. 9464, effective May 11, 1972.) 21 CFR.

N-Octylbicycloheptene dicarboximide.

Section 121.1205 The food additive *N*-octylbicycloheptene dicarboximide may be safely used in accordance with the following prescribed conditions:

(a) It is used in combination with piperonyl butoxide and pyrethrins for insect control in food-processing and food-storage areas, provided that the food is removed or covered prior to such use.

(b) Residues in food resulting from the use described in paragraph (a) of this section shall not exceed 10 parts per million of *N*-octylbicycloheptene dicarboximide, 10 parts per million of piperonyl butoxide, and 1 part per million of pyrethrins.

(c) To assure safe use of the additive, its label and labeling shall conform to that registered with the U.S. Department of Agriculture and it shall be used in accordance with such label and labeling.

(As added, 32 F.R. 7458, effective May 19, 1967.) 21 CFR.

These tolerances are based, in part, on residue data obtained with the analytical procedures described above. Some of these established tolerances are quite low. It should be noted that the magnitude of the tolerance is not necessarily, in many cases, based on the toxicity of the chemical under consideration but rather on the magnitude of the residue remaining consistent with good agricultural and sanitary practices.

VI. Summary

Elliott *et al.* (1972), in their study on the metabolic fate of pyrethrins I and II and allethrin administered orally to rats, indicate that some of the ingested pyrethroids are excreted in unmetabolized form and that the metabolites of pyrethrins are likely to be less toxic than the pyrethrins themselves. They point out that it is improbable that pyrethrins would be concentrated when passing through the food chains or that they will persist unmetabolized in the mammalian body.

Pyrethrins may be used safely to control insects infesting food crops and food in storage and processing plants without significant contamination of the environment. Such use is of no hazard to humans and animals at the established tolerance levels.

References

Bruce, W. N. (1967). Detector cell for measuring picogram quantities of organophosphorus insecticides, pyrethrin synergists and other compounds by gas chromatography, *J. Agr. Food Chem.* **15**, 178–181.

Elliott, M., Janes, N. F., Kimmel, E. C., and Casida, J. E. (1972). Metabolic fate of pyrethrin I, pyrethrin II, and allethrin administered orally to rats. *J. Agr. Food Chem.* **20**, 300–313.

Gunther, F. A. (1962). Instrumentation in pesticide residue determinations. *Advan. Pest Contr. Res.* **5**, 191–206.

Mills, P. A. (1961). Collaborative study of certain chlorinated pesticides in dairy products. *J. Ass. Offic. Agr. Chem.* **44**, 171–177.

Moffitt, R. A. (1963). Residue analysis in the dairy industry. *In* "Analytical Methods for Pesticides, Plant Growth Regulators, and Food Additives" (G. Zweig, ed.), Vol. 1, pp. 545–570. Academic Press, New York.

Moore, J. B. (1970). Terminal residues of pyrethrin-type insecticides and their synergists in foodstuffs. *Residue Rev.* **33**, 87–100.

O'Donnel, A. E., Neil, M. M., Weiss, F. E., Bann, A. M., DeCino, T. J., and Lau, S. C. (1954). Pesticide residues: Chemical determination of aldrin in crop tissues. *J. Agr. Food Chem.* **2**, 573–580.

Chapter 17

Advantages and Disadvantages of Pyrethrum

EMIL M. MRAK

I. Introduction

Today, in dealing with pesticides, we are burdened with emotions, pressures, court instructions, the necessity of making quick decisions—sometimes too hasty and even faulty, as a result of the absence of sound and reliable information. However, so often there is information which unfortunately is tucked away in the files of research workers, and particularly those in industry. This volume brings to light such unpublished information, which is sorely needed, whether it be favorable or unfavorable to pyrethrum. In fact, this volume brings both the published and unpublished information about pyrethrum up to date in readily available form for the first time in almost 30 years. A single multiauthor treatise covering all aspects of a single pesticide is a rarity indeed. As such, it will be very useful to those involved in decision making with respect to the role and place of pyrethrum in the spectrum of insecticides. We are now in a position to evaluate the benefits of pyrethrum relative to its disadvantages.

II. Production, Chemistry, and Biochemistry

Pyrethrum flowers are produced in Kenya, Tanzania, Rwanda, several other countries in Africa, Ecuador, and elsewhere, many of which nations are economically developing. As a matter of fact, pyrethrum is extremely important to the economy of Kenya and certain other areas, for it is one of their most important exports, if not the most important. Thousands of small growers—about 85,000 in Kenya alone—are involved in the production of pyrethrum flowers. Kenya has developed an effective central control for quality and quantity. This is important, for it has enabled the production of a better quality pyrethrum at a lower cost. Since the United States is the most important purchaser of pyrethrum extract, price and quality are of great concern to us. It is important, too, that the development of this product in Kenya has encouraged a philosophy indicating the need for various aspects of agricultural research. A wide-spectrum research program has evolved in Kenya, a development that is good not only for pyrethrum production but for all of agriculture.

It is apparent that we know a great deal about the chemistry and biochemistry of pyrethrum. It is rapidly biodegraded in mammals by established pathways. Although the levels in water, air, and soil are not known, little is expected to persist because of its photolabile and biodegradable nature. We know a little about how it acts but there is still much to learn. For example, we do not know whether it acts at only a single site or at several targets in the nervous system to repel, knock down, or kill insects. Nevertheless, I only wish we knew as much about all other pesticide chemicals as we do about pyrethrum. This does not mean, however, that we should discontinue research on this important material but rather that we should encourage more.

III. Uses

Pyrethrum has proved to be an excellent material for protection against mosquitos and other insects that may have found their way into airplanes. It is effective at very low dosage when applied from the air for control of mosquitos, tsetse flies, and tabanids and when applied with ground equipment as a mosquito adulticide. When used as smoke in small houses in some countries, it has proved to be an excellent repellent and knockdown agent. Mosquitos stop feeding almost immediately. Other insects are also repelled by pyrethrum and succumb to its knockdown effect. It is very useful in homes, barns, and similar structures for the control of various types of flies, mosquitos, gnats, ticks, and so on. Likewise it is very useful in the control of insect pests associated with households and food commodity storage and processing facilities. A great deal of it is used for these purposes.

Prior to World War II pyrethrum was extensively used in control of agricultural and horticultural insect pests and this use area may become prominent once again with improved formulations and methods of application. Recent experiments conducted in forest areas indicate promise in the control of certain defoliating forest insects. Studies in the Rocky Mountains of Colorado establish that it does not harm birds or fish even when they eat insects knocked down by pyrethrum. Pyrethrum does not harm wildlife, including birds, fish, and larger animals, under normal use conditions. Likewise, it does not harm domestic animals and has never been known to kill a pet or a farm animal.

We can expect that any population of an insect pest suitably pressured with pyrethrum for many generations will show some resistance although it may be minimal. However, it is likely that the level of resistance will not significantly impair the effectiveness of pyrethrum in control. There are some indications and speculations about why insects do not rapidly develop resistance to pyrethrum, but additional studies in this area are needed.

IV. Safety

The general impression from the available information on pyrethrum is that no health hazards are associated with it. It does not depress cholinesterase or induce other harmful actions on large animals. It deteriorates so rapidly there is no residue—so why worry about it?

From the standpoint of the repression of growth of small animals, it requires a large dose indeed—5000 ppm. Yet, in subsequent generations, there is no evidence of effect on growth or reproduction. When pyrethrum is fed to rabbits by capsule at 90 mg/kg/day during the gestation period, no classical teratogenic effects are produced and this dose is an extremely large one, and beyond the realm of reason. There is no evidence whatsoever of pyrethrum inducing carcinogenesis or mutagenesis; however, carcinogenesis is an area that needs more study.

There has been some concern about possible allergenic properties of pyrethrum. The unpurified materials sometimes induce the same dermatitis responses as ragweed. These, however, disappear rapidly when the allergen is removed during normal manufacturing procedures. Observations made over a period of 18 years, in a factory manufacturing and selling pyrethrins-containing products, showed essentially no cases with evidence of allergies, such as asthma or dermatitis. They were rare indeed.

Interaction with other chemicals is an extremely interesting problem for research and more should be done in this area. Pyrethrum has been studied in combination with many other chemicals, particularly alcohol, aspirin, caffeine, nicotine, several synergists, and DDT. In some instances, there is a weight reduction when large amounts are fed in conjunction with other chemicals, but the general conclusion is that there are no real serious effects. Studies made on the effects on liver, kidneys, gonads, heart, and other tissues in general show no ill effects. Likewise, there are no marked pathological effects. The effects of pyrethrum when used with sulfoxide, caffeine, and alcohol appear to be somewhat harmful. On the other hand, when pyrethrum is combined with piperonyl butoxide, a synergist in aerosol sprays containing pyrethrum, the harmful effects of the synergist are negligible as compared to the chemicals when used alone.

There are extensive records on human exposure and response to pyrethrum used as an insecticide or vermifuge, and these establish the negligible hazard to the users of pyrethrum products.

Pyrethrum is safe to use on food plants, and the evidence supporting this comes from various sources. It is permitted for use in the storage of grains, fruits, and foods of this type. There is a tolerance value for milk because of its use on animals or perhaps the possibility of them eating stored grains.

There is some question, however, about continuous spraying in food service facilities, such as restaurants and hospitals, because of the possibility of contaminating exposed foods. There is also some concern about continuous inhalation of pyrethrum, the synergist piperonyl butoxide, and fluorinated hydrocarbons by workers. In view of the strong indications of safety insofar as pyrethrum is concerned and its other approved uses, one cannot but wonder why there is concern about its use in feeding establishments. In regard to the possible contamination of food, however, the problem seems to be one of a technicality because no tolerance has been established for restaurant foods. Furthermore, if it is established, it will depend on the local agencies, such as they may be, to enforce rather than the Federal Food and Drug Administration. There are over 500,000 food service facilities in the United States and the Federal Food and Drug Administration is just not equipped to police them. Therefore, this is a technicality, for as the laws are now written, they do not permit the use of judgement, even though there is every indication that such contamination would be completely harmless.

V. Disadvantages

The disadvantages of pyrethrum are few indeed. Pyrethrum is not a persistent chemical. It is destroyed by light and in addition is biodegradable. This is a favorable point on the one hand, but on the other it is a disadvantage in that it can only be used as a contact insecticide. For knockdown, insects must have direct contact when sprayed. Because pyrethrum is not a persistent material, the spraying of surfaces cannot be relied on for control as is the case with DDT.

In spite of the fact we know so much about pyrethrum, its formulations, and its great advantages, there is still the need for more research, and this should indeed be encouraged. We are still lacking in knowledge with respect to application technology, especially when used on farms and in forests. The allergic responses, once a problem, seem no longer to be of concern, especially when the purified material is used. There are problems in periodic use in food establishments because of legal technicalities. There is some question about the carrying agents and formulation constituents. These points need further study.

VI. Balance of Advantages Versus Disadvantages

Without a doubt, the benefits of pyrethrum certainly outrank its disadvantages. One of the earliest insecticides, it is a tried and true material from the historical point of view. Pyrethrum controls certain insects as effectively and more safely than most of the currently used synthetic organic insecticides. There is no evidence of harm to humans, domestic animals, or wildlife, when used as directed. It is hoped that this volume reviewing the available information on pyrethrum will encourage continuing research to find improved methods for using this insecticide and to establish the degree of safety of pyrethrum-containing products.

Author Index

Numbers in italics refer to the pages on which the complete references are listed.

Y

Z

Subject Index

H

I

T

U

Y

Z